Y0-BTD-386

PERSPECTIVES IN URBAN ENTOMOLOGY

ACADEMIC PRESS RAPID MANUSCRIPT REPRODUCTION

PERSPECTIVES IN URBAN ENTOMOLOGY

EDITED BY

G. W. Frankie

Department of Entomological Sciences
University of California
Berkeley, California

C. S. Koehler

Cooperative Extension
University of California
Berkeley, California

ACADEMIC PRESS New York San Francisco London 1978
A Subsidiary of Harcourt Brace Jovanovich, Publishers

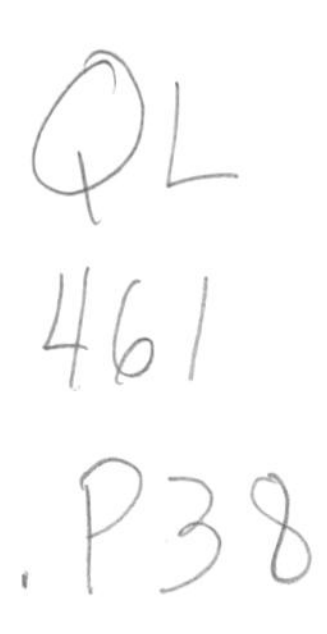

ACADEMIC PRESS, INC.
111 Fifth Avenue, New York, New York 10003

United Kingdom Edition published by
ACADEMIC PRESS, INC. (LONDON) LTD.
24/28 Oval Road, London NW1 7DX

Library of Congress Cataloging in Publication Data

Main entry under title:

Perspectives in urban entomology.

"An outgrowth of a symposium . . . held during the XV International Congress of Entomology, 1976, in Washington, D.C."
Includes bibliographical references.
1. Insects—Congresses. 2. Insect control—Congresses. 3. Urban fauna—Congresses.
I. Frankie, G. W. II. Koehler, Carlton S.
III. International Congress of Entomology, 15th, Washington, D. C., 1976.
QL461.P38 595.7'05'26 78-6746
ISBN 0-12-265250-9

PRINTED IN THE UNITED STATES OF AMERICA

CONTENTS

LIST OF CONTRIBUTORS

Numbers in parentheses indicate pages on which authors' contributions begin.

JOHN T. AMBROSE (187), Department of Entomology, North Carolina State University, Raleigh, North Carolina 27607

GARY W. BENNETT (409), Department of Entomology, Purdue University, West Lafayette, Indiana 47907

MICHAEL BURGETT (187), Department of Entomology, Oregon State University, Corvallis, Oregon 97331

DEWEY M. CARON (187), Department of Entomology, University of Maryland, College Park, Maryland 20740

HARRY G. DAVIS (163), Yakima Agricultural Research Laboratory, Agricultural Research Service, USDA, Yakima, Washington 98902

WALTER EBELING (221), Department of Entomology, University of California, Los Angeles, California 90024

L. E. EHLER (349), Department of Entomology, University of California, Davis, California 95616

GORDON W. FRANKIE (249, 267, 359), Department of Entomological Sciences, University of California, Berkeley, California 94720

R. A. FRENCH (31), Rothamsted Experimental Station, Harpenden, Hertfordshire, England, AL5 2JQ

MICHAEL J. GAYLOR (267), Department of Entomology and Zoology, Auburn University, Auburn, Alabama 36830

LAWRENCE E. GILBERT (1), Department of Zoology, University of Texas, Austin, Texas 78712

ALAN I. KAPLAN (311), Department of Entomological Sciences, Division of Biological Control, University of California, Berkeley, California 94720

GERALD N. LANIER (295), State University College of Environmental Science and Forestry, Syracuse, New York 13210

HANNA LEVENSON (359), Langley Porter Institute, University of California Medical School, San Francisco, California 94143

RICHARD W. MERRITT (125), Department of Entomology, Michigan State University, East Lansing, Michigan 48824

DAVID L. MORGAN (267), Texas Agricultural Experiment Station, Texas A&M University Research and Extension Center, Dallas, Texas 75252

BERNARD C. NELSON (87), Vector and Waste Management Section, California Department of Health, Berkeley, California 94704

H. D. NEWSON (125), Department of Entomology, Michigan State University, East Lansing, Michigan 48824

HELGA OLKOWSKI (311), Department of Entomological Sciences, Division of Biological Control, University of California, Berkeley, California 94720

WILLIAM OLKOWSKI (311), Department of Entomological Sciences, Division of Biological Control, University of California, Berkeley, California 94720

DENIS F. OWEN (13), Department of Biology, Oxford Polytechnic, Headington, Oxford, England OX3 OBP

GARY L. PIPER (249), Department of Entomology, Texas A&M University, College Station, Texas 77843

FREDERICK W. PLAPP, JR. (401), Department of Entomology, Texas A&M University, College Station, Texas 77843

J. D. SHORTHOUSE (67), Department of Biology, Laurentian University, Sudbury, Ontario, Canada, P3E 2C6

MICHAEL C. SINGER (1), Department of Zoology, University of Texas, Austin, Texas 78712

L. R. TAYLOR (31), Rothamsted Experimental Station, Harpenden, Hertfordshire, England, AL5 2JQ

ROBERT VAN DEN BOSCH (311), Department of Entomological Sciences, Division of Biological Control, University of California, Berkeley, California 94720

I. P. WOIWOD (31), Rothamsted Experimental Station, Harpenden, Hertfordshire, England, AL5 2JQ

FOREWORD: THE URBAN HABITAT

Urbanized areas are, by definition, environments dominated by the actions of man. Other organisms share urban habitats with human beings for a variety of reasons. Some are nurtured actively by people, because they give aesthetic pleasure or companionship or economic return—street trees, ornamental shrubs and herbs, and domestic pets are examples. Some wild creatures endure because the indoor or outdoor environmental characteristics of cities are within their ranges of environmental tolerance as long as no active human effort is expended to extinguish them, native plants on remnant undisturbed sites, migratory waterfowl. Others—weeds, pests, and parasites of many kinds—thrive even though unwanted and actively suppressed, because their own requirements are met by the environments people create. The quantity of information available concerning the nonhuman biota of urban areas rather closely reflects the perceived importance of several organisms to people. Species of economic significance have received the most attention.

Scholars in many disciplines have begun to study urbanized areas as habitats for humans and for other organisms. The growing literature is scattered widely. Urbanists in many fields will find the contributions by entomologists in this collection of value to their understanding of urban ecosystems. What insects can dwell in cities, and how to control those which become excessively abundant—without contaminating the rest of the environment with toxic chemicals—are questions of interest to contemporary society generally.

Several generalizations can be made concerning typical aspects of the urban habitat, although current knowledge is far from complete. Recurrent patterns of topography, soils, climate, vegetation, and biota are associated with urbanized areas. Exceptions to the generalizations also may be worth noting.

The landforms present in an urban region greatly affect the spatial arrangement of human activity that is undertaken there. Modern industrial societies can transform slopes and drainage radically if there is an economic or political incentive to do so. But the preurban topography almost everywhere influences the configuration of land that is built up first and of the intervening islands of more nearly natural vegetation associated with cliffsides, watercourses, and marshes. Over time such areas may be reduced by continuing urban growth, if they are not protected by effective political land-use controls. The trend is toward ever smaller islands of urban land not given over to intensive human use.

When land is urbanized, its soils are altered radically. Erosion may precede urbanization on land that is farmed or cleared for fuel or timber. During urban construction, soils are drained and their structure is disturbed. Substantial erosion is typical. After construction the erosion slows. Hard pavements and buildings seal off great expanses of soil from the flux of energy, moisture, and biotic activity, and the habitat is essentially lost to most soil-dwelling organisms. Where soils are not covered, they may be packed hard by the constant passage of people or machines. In suburban areas around modern American or European cities, soils may be preserved beneath lawns and their biota may be stimulated by water and nutrients from septic tanks. Suburban soils may be unlike the natural soils of a region. Their structure may be altered greatly during development or may be changed by residents to accommodate cultivated plants. They may be treated with fertilizer and insecticides but are unlikely to accumulate leaf litter and its complex fauna.

Urban climate has two basic characteristics by which it differs from the climate of a city's hinterland. The first of these is the distinctive urban heat-water balance; the second is the air pollution endemic to the modern urban-industrial complex.

Cities outside the arid regions are warmer and less humid than their rural surroundings. They exhibit extensive stone and asphalt surfaces, and they require massive combustion of fuels. Hence cities have measurably warmer air than their surroundings, and the intensity of the difference tends to increase with the size of the urban agglomeration. Consequently, when the warm urban air rises, a mild local breeze blows inward toward the city from all compass directions when it is not overpowered by regional winds. The availability of the higher recorded precipitation in urban areas, however, is probably more than offset by rapid artificial drainage. Snow covers the ground for a shorter period than in surrounding rural areas. In desert or semidesert regions, cities may be oases where irrigation maintains more vegetation and a moister atmosphere than in the surrounding countryside. Indoors, of course, the range of variation in temperature and humidity generally is less than outdoors. Indoor climate is controlled to suit the preferences of the human inhabitants.

Air pollution is a much studied aspect of the urban climate. Emissions from stationary sources and from vehicles are greater than can be dispersed rapidly most of the time in many North American and European industrial cities. Thus concentrations of particulate matter and gases may reach levels unhealthy to people and other organisms. Spectacular episodes may result when stagnant air persists for a number of days. General atmospheric circulation patterns dictate the general frequency of stable, subsiding air masses; local topographic conditions also play a major role in the actual experience of air pollution incidents. The release of particulates into the atmosphere has been controlled during recent decades, and progress is notable in cities with extensive heavy industry (e.g., Pittsburgh, Pennsylvania). Gaseous pollutants are less tractable generally than particulate matter, and those produced by motor vehicles are especially costly to control. Urban pollutant concentrations are measurable in the atmosphere for tens and hundreds of miles downwind from urban areas.

Vegetation in urban areas typically differs from that of surrounding regions in quantity, composition, and management practices. In cities, vegetation is consigned to the open space not occupied by buildings or hard surfaces. Near the center of an urban region most open space is publicly owned; in the suburbs it may be owned by many private individuals. Urbanized areas may have more forested land than surrounding agricultural regions from which forests have been cleared, as in eastern North America. Alternatively, urban areas may support more trees than the countryside in generally treeless regions such as the midlatitude grasslands or deserts. Vegetation beneath the tree canopy typically is less abundant in the managed landscapes of urban areas than in unmanaged ecosystems.

The kinds of plants found in cities may be quite distinct from those of nonurban regions. Relatively few native plant communities survive in urban areas; they are replaced by cultigens and weeds. The potential range of cultivated plants is vast, as demonstrated by the immense diversity of species displayed in botanic gardens and arboreta. But the actual diversity of commonly cultivated plants actually is far smaller than the potential, at least in American cities for which data are available. For trees, shrubs, and herbs a handful of species accounts for the overwhelming preponderance of individual plants.

Management of urban vegetation is fragmented among numerous agencies and private landowners, unlike forest, ranch, or farm regions where hundreds or thousands of acres may be under a single management. This means that the patches of vegetation across an urban region may be managed for diverse purposes and goals. One manager may control weeds and insects closely; another may allow both to develop without interference. In cities of the United States the central reason for vegetation is ornamental; some European cities, however, derive economic products from urban forests. The extent of human interference in biological processes may vary sharply over short distances, even in urban vegetation that presents a similar physiognomic appearance.

Animals that inhabit cities are those able to adjust their behavior patterns to human activity, to utilize patches of open or woodland-edge habitat, to avoid recognizable competition with people, or to attract human appreciation and esteem. Others, especially some insects, are able to thrive inside buildings and to tap man's own food supply surreptitiously. The diversity of wild species that can survive in urbanized areas may be surprising to urbanites who do not customarily study natural history. One compilation for London suggests that a substantial proportion of the British fauna can persist in the urban vicinity (Table I). There is no reason to regard Table I as an exhaustive list; it merely

Table I. The Biota of Contemporary London[a]

Kind	Number of species: Seen in London	Number of species: Seen in the United Kingdom	London species as percentage of United Kingdom species
Higher Plants	1835	3000	61
Insects			
Hemiptera-Homoptera (bugs)	317	390	82
Coleoptera (beetles)	248	3700	7
Macro Lepidoptera (moths, butterflies)	728	930	78
Diptera (true flies)	2300	5200	44
Fishes (fresh and brackish water)	33	45	73
Amphibians	8	12	66
Reptiles	6	10	60
Birds	203	301	66
Terrestrial mammals			
Insectivora (shrews, moles, hedgehogs)	5	6	83
Chiroptera (bats)	10	15	66
Lagomorpha (hares, rabbits)	2	3	66
Rodentia (squirrels, voles, rats, mice)	8	17	47
Carnivora	4	11	36
Artiodactyla (deer)	3	9	33

[a]Compiled by Gill and Bonnett (1973) for London as defined by a circle of 32 km radius centered on St. Paul's Cathedral (about 3200 km^2).

reflects existing information on one urban region that has received a substantial amount of attention from local naturalists.

The biomass of urban animals in general can be hypothesized to be less than that of rural nature preserves, because the biomass of vegetation is relatively small in urban areas. Yet particular animals may become more abundant in urban areas than elsewhere. House sparrows, rock doves, rats, and feral dogs are a few examples. Formerly rare organisms may become more abundant as a result of urban and industrial growth. For example, after air pollution eliminated the lichens that camouflaged light-colored British moths from predators, the formerly rare, dark-colored mutants became abundant in the countryside downwind from heavy industrial districts.

It is up to entomologists to promote public understanding of the diverse and complex world of insects that persists in urban areas. At present few urbanites know or want to know about the insects that share their environment other than the simplest way to eradicate noxious pests. Urbanites understandably are reluctant to share their homes with pests or to see their scarce vegetation defoliated. But the simplest chemical means to exterminate a pest may affect many species other than the target, and may contaminate the human environment as well. As the scientific understanding of insects and their management increases, urbanites can expect better methods to restrain overabundant insects, yet encourage those which provide economic benefit or aesthetic pleasure as part of the changing seasons. Entomologists must call upon other scientists to provide data useful to their work with insects, so that all can contribute ultimately to the survival of man in the urban milieu.

Key Sources on Urban Environment

Andresen, J. W. (ed.) (1977). "Trees and Forests for Human Settlements." Centre for Urban Forestry Studies, University of Toronto, Toronto, Ontario, Canada.

Detwyler, T. R., and Marcus, M. G. (eds.) (1972). "Urbanization and Environment: The Physical Geography of the City." Duxbury Press, Belmont, California.

Frankie, G. W., and Ehler, L. E. (1978). Ecology of insects in urban environments. *Annu. Rev. Entomol. 23,* 367–387.

Gill, D., and Bonnett, P. A. (1973). "Nature in the Urban Landscape." York Press, Baltimore, Maryland.

Hay, C. J. (1977). Bibliography on Arthropoda and Air Pollution. USDA Forest Service, General Tech. Rep. NE-34.

Noyes, J. H., and Progulske, D. R. (eds.) (1974). "Wildlife in an Urbanizing Environment." Planning and Resource Development Series 28. Cooperative Ext. Serv., University of Massachusetts, Amherst, Massachusetts.

Santamour, F. S., Jr., Gerhold, H. D., and Little, S. (eds.) (1976). "Better Trees for Metropolitan Landscapes." USDA Forest Service General Technical Report NE-22.

Schmid, J. A. (1974). The environmental impact of urbanization. *In* "Perspectives on Environment" (I. R. Manners and M. W. Mikesell, eds.), pp. 213–251. Association of American Geographers, Washington, D.C.

Schmid, J. A. (1975). "Urban Vegetation: A Review and Chicago Case Study." University of Chicago, Department of Geography, Research Paper 161.

James A. Schmid
Jack McCormick & Associates
Berwyn, Pennsylvania 19312

PREFACE

Much has been said in recent years about the emerging field of urban entomology, yet there is less than complete agreement as to what is meant by the term. In conceptual terms, urban entomology should be antithetical to rural entomology but in strict terms it is not, as some chapters in this volume will verify. To most persons, urban entomology simply refers to the study of insects, including their management, in urban environments. Whatever definition is used, however—and several chapter authors have offered their perspectives on this point for their particular sphere of interest—most who have had contact with the field would agree that in no other area of entomology is man more involved with insects, and insects with man, in such a wide variety of ways. In this connection, at one extreme are those persons who recognize and appreciate the aesthetic value of selected insect species, while at the other end of the spectrum are those who fear the presence, or even the threat of the presence, of any insect. The opinion of the majority, however, would likely fall between these extremes.

The question of why urban entomology is finally emerging is an interesting one. For years entomologists have given attention to the applied problems of insects and associated organisms as they impact on man in urban environments. The early text and reference books on shade tree insects by Felt[1] and Herrick[2] certainly must be considered contributions to urban entomology although the term was not used then. Many of the contributions to our knowledge of pests of medical importance similarly fall into this category.

Very possibly economics is the root cause of the recognition urban entomology is accorded today. By virtue of employment of most of our entomologists in the public sector, with support provided from tax dollars, contributing service and information to those who provide this support can only help to preserve this relationship. In most developed countries such a small proportion of the population is now directly engaged in agricultural production that the tax support base resides principally in the urban sector. An associated factor contributing to the emergence of urban entomology is the ecology and environmental movement. People want to know more about the plants and animals

[1]Felt, E. P. (1924). "Manual of Tree and Shrub Insects." Macmillan, New York.
[2]Herrick, G. W. (1935). "Insect Enemies of Shade Trees." Comstock Publ., Ithaca, New York.

around them, and about the various options for improving or protecting their surroundings. Quite logically, insects and related organisms have been newly discovered by this more perceptive urban populace.

Quite apart from the question of why urban entomology is today gaining new prominence, many other questions of a practical and philosophical nature deserve attention. Do concepts, principles, and theory have a place in this field? Should urban entomologists be expected to make basic as well as practical contributions to science? What boundaries should limit the kinds of activities and problems with which urban entomologists become involved? Where is the "pulse" of the clientele group we are attempting to serve, and if found, should lay involvement be invited to participate in setting research or other priorities for the urban entomologist? What kinds of collaborative arrangements, if any, should the urban entomologist seek with those in other sciences, including the social sciences? How should the prospective urban entomologist be trained? How can the urban entomologist best communicate research findings to the general public?

In this volume, 17 chapters provide insight to some of these and other relevant questions. The contributions resulted largely as an outgrowth of a symposium entitled Ecology and Management of Insect Populations in Urban Environments, held during the Fifteenth International Congress of Entomology, 1976, in Washington, D.C. The contributors were requested to provide a broad account of their respective topics. Exploration of appropriate concepts and principles and speculation on future trends were encouraged. Each was requested to emphasize the relationship of man to his particular entomological topic. The topics represent the diverse characteristics of urban entomology; the book is not a catalog of all relevant subject areas. Examples of additional topics of interest to entomologists working in urban environments include endangered insect species, entomophobia and delusions of parasitosis, and insects associated with turfgrass and soil, with food processing plants and dining establishments, with stored products, and with urban aquatic habitats.

We fully expect additional coverage of the subjects of which urban entomology is composed to appear in the years ahead. It is hoped that the readers of these chapters will gain insight to the state of the science as of this year. For those readers who are or will become urban entomologists we earnestly seek answers to the questions raised earlier and to new questions yet to be asked.

We would like to acknowledge Jutta Frankie who typed and assisted in the copyediting of this volume.

The Editors

PERSPECTIVES IN URBAN ENTOMOLOGY

ECOLOGY OF BUTTERFLIES IN THE URBS AND SUBURBS

Michael C. Singer
Lawrence E. Gilbert

Department of Zoology
University of Texas
Austin, Texas

Information on the ecology of butterflies in urban environments is sparse, and we have broadened our scope to discuss the impact on butterflies of man's various modifications of the landscape, including urbanization. We shall attempt to describe in a qualitative way the interactions between regional land use and climate as they affect different butterfly species.

I. EFFECTS OF MAN ON ECOLOGICAL REQUIREMENTS OF BUTTERFLIES

The ecological needs of butterflies are not well understood. For many species one can attempt to define the param-

ISBN 0-12-265250-9

eters of a suitable habitat in terms of climatic conditions and such resources as host plants of larvae, nectar supply for adults, sites for pupation or diapause, etc., and yet still one finds many habitats which seem to meet these criteria but will not support a population of the species concerned. In consequence attempts to manipulate, say, a nature reserve to render it suitable for some endangered butterfly species stand a rather low chance of success unless the species has occurred at that site in the past, and, even in such cases, may prove problematical (e.g., *Lycaena dispar* Haworth in England - Ford, 1945; Duffey, 1968).

A further consequence of our lack of understanding in this area is that we cannot always specify what aspect of an urban environment renders it unsuitable for a butterfly species. As an example, species of hackberry butterfly (*Asterocampa* spp.) are relatively rare away from natural or seminatural vegetation in central Texas, yet the larval hosts (*Celtis* spp.) grow abundantly as "weedy" native trees in the yards of Austin. Adult *Asterocampa* are known to be opportunistic in their feeding habits, utilizing flowers, fruit, sap flows and exudates from fungus-infected plants, some of which are available in suburban areas for most of the year.

Some sedentary butterflies require resources for both adults and larvae to occur in close proximity. For example, members of some populations of the California checkerspot, *Euphydryas editha* Bsd. in the San Francisco region require juxtaposition of nectar sources, diapause sites and two different hosts for the larval stage (Singer, 1972). Nowhere have these resources remained intact as land has been developed, and many local populations of the butterfly have become extinct. However, in other parts of California such as San Diego some natural populations of *E. editha* occupy temporary, early succession habitats and require only one larval host plant. In one area the required assemblage of larval host and nectar-providing plants appears to be maintained solely by human disturbance (Gilbert and Singer, 1973). In such areas the adult insects have evolved a greater tendency to disperse and the disappearance of habitats does not eliminate the insect from the region, since new habitats which appear can be colonized rapidly.

In contrast, other species typically commute between nectar sources and larval hosts or may traverse large tracts, constantly relocating these resources. Such species (e.g., many *Pieris* and *Papilio*, and *Danaus plexippus* L.) need to be efficient at finding mates, nectar, and oviposition sites. They require their larval hosts only just long enough for completion of a single generation. Evidently, they are better preadapted to urban life than even the San Diego race of *E. editha* which still requires juxtaposition of adult and larval resources, with habitat patches that last for at least several years.

Thus, patterns of movement of butterflies and their resource requirements are important as determinants of their response to habitat disturbance. The existing literature on these topics has been reviewed by Gilbert and Singer (1975). The temporal pattern of resource availability is also important, and one problem with some urban habitats may be that larval and adult resources are not available in the correct sequence at the "right" times of year. Human activity may either prolong the availability of a resource or restrict it. An example of the first is given by Owen (1971). He describes how the trimming of shrubs in African gardens provides year-round new growth and a more constant supply of food for lycaenid larvae than would be obtained in the surrounding forests. Awadallah *et al.* (1970) give details of the life cycle of an Egyptian lycaenid, *Virachola livia* Klug., which can breed throughout the year by switching between five different hosts, some of which are crop plants. This may be a general phenomenon in Mediterranean climates, where gardening and agricultural irrigation reduce the effects of the dry season and may either allow more generations per year of insects already present or may render habitats suitable for introduced or migratory species. Examples in California are skippers, which as larvae feed on lawn grasses; *Colias* spp. on alfalfa and *Agraulis vanillae* L. which as a larva feeds on *Passiflora* growing in gardens and as an adult utilizes nectar also supplied by man in midsummer when neither resource is present in the surrounding habitats. The same phenomenon (i.e., the supplementation and prolongation of butterfly resources by human activity) has been observed by one of us (LEG) in the case of *Dione juno* Cramer at a comparable southern latitude in Lima, Peru.

Many of man's environmental manipulations restrict the length of time for which butterfly resources are available. Habitat patches may endure for only a few seasons; grass-cutting, burning, herbicide application, hedge-clipping or tilling may shorten the time within each season that larval hosts can be utilized. Shortened life of habitat patches will give rise to selection pressure for greater mobility, while shortening of the effective growing season will produce selection for shorter generation time to the extent that individual larvae fail to mature before their host-plant is removed. Mowing, etc., may have a less severe effect in tropical climates, where some proportion of the insects will usually be not only in the adult stage but also alert and able to escape.

As well as influencing the timing of availability of resources for butterflies, man also destroys and creates resources. Resource destruction occurs largely as a consequence of total habitat destruction (or application of selective herbicides). Heath (1974) regards this process (e.g., draining of marshes) as responsible for most changes in range and extinction of British Lepidoptera in the past 100 years, though climatic factors may also have some influence (Ford, 1954; other references in Gilbert and Singer, 1975). Habitat destruction is also held responsible for the extinction of the "xerces blue," *Glaucopsyche xerces* Bsd., which was endemic to the San Francisco Bay region of California (Pyle, 1976). The removal of rain forests from many parts of the world has likewise reduced butterfly faunas (Brown, 1976). Short of complete destruction, the manicuring of potential habitats such as road margins and stream banks, and the reduction of plant species diversity in commercially grown forests, remove plant species necessary for butterfly survival and affect the availability of suitable sites for diapausing stages. Many species in semi-arid South Texas rely on stream bottom refuge areas for persisting through both the regular midsummer drought and periodic dry years. Even dry stream beds are important as roosting areas (e.g., for *Battus philenor* L., *Danaus gilippus* Cramer) as adult diapause sites (e.g., for *Kricogonia* and *Libythea*; Gilbert, unpubl. observations) and as sources of flowers and new growth for larvae long after most of these resources have disappeared from the region generally (e.g., for *Asterocampa* and *Libythea*). In extremely dry periods this assemblage of species retracts towards major river bottoms, while in wetter periods it expands into the brushlands. Since dry creek bottoms are often the first areas where brush is removed because of their greater potential for grass production, brush control programs are threatening the pattern of existence of these species, and damming of rivers may have similar harsh effects.

On the opposite side of the coin, man also provides new resources for butterflies. There are several recorded instances of butterflies including in their host ranges or even preferring introduced plants. One example is the pierid *Anthocaris cardamines* L. which is reported to oviposit on a crucifer planted in British gardens (Owen, Chapt. 2, this volume), an event which seems to be enabling this typically rural species to become at least partly suburban. Owen (1971) also cites *Papilio demodocus* Esp. in tropical Africa as having switched host preference almost entirely to planted *Citrus* and only rarely utilizing the original hosts - wild species of Rutaceae.

New habitats are constantly being created for butterflies specifically adapted to secondary successional stages. There appear to be many more such species in the tropics than in the temperate zones. Owen (1971) used three trapping methods to estimate butterfly species diversity in a tropical garden and found more species than would occur in this region in either natural forest or grassland habitats. Many of these species completed their life cycle in the garden. Owen noted that *Charaxes* spp. and satyrines in particular were more diverse in his garden than elsewhere. Some neotropical satyrines also seem to be especially adapted to areas disturbed by man (Singer and Ehrlich, unpub. observations).

One further example of new resource provision by man is the building of outhouses, garages, etc., which are used as diapause sites by adult butterflies (such as *Aglais urticae* L. in Europe). The species that use these man-made structures derive considerable protection from predators and weather but suffer some risk in becoming trapped on emergence from hibernation.

II. EFFECTS OF HUMAN ACTIVITY ON PREDATORS OF BUTTERFLIES

Feeding the birds in winter is a widespread activity in Britain. As a result, bird mortality is reduced (at a time when natural food resources are at their lowest point), and higher year-round populations of birds occur in urban areas than would otherwise exist. This effect may underlie Baker's (1970) finding that birds were the major predators of eggs and larvae of *Pieris rapae* L. in his British garden, while in agricultural habitats arthropod predation is paramount.

Cereal growing in the North Temperate Zone often produces a large biomass of aphids just prior to harvest. In some years in southwest Britain (e.g., 1976), coccinellid beetles are able to complete their first generation on aphids associated with the cereals, emerge as adults, and migrate in search of food. During the first week of July, 1976, one of us (MCS) observed *Coccinella septempunctata* L., which had been generated in the agroecosystem, descending en masse on every other type of habitat, including thick woodland, where they were foraging over every plant surface for small, soft-bodied insects. Any butterfly species which happened to be present at this time only as first- and second-instar larvae would have run severe risk of population extinction by these predators, no matter how apparently stable its habitat may have been.

In East Texas, the mowing of road margins and pipeline tracks has created grassland habitats suitable for the nests of the imported fire ant, *Solenopsis invicta* Buren, which is a voracious predator, foraging up to 200 m into adjacent woodland (Gilbert, personal observations). It should be of concern to conservationists that the impact of *S. invicta* on ground-level arthropods (including some butterfly larvae) seems to be most extreme within the bounds of state parks and state forests where mirex application was prevented (Hooper, unpub. observations).

Effects of parasitoids on butterflies in urban environments have not been assessed and research in this area is badly needed. Hessel (1976) found high densities of the hairstreak

Atlides halesus Cr., in the Sacramento Valley of California and attributed this to escape from parasites, resulting from locally high rates of pesticide application. In general, *A. halesus* may be more successful in close proximity to man. It seems more conspicuous in the city of Austin (Texas) than in surrounding natural habitats; however, its host (mistletoe) is common in all habitats in south central Texas. A more constant supply of flowers for the adult in the city may be a major factor for this observed difference.

One of us (LEG) has observed intense predation on larvae of *Papilio polyxenes* F. in Austin by the wasp *Polistes exclamans*, a species which may be assisted in urban areas by the provision of abundant nest sites on buildings. *Polistes* and other hymenopteran predators and parasites of early butterfly stages may also be supported by man's provision of adult nectar in the form of *Ligustrum* and other flowering shrubs of yards.

III. EVOLUTIONARY EFFECTS OF HUMAN ACTIVITY

The evolution of dark wings by *Biston betularia* L. in response to pollution-darkened substrate, or the evolution of DDT resistance in many insects are relatively simple changes genetically. In contrast, the evolution of several radically new life history tactics by a butterfly population faced with major habitat disruption would necessarily involve many genes changing simultaneously in the appropriate direction. As a consequence, species with populations that are small, sedentary, and which have complex resource requirements should rapidly become extinct as habitats are drastically altered. Conversely, species with populations which have evolved to exploit disturbed, temporary habitats in nature should not be so drastically affected by habitat manipulation as would ecologically more specialized species.

As would be expected from genetic considerations, the only known or suspected cases of evolutionary change in butterflies resulting from human activity involve habitat changes expressed one dimension at a time. For example, *Apatura* in Britain is thought to have altered its adult feeding behavior over the last century (Heslop *et al.*, 1964). The selective agent in this instance appears to be increased bird predation of adults feeding exposed on fecal matter, but the use of bait by collectors of this highly prized insect may have played a selective role as well. Incidentally, the increase in bird predation in the woodland habitat of *Apatura* probably resulted from a decline in the control of magpies and jays by game keepers.

A second case involves a possible adaptation by a butterfly to the progressive reduction of its area of natural habitat. Dempster *et al.* (1976) have shown a gradual decrease in the wing length of *Papilio machaon* L. that is correlated with the shrinkage and increasing isolation of its habitat. If, as these authors suggest, reduction in wing length reduces dispersal by individuals, we have been observing an evolutionary reduction in dispersal ability, which is a well-known phenomenon among birds, plants, and insects on oceanic islands.

A third probable instance of evolution by a butterfly in response to human activity is the shift of a riodinid butterfly from unknown native hosts to *Eucalyptus* during the first 75 years of its planting in Brazil (Anon., 1968). In this case it may be the increasing "target size" of the new resource as much as the shrinkage of natural habitat which sets the stage for the evolution of a new host preference. Gilbert (1978) discusses in considerable detail how this kind of evolutionary change in butterflies may occur.

None of these presumed cases of evolution resulting from human alteration of butterfly habitats is exempt from alternative explanations and all deserve additional study. More generally we need much additional information on the nature of genetic differences which exist between geographically and ecologically distinct populations of particular butterfly species (see Ehrlich *et al.*, 1975). We have good circumstancial evidence that interpopulation differences in *Euphydryas editha* are at least partly genetic (Gilbert and Singer, 1973) and egg size, clutch size, adult body weight have evolved to suit local circumstance (Singer and Gilbert, unpub. observations). Better understanding of the selective forces involved in such intraspecific differentiation of life history traits of butterflies will give us a basis for predicting the way that particular habitat alterations will affect particular butterfly species.

IV. DISCUSSION AND CONCLUSION

A central but ignored problem of urban ecology involves explaining why particular organisms are absent. We prefer to concentrate on explaining why certain species manage to survive. But the more troublesome question of absence involves a level of knowledge of the natural history of a species and an understanding of the nature of the habitat, which has been obtained in few cases.

Interest in insects that inhabit urban areas has been directed either to control of species which are commonly present or (rarely) to helping species persist that usually are not present. To be on secure footing as a scientific discipline, urban entomology should eventually be based on theory which is sufficiently robust to answer basic questions of both commonness and rareness.

The necessary conceptual framework is provided by Southwood (1976) who used broad terms to show how habitat acts as the "templet" against which the life history strategy of an insect will evolve. Southwood classifies habitats according to the following characteristics:

(1) *Durational stability*, the length of time the particular habitat type remains in a particular geographic location.

(2) *Temporal variability*, the extent to which the carrying capacity of the habitat varies during the "duration."

(3) *Spatial heterogeneity*, the degree of spatial patchiness of the habitat.

Obviously, "urban environment" is no more meaningful than "forest environment" in rigorously describing the actual habitat as "seen" by an insect of a particular life history type. Regional and local climate, the way surrounding land is managed, the horticultural habits and affluence of the populus, and the size and shape of urban areas are just a few obvious variables that have to be considered when classifying urban habitats in Southwood's general terms.

To make things more complex, how a particular site will be classified will depend upon the life history and resource needs of the insect chosen for study. A phytophagous insect like a butterfly may only view native vegetation as suitable habitat. Thus, an urban center would be viewed as a series of habitat islands, the carrying capacity of each dependent on host plant density and local mowing or herbicide schedules, with average durational stability dependent upon rates of real estate development. In contrast, a house-dwelling species of cockroach would see the same urban area as an effectively continuous habitat with micro-site variation in carrying capacity (depending on food and garbage handling habits of families), but with great durational stability. On the other hand, the same species of cockroach 50 km from the city might view a trailer park as the only suitable habitat available for a considerable distance.

These considerations lead to the inevitable conclusion that to understand how a particular episode of urbanization will affect insects, such as butterflies dependent on natural vegetation, we must understand the detailed resource requirements of each species and how urbanization alters the spatial and temporal patterning of these resources. Basic ecological studies of a species over an adequate sample of its natural habitat(s) provides the information required to develop initial management procedures for conserving this biological heritage. However, we have given examples above of cases where appropriate larval and adult resources are present but a butterfly is still absent or rare. Such cases should receive special attention from urban entomologists since in essence some sort of rather effective biological control is occurring. Elucidation of that control may have general implications for the control of similar insects which are pests of horticultural plants.

We further suggest that the study of urban butterflies be applied to the problem of managing urban remnants of natural vegetation. Because butterflies are sensitive yet conspicuous indicators of particular host plants as well as other complex and less tangible ecological requirements, designing management procedures which would maintain diverse assemblages of native butterflies along streams, in parks, along freeways, etc., would, no doubt, simultaneously maintain many other native insects and plants (see Gilbert, 1977). In particular, it might be possible to effect changes in many haphazard habitat management practices currently carried out by city and county governments such that faunal and floral diversity is maximized without altering the short-term "usefulness" of the area being managed.

POSTSCRIPT

Gross ignorance (even fear) of insects and native plants ("weeds") by the general public, including government officials, stands in the way of the kind of habitat management which would allow phytophagous insects, having complex life cycles, to persist in the face of urbanization. Particularly in the U.S. where we can afford it, there is a level of maintained habitat alteration involving private as well as municipal lands which pathologically exceeds the human need for a safe, predictable environment. Indeed, "the ecology of urban butterflies" is a nonsubject in many urban areas of the U.S. - unless one wishes to study the effects of nitrous oxides on migrating monarchs.

We feel that it is in the long-term interest of our species to hang on to any vestige of primeval diversity. As urban areas expand, these last remnants of nature will be increasingly surrounded by urbs, suburbs and the agricultural areas which support them. It is thus important to immediately begin to save urban patches of natural vegetation from arbitrary destruction so that urban ecology can develop beyond the study of cockroaches, termites, and other pests.

The most rapid way to change public policy toward natural habitats within urban areas is to educate the public concerning the requirements of aesthetically pleasing organisms such as butterflies. A notable attempt to educate the public on the enhancement of urban lawns for butterflies is the article and host plant chart by Julian P. Donahue of the Los Angeles County Museum (see Donahue, 1976). See also Owen (1975).

Finally, there are now serious attempts being made to save endangered butterfly species through habitat preservation. Since most cases of extinction are related to the loss of natural habitat to urban expansion, those concerned with preserving butterfly species must quickly discover the minimal ecological conditions which will both support a particular butterfly and be compatible with the certainty of urban expansion. The Xerces Society and its journal, *Atala*, is the most significant source of information on current nationwide attempts to preserve urban butterfly diversity. *Atala* may be obtained by writing to Jo Brewer, 300 Islington Road, Auburndale, MA 02166, U.S.A. *Atala* is rapidly becoming a rich source of information on the ecology of urban butterflies and should be read by all concerned with this topic.

REFERENCES

Anonymous. 1968. Report of Servicio Forestal da Cia Siderugica Belgo - Mineira, July 1968 (Brazil) (mimeo).

Awadallah, A. M., A. K. Azab, and A. K. M. El-Nahal. 1970. Studies on the pomegranate butterfly, *Virachola livia* (Klug.). *Bull. Soc. Entomol. Egypt 54*: 545-567.

Baker, R. R. 1970. Bird predation as a selective pressure on immature stages of the cabbage butterflies, *Pieris rapae* and *P. brassicae*. *J. Zool. 162*: 43-59.

Brown, K. S. 1976. Geographical patterns of evolution in neotropical forest lepidoptera. *In* Biogeographie et evolution en amerique tropicale (H. Descimon, ed.). Publ. du Laboratorie de Zool. de l'Ecole Normal Superieure No. 9, Paris, pp. 118-160.

Dempster, J. P., M. L. King, and K. H. Lakhani. 1976. The status of the swallowtail butterfly in Britain. *Ecol. Entomol. 1*: 71-84.

Donahue, J. P. 1976. Take a butterfly to lunch. A guide to butterfly gardening in Los Angeles. *Terra 14*: 3-12.

Duffey, E. 1968. Ecological studies on the large butterfly *Lycaena dispar* Haw. at Woodwalton Fen National Nature Reserve, Huntingtonshire. *J. Appl. Ecol. 5*: 69-96.

Ehrlich, P. R., R. R. White, M. C. Singer, S. W. McKechnie, and L. E. Gilbert. 1975. Checkerspot butterflies: A historical perspective. *Science 188*: 221-228.

Ford, E. B. 1945. The butterflies. Collins, London.

Gilbert, L. E. 1977. The role of insect-plant coevolution in the organization of ecosystems. *In* Comportement des insectes et milieu trophique (V. Labyrie, ed.). C.N.R.S. Paris, pp. 399-413.

Gilbert, L. E. 1978. Development of theory in the analysis of insect-plant interactions. *In* Analysis of ecological systems (D.J. Horn, R.D. Mitchell, and G.R. Stairs, eds.). Ohio State University Press, Columbus.

Gilbert, L. E., and M. C. Singer. 1973. Dispersal and gene flow in a butterfly species. *Amer. Nat. 953*: 58-72.
Gilbert, L. E., and M. C. Singer. 1975. Butterfly ecology. *Ann. Rev. Ecol. Syst. 6*: 365-397.
Heath, J. 1974. A century of change in the Lepidoptera. *In* The changing flora and fauna of Britain (D. Hawksworth, ed.). Syst. Assoc. Spec. Vol. No. 6, Academic Press.
Heslop, I. R. P., G. E. Hyde, and R. E. Stockley. 1964. Notes and views of the purple emperor. Southern Publ. Co. Ltd., Brighton.
Hessel, S. A. 1976. A preliminary scan of rare and endangered neartic moths. *Atala 4*: 19-21.
Owen, D. F. 1971. Species diversity in butterflies in a tropical garden. *Biol. Conserv. 3*: 191-198.
Owen, D. F. 1975. Suburban gardens: Englands most important nature reserve? *Environ. Conserv. 2*: 53-59.
Pyle, R. M. 1976. Conservation of the Lepidoptera in the United States. *Biol. Conserv. 9*: 55-75.
Singer, M. C. 1972. Complex components of habitat suitability within a butterfly colony. *Science 173*: 75-77.
Southwood, T. R. E. 1976. Bionomic strategies and population parameters. *In* Theoretical ecological principles and applications (R. May, ed.). Saunders, pp. 26-48.

INSECT DIVERSITY IN AN ENGLISH SUBURBAN GARDEN

Denis F. Owen

Department of Biology
Oxford Polytechnic
Headington, Oxford, England

I. THE SUBURBAN GARDEN SCENE

The English countryside with its small fields, hedges and tiny patches of woodland is not "natural" in the sense normally understood. Centuries ago the land was covered by oak forest, but this has now almost gone, and instead the landscape is man-made and designed for food production, for industry and for transport. The need for houses, towns and roads is accepted, and there is seemingly no end to efforts at changing and modifying the countryside to suit ever-increasing demands.

Most English people live in urban or suburban areas, and their only regular contact with nature is through gardening. No other outdoor activity is as popular. Millions spend time and money looking after their own patch of land; indeed, gar-

ISBN 0-12-265250-9

Fig. 1. A view of part of the Leicester garden.

to nectar seeking insects and this no doubt accounts for the abundance and diversity found to be present in these groups. The overall impression is of a small area of land with an incredible variety of plants, more than would be found in a natural area of comparable size.

The contrived plant diversity typical of English gardens is further augmented in mine by an avoidance of pure stands, even of vegetables. Food crops are grown intermixed with flowers in an attempt to minimize outbreaks of pests by maintaining a high diversity of species of plant feeding insects. Few species of insects are really common and there are rarely outbreaks of pests. Hence, I have found it unnecessary to use pesticides, and nor do I use herbicides and commercial fertilizers. Every effort is made to maintain a complete ground cover of vegetation and many plants regarded as weeds by most gardeners are tolerated.

III. TWO IMPORTANT FOOD WEBS

A garden as varied as mine might be expected to support a variety of food webs in such complexity that to sort them out would require a lifetime. But even casual observations indicated that of all the varied food webs present, two are outstanding in terms of number and variety of insects associated with them.

The first and most obvious is the association of aphids with the variety of species of plants. Many (but not all) aphids are host specific to an extent that almost every species (or group of related species) of plant supports a dependent species of aphid. Other aphids exploit a wide variety of plant species, some of them having become adjusted to feeding on exotic plants that have no close relatives in the native British flora. Every spring there is a buildup in aphid numbers, the magnitude of which depends primarily on the weather: fine, warm springs being better for aphids than cool, damp ones. Aphids are exploited by two main (there are others) groups of predators: the larvae of hoverflies (Syrphidae), and the larvae of ladybirds (Coccinellidae) which, in turn, means that adults of both hoverflies and ladybirds are abundant in the garden. Hoverfly larvae are exploited by parasitic insects, including the Ichneumonidae, and this is one reason for the extraordinary diversity of these insects now known to occur in the garden. I have not thus far investigated the parasites of ladybirds, but they exist, and I predict that they too are abundant and diverse.

The second food web, also intimately associated with plant diversity, is the variety of species of caterpillars of the Lepidoptera, particularly the families Noctuidae, Geometridae, and Pyralidae. Here, as with aphids, there are species that range widely in their choice of food, but the majority are restricted to a relatively narrow selection of plant species. Many moth caterpillars have become adjusted to introduced plants and are nowadays rarely found on what are presumed to be their "natural" food plants. The caterpillars are exploited by several groups of parasitic insects, including the Tachinidae and the Ichneumonidae. I have not investigated the Tachinidae but the diversity of Ichneumonidae known must at least in part be determined by the diversity of Lepidoptera in this and adjacent gardens, and by the variety of ecological niches available to caterpillars.

IV. MONITORING THE INSECT FAUNA

The provisional identification of two important food webs enabled me to decide which insect groups should be monitored in order to obtain some estimate of the total diversity present. When monitoring began in 1972, it soon became obvious that it would be necessary to severely restrict the number of groups to be studied in detail; there were simply too many kinds of insects. I therefore chose certain groups because 1) they are part of the main food webs, 2) identification to species was possible, and 3) they are especially suitable for the type of quantitative sampling I had in mind.

Several different sampling techniques have been tried. The one that has consistently provided the best information is the Malaise trap (named after its inventor); an open-sided, tent-like construction of fine netting with an internal baffle of netting, supported by poles and strings. Flying insects that wander into the trap tend to fly upwards on encountering the central baffle, and eventually fall into a jar containing 70% alcohol attached at the apex. The trap's suitability for sampling flying insects depends on two features: no attractant is used, the only insects caught being those that fly into it of their own accord; and it can be operated continuously in all weather throughout the year. All insects that fall into the jar are of course killed, but the effect of the trap on the garden fauna is negligible because it samples an area of only 2.6 m^2 to a height of 1.1 m.

I also used methods that will be more familiar to collectors, including a mercury vapor light trap, and a special trap baited with rotten and fermented fruit. Records obtained from these and other techniques were supplemented by collecting with an ordinary insect net and, in the case of butterflies, by capturing, marking, and releasing.

Although I would be the first to admit that there is no such thing as a "normal" year, the weather for the first three seasons (1972 to 1974) was typical of what might be expected in central England: variable, with rain and sunshine, not especially hot, and rarely cool. The 1975 season was much drier than usual, and then in 1976 there was a severe drought. Indeed, the summer of 1976 was the longest and hottest that most could remember and, some said, the driest for at least 250 years. These two unusual summers were something of a bonus to my monitoring of garden insects for I had a run of three normal years followed by two exceptional ones, an event that is generally appreciated by an ecologist. The Malaise trap records showed how insects reponded to the drought and provided evidence of movements into the garden from the surrounding countryside.

V. DIVERSITY IN SELECTED GROUPS

A. Syrphidae

About 250 species of Syrphidae occur in Britain and 83 (about a third of them) have thus far been found in the Leicester garden. In 1972 to 1976, 16,782 individuals were caught in the Malaise trap. This figure approximates the number flying over 2.6 m^2 of garden and means that over four million hoverflies entered the area of the garden in the five years, a necessarily conservative estimate as the trap captures only those individuals flying less than 1.1 m above the ground.

Table I shows the number of individuals and species caught each year and the number of species added in each successive year. From these figures it seems possible that another five years of sampling will bring the total of individuals to over thirty thousand and of species to about a hundred.

Adult hoverflies occur in the garden from April to November and in four of the five years there was a peak in abundance in August. The exceptional numbers recorded in 1975 were the result of an immigration in August of three species, *Syrphus corollae* F., *S. balteatus* Deg., and *S. ribesii* L., all with aphid-feeding larvae. Presumably, they came in from the countryside. The fine weather that year allowed for a massive buildup of aphids on field crops and a consequent rise in the numbers of aphid predators, including the larvae of hoverflies. It is likely that the food supply was suddenly exhausted and there was nowhere for the females, which made up the bulk of the immigrants, to oviposit. There may also have been a shortage of nectar and the insects were thus forced to move into gardens where flowers were plentiful. The events of 1975 were repeated in 1976, but four weeks earlier in July, and thus both the 1975 and 1976 samples were distorted by exceptional numbers of a few species for a short period of time. In both years,

TABLE I Number of Individuals and Species of Adult Hoverflies Collected in a Malaise Trap in the Garden during Five Consecutive Years

Year	*Individuals*	*Species*[a]	*Species unrecorded in previous years*	*Running total of species*
1972	*1,339*	*47*	-	*47*
1973	*3,164*	*62*	*23*	*70*
1974	*2,205*	*54*	*3*	*73*
1975	*6,363*	*55*	*4*	*77*
1976	*3,411*	*55*	*4*	*81*

[a]*Two species not recorded in the Malaise trap were collected in the garden, one in 1971, the other in 1973 and subsequently. This brings the total species to 83.*

the warm summer was responsible for the chain reaction of events that occurred.

The life histories of many of the British species of hoverflies are known, at least in outline, and it is thus possible to divide the Malaise trap sample according to the trophic position of the larvae. Results for 1972 to 1975 are shown in Table II where I recognize five feeding categories: one group of secondary consumers, one group of primary consumers, and three of primary to nth order decomposers. The aphid-feeding secondary consumers dominate the sample with nearly 82% of the individuals and 65% of the species. Thus most garden hoverflies, viewed in terms of individuals and species, are dependent as larvae on aphids and as adults on nectar. Exactly how the species divide up their resources and to what extent there is ecological segregation between them is a matter for further investigation.

Most of the individuals and species in the primary consumer group are dependent as larvae on living bulbs. The remaining three groups are all decomposers of organic material; by far the most important being species of *Eristalis* whose larvae are associated with decaying plant material in damp or even aquatic sites. Few of these breed in the garden and the adults are abundant because they are attracted to garden flowers.

TABLE II Trophic Position of the Larvae of Hoverflies Trapped in the Garden during Four Seasons, 1972 to 1975

Food source of larvae	*Trophic position*	*Individuals*	*Species*
Aphids and other Homoptera	*Secondary consumers*	*10,943*	*50*
Aquatic vegetation, stems, roots, bulbs	*Primary consumers*	*945*	*7*
Tree sap, rotting wood	*Primary to nth order decomposers*	*5*	*4*
Manure and dead vegetation, often in stagnant water	"	*1,466*	*14*
Scavengers in wasp nests	"	*12*	*2*
	Total	*13,371*	*77*

B. Ichneumonidae

Ichneumons as a group are abundant in the garden but there are apparently few common species, and there is no evidence that numbers can suddenly rise, as with hoverflies. All species are parasitic on the larvae of other insects, especially the Lepidoptera and the Diptera. Identification to species is difficult and results (all from the Malaise trap) are available for only the first three years. Thus far 529 species have been identified, more than a quarter of the 2,000 species known from Britain; but included among them are species never previously reported from Britain and one or two that are apparently new to science. Evidently, ichneumons are rather poorly known as it is possible to turn up "new" or "rare" species in a country like Britain where knowledge of the fauna is generally assumed to be in an advanced state.

Most impressive is the extraordinary diversity of species, shown in Table III where results of trapping for the first two years are summarized. No species is common and many are rare; in each year 117 species were recorded once only which alone

TABLE III Massive Species Diversity in Ichneumonidae in the Garden

	1972	*1973*
Individuals	*2,495*	*3,950*
Species	*322*	*354*
Average number individuals/species	*7.8*	*11.2*
Species taken once only	*117*	*117*
Commonest species	Helictes *sp.*	Charitopes *sp.*
Commonest species as percent of total	*3.2*	*6.6*

attests to the rarity of most of them. If ichneumons reflect the diversity of species present in other groups of parasitic insects, including other parasitic Hymenoptera, it would appear that there are substantially more species of parasites than of hosts; most plant feeders must be utilized by several species of parasite.

C. Serphidae (Proctotrupidae)

These tiny insects are parasitic on a variety of other insects, particularly Coleoptera. Little is known about them and there is no doubt that the 32 species reported for Britain is an underestimate. The Malaise trap produced 20 species. Nothing is known about what they are doing in the garden nor of their ecological relationships with other insects. But the number of species found suggests that they are well established.

D. Bumblebees and Cuckoo Bees (Apidae)

Bumblebees, *Bombus* spp., are common and conspicuous flower visitors. These and their "mimics," the cuckoo bees, *Psithyrus* spp. (which, incidentally, have never been seen alive in the garden) belong to the Apidae, other genera of which are also numerous but have not been investigated in quantitative terms.

Nineteen species of *Bombus* and six of *Psithyrus* are recorded from Britain; the Malaise trap has thus far yielded eight species of *Bombus* and two of *Psithyrus*. Numbers of each species trapped in 1975 and 1976 are given in Table IV. The two years produced similar totals but the relative frequency of the species varied, as shown.

E. Coccinellidae

Only seven of the 42 species of ladybirds known from Britain have been found in the garden. All except *Thea 22-punctata* (L.) (which apparently feeds on mildew) are predators of aphids, and their fluctuations in numbers are similar to those observed in the aphid-feeding hoverflies. The relatively low number of species compared to other groups examined may arise because ladybirds are relatively sedentary and tend to be confined to specific habitats: only three of the British species, *Coccinella 7-punctata* L., *C. 11-punctata* L., and *Propylea 14-punctata* (L.) seem able to undertake substantial movements and to vary markedly in numbers from year to year.

In England, *Adalia bipunctata* (L.) is the common garden ladybird, familiar to everyone. The Malaise trap records (Table V) indicate that it occurs commonly every year with relatively small fluctuations in numbers. If, as I have suggested earlier, 1972 to 1974 are taken as normal years, it emerges that *A. bipunctata* is usually the only common ladybird and that the remaining species are relatively infrequent. In early August 1975, during dry, warm weather, there was an influx of *C. 7-punctata* and *C. 11-punctata* from the surrounding countryside that coincided with the arrival of large numbers of the three aphid-feeding hoverflies. These individuals remained and hibernated and in the spring of 1976 bred in the garden, which they had not in previous years. Later in 1976 there were very large numbers of *C. 7-punctata* which, in turn, were reinforced by further immigrants from the countryside, while *C. 11-punctata* remained at about the same (high) level as in 1975. *P. 14-punctata* is usually associated with woods or woody vegetation. As the drought developed in 1976 large numbers arrived in the garden so that by the end of the year this species, which had previously been relatively infrequent, shared second place with *A. bipunctata*.

Trapping results suggest that *C. 7-punctata* was the opportunist, the ladybird that was best able to exploit the superabundance of aphids that occurred in 1975 and 1976. *C. 11-punctata* and *P. 14-punctata* were significant runners-up, but *A. bipunctata* was apparently unable to exploit the changed circumstances produced by two dry summers. There remains the intriguing possibility that these ladybirds compete with each

TABLE IV Frequency of Bumblebees and Cuckoo Bees in a Malaise Trap in the Garden during Two Seasons

Species	*1975*	*1976*
Bombus agrorum *(F.)*	*163*	*196*
B. ruderarius *(Müller)*	*28*	*13*
B. lapidarius *(L.)*	*9*	*41*
B. terrestris *(L.)*	*93*	*47*
B. ruderatus *(F.)*	*5*	*6*
B. lucorum *(L.)*	*28*	*24*
B. pratorum *(L.)*	*48*	*103*
B. hortorum *(L.)*	*90*	*44*
Psithyrus sylvestris *(Lepeletier)*	*1*	*1*
P. vestalis *(Geoffroy)*	-	*1*
Total	*465*	*476*

TABLE V Fluctuations in the Frequency of Ladybirds in a Malaise Trap in the Garden during Five Successive Years

Species	*1972*	*1973*	*1974*	*1975*	*1976*	*Total*
Adalia bipunctata *(L.)*	*168*	*324*	*155*	*424*	*346*	*1,417*
A. 10-punctata *(L.)*	*4*	*1*	*4*	*9*	*25*	*43*
Coccinella 7-punctata *L.*	*4*	-	-	*121*	*562*	*687*
C. 11-punctata *L.*	*2*	-	*1*	*108*	*97*	*208*
Thea 22-punctata *(L.)*	*1*	*16*	-	*14*	*39*	*70*
Propylea 14-punctata *(L.)*	*15*	*8*	*8*	*38*	*346*	*415*
Calvia 14-guttata *(L.)*	-	-	-	-	*1*	*1*
Total	*194*	*349*	*168*	*714*	*1,416*	*2,841*

other for food; if they do, it appears that *A. bipunctata* is a loser during warm weather but a winner in "normal" summers.

F. Butterflies

Only about 70 species of butterflies occur in Britain; the exact number depends on which of the several vagrants are counted as genuinely British. Twenty-one (belonging to five families) have been found in the garden, more species than are known from any of the local nature reserves.

For most butterflies the garden is a refuelling station; they are particularly attracted to the flowers of *Buddleia davidii* Franch. (Loganiaceae), and are regular visitors to about 10 other species, mostly Compositae. The three species of *Pieris* and *Anthocharis cardamines* (L.) (all Pieridae) are the only ones known to have bred in the garden, although a female *Maniola jurtina* (L.) (Satyridae) was once seen laying eggs on the closely cropped grass of a lawn.

Butterflies were caught with an ordinary insect net, marked on the wing with a spot of colored ink, and released. Captures were made without bias for either rare or common species and the results for six seasons (1971 is included) are given in Table VI. There were three very common species, *Pieris brassicae* (L.), *P. rapae* (L.) (both regarded as pests of cabbages), and *Aglais urticae* (L.); three moderately common species, *Pieris napi* (L.), *Vanessa atalanta* (L.), and *Inachis io* (L.); while the remaining 15 species were relatively infrequent, six being recorded once only.

In Table VII, the species and individuals marked and released are grouped according to the food plants of their caterpillars. The vast majority of individuals were dependent on Cruciferae (chiefly cultivated species and varieties) and Urticaceae, almost exclusively *Urtica dioica* L., an abundant weed of cultivation. Seven species were dependent on grasses, (Gramineae), but the number of individuals involved was small, as it was for the remaining six species which feed on a variety of other plant families.

The most remarkable result of marking and releasing is the extraordinarily large number of individuals that came into the garden. *Pieris* spp. excepted, butterflies as a group are popular and most people encourage them, although few would appreciate just how many enter a garden during a year. From the point of view of conservation, gardens must be regarded primarily as refuelling stations for adults, and I envisage suburbia as supporting a vast, highly mobile community of species that have bred in the countryside.

TABLE VI Butterflies Marked and Released in the Garden, 1971 to 1976

Families/species	*Individuals*
Hesperiidae	
Thymelicus sylvestris *(Poda)*	5
Ochlodes venata *(Bremer and Grey)*	2
Pieridae	
Gonepteryx rhamni *(L.)*	37
Pieris brassicae *(L.)*	1,608
P. rapae *(L.)*	5,135
P. napi *(L.)*	572
Anthocharis cardamines *(L.)*	22
Lycaenidae	
Strymonidia w-album *(Knoch)*	1
Lycaena phlaeas *(L.)*	14
Polyommatus icarus *(Rottenburg)*	1
Nymphalidae	
Vanessa atalanta *(L.)*	304
V. cardui *(L.)*	11
Aglais urticae *(L.)*	2,382
Inachis io *(L.)*	591
Polygonia c-album *(L.)*	3
Argynnis paphia *(L.)*	1
Satyridae	
Lasiommata megera *(L.)*	64
Melanargia galathea *(L.)*	1
Pyronia tithonus *(L.)*	1
Maniola jurtina *(L.)*	72
Coenonympha pamphilus *(L.)*	1
Total	10,828

TABLE VII Food Plants of Caterpillars of Butterflies Marked and Released in the Garden

Food plant of caterpillars	*Species*	*Individuals*
Cruciferae	*4*	*7,337*
Urticaceae	*4*	*3,280*
Gramineae	*7*	*146*
Other	*6*	*65*

VI. ADJUSTMENT TO THE GARDEN ENVIRONMENT

A key to understanding adjustment by insects to the contrived vegetational diversity of the garden environment must lie in the relationship between the primary consumers and the plants they eat. The insects involved fall into two categories: those that extract liquid food from plant tissue, chiefly Homoptera, and mostly aphids; and those that chew, chiefly the caterpillars of Lepidoptera, but including also the larvae of sawflies (Symphyta) and many Diptera and Coleoptera. To these must be added the numerous nectar and pollen feeders, all of them adult insects, and belonging mainly to the Lepidoptera, Hymenoptera, Diptera, and Coleoptera. To what extent have these insects become adjusted to an environment dominated by introduced species of plants and by native plants grown in assemblages that would never be encountered in nature? If, as I believe, there has been a rapid adjustment to new food plants and to new assemblages of long-standing food plants, the diversity of the dependent predators and parasites of the insects involved is easily explained. Thus, if it can be shown that a substantial proportion of the Lepidoptera found in gardens feed as caterpillars on introduced plants or on native plants grown "out of place," then the presence of 529 species of Ichneumonidae in the Leicester garden is at least partly explained.

Some primary consumers are known to utilize a wide range of plants. For example, the spittlebug, *Philaenus spumarius* (L.) (Cercopidae) has been recorded in the Leicester garden on 62 species, most of them nonnative. Indeed, the spittlebug seems to exploit almost anything available, avoiding only the very woody, and, oddly enough, the Cruciferae. Another wide ranging plant feeder is the caterpillar of the moth *Phlogophora meticulosa* L. (Noctuidae) which seems to feed on the leaves of most garden plants, including such a varied array as cabbage, lettuce, and parsley.

The collection of food plant records is dependent on the more or less chance discovery of an insect ovipositing or of a larva feeding on the plant; nevertheless records from the Leicester garden are beginning to accumulate. Many primary consumers are indeed restricted to a narrow range of plants and many have become adjusted to introduced species. Thus, the butterfly *Anthocharis cardamines* (L.) (Pieridae) lays eggs and the caterpillars feed on green seed pods of *Arabis albida* Stev. (Cruciferae), a common rock garden plant introduced from south-east Europe. *A. cardamines* was at one time considered as endangered because its caterpillars feed on roadside Cruciferae which until recently suffered from the application of herbicides as a means of controlling roadside vegetation. The rather recent spread of the butterfly into gardens is welcomed by conservationists, and now that it is known to breed on a garden plant its future seems assured. Caterpillars of the moth *Gymnoscelis pumilata* (Hübner) (Geometridae) have been found feeding on the flower heads of *Buddleia davidii* Franch., an attractive shrub introduced from China, and a member of the Loganiaceae, a family unrepresented in the native British flora.

Even more striking are species that have invaded Britain and adjusted themselves to the garden environment to the virtual exclusion of other habitats. There are several examples, the best known being the moths *Hadena compta* (Schiff.) and *Polychrisia moneta* (F.) (both Noctuidae), which during the last 100 years have colonized gardens throughout much of the country and whose caterpillars feed on cultivated *Dianthus* (Caryophyllaceae) and *Delphinium* (Ranunculaceae), respectively.

Dependence on garden plants by lepidopteran caterpillars, and to some extent by other primary consumers, is thus well established. The next step is to produce a description of the resulting food webs. Some of the admittedly preliminary results are quite bizarre. For example, one of the ichneumons collected in the Malaise trap was found to be *Hyposoter singularis* Schmiedeknecht, a species hitherto unrecorded in Britain and previously known from localities as far apart as Germany and Japan. In 1974 this species was reared from caterpillars of *Abraxas grossulariata* (L.) (Geometridae) feeding on *Ribes sanguineum* (Pursh.) (Grossulariaceae), a plant that originates in western North America. Here, then, is part of a food web associated with the man-made garden ecosystem, the investigation of which led to the discovery of an insect new to Britain.

VII. AN UNDERESTIMATED NATURE RESERVE

In 1941, Frank E. Lutz published a book[1] which he called "A Lot of Insects: Entomology in a Suburban Garden." The book is long out of print, and I find that very few have even heard of it. Lutz's "lot" was his yard in the suburbs of New York City. He was at that time an entomologist at the American Museum of Natural History, and he describes how he tried to persuade the Director of the Museum to hire more entomologists on the grounds that there are far more species of insects than all other groups put together. The discussion is reputed to have led to a bargain (which in the event was never made) that the Museum would raise Lutz's salary by 10 dollars a year for every species above 500 found in his yard, and reduced by the same amount for every species short of 500. Lutz found 1,402 species and the book describes how this was done, and also much about the ecology and behavior of the various groups. A good deal of the book is anecdotal but there are lists of species which themselves confirm the diversity that must have been present. He is remarkably short of species in some groups: only 167 Hymenoptera, but he did not have a Malaise trap and so collected rather few of the small parasitic species. Nevertheless, I think that Lutz must be rated as among the first to appreciate that gardens are interesting and contain an extraordinary variety of species assembled together in complex food webs.

It is sometimes said that the conservation movement neglects the vast majority of species of organisms because, apart from butterflies, insects are ignored. There is much truth in this; it is equally true that our most familiar environment, the garden, has been neglected; and, I therefore believe that the time has come to find out more about it. Gardens are valued for the crops and flowers they produce; many also appreciate the birds and other conspicuous creatures that visit them. But the insects, whether primary or secondary consumers or decomposers and which are so essential to the economy of the garden are either destroyed or ignored. In England, as in all industrial countries, the garden environment is in no sense threatened; indeed it is expanding, and the insect life it supports is thus of great interest now that many natural areas are disappearing. The richness of gardens depends on two attributes: the diversity of plants and the extent to which they are exploited by insects; and the incredible patchiness of the environment, repeated kilometer after kilometer, and which, as I have shown, supports as many if not more kinds of insects

---

[1] *G. P. Putnam's Sons, New York.*

than occur in the countryside, especially now that much of it has been converted to monotonous monocultures.

FURTHER READING ON THE LEICESTER GARDEN

Owen, D. F. 1975. Estimating the abundance and diversity of butterflies. *Biol. Conserv.* 8: 173-183.

Owen, D. F. 1976. Ladybird, ladybird, fly away home. *New Scientist* 71: 686-687.

Owen, D. F. 1976. Conservation of butterflies in garden habitats. *Environ. Conserv.* 3: 285-290.

Owen, D. F., and J. Owen. 1974. Species diversity in temperate and tropical Ichneumonidae. *Nature*, (London) 249: 583-584.

Owen, J., and D. F. Owen. 1975. Suburban gardens: England's most important nature reserve? *Environ. Conserv.* 2: 53-59.

THE ROTHAMSTED INSECT SURVEY AND THE URBANIZATION OF LAND IN GREAT BRITAIN

L. R. Taylor
R. A. French
I. P. Woiwod

Rothamsted Experimental Station
Harpenden, Hertfordshire, England

I. INTRODUCTION

With increasing urban development and intensification of agriculture, there has been a revival of interest in the natural fauna and its changes in Britain, as elsewhere, but claims made about the accelerating impact of human activities upon the fauna usually remain qualitative or confined to rare or striking species. The task of monitoring changes in the general level of the fauna, quantitatively on a national scale, is insurmountable for many groups of animals because of the difficulties of sampling over a sufficiently large area. An added problem is to find taxa with enough species to measure changes in the species content of the fauna, as well as in the number of individuals, for it is the loss of species from the fauna

ISBN 0-12-265250-9

that excites most concern. Insects are an obvious choice because they form a large and important part of the total fauna, as predators and prey for other animals and as potential competitors with man for food, as well as having their own intrinsic interest. What happens to insects is thus of interest to agriculturalists as well as to conservationists.

Our concern is primarily with agricultural pests (Taylor, 1973), the ultimate objective being to understand the effects of changing agricultural practice on the structure of the insect fauna and hence to anticipate new pest problems. However, agriculture is seen as just one stage in man's increasing domination of the land and so it is to the whole sequence of change that we look for enlightenment about the population processes involved in pest creation, and the prospects for its prediction. Population dynamics theory cannot yet be used to forecast the effects on pest status of changing the environment because fundamental work has concentrated on numerical change in populations, whereas these problems are partly spatial. They are concerned with what controls the distribution of insects as well as what affects population density. Information on the rate of change of distribution of insect populations is meager. Some species, many aphids for example, are compulsive nomads and periodic movements result in the complete redistribution of the population so that the only relevant measure of its size is one made over a geographically large area (Taylor, 1977). At the other extreme there are insect species so insular that each population center has been known for generations and these are especially familiar in the Lepidoptera whose natural history has been studied for nearly 200 years in Britain.

For this reason we regard the urban insect population, like the agricultural insect population, as an integral part of the whole insect community, with a potential for continuous interchange within, and perhaps continued renewal from a wider ecological system.

In the first half of this paper we have attempted to explain our approach to the problem encountered in establishing a monitoring survey for insects, the principles we recognize, and the hypotheses we can see developing. The second half illustrates briefly some relevant results now emerging from the sampling system which was designed to produce data that could be used to analyze many different processes, including the effects of urbanization on populations.

II. MONITORING INSECT POPULATIONS

A. The Insects Sampled

Because of their large numbers, divergent habits and the difficulties of identification, not all insect taxa are equally easy to monitor and a recording system must be a compromise between the ideal and the feasible. The issue resolves itself into facing the technical and economic problems of measuring the distribution over Great Britain of sufficient species quickly and sensitively enough to register the changes as they occur. Ground sampling of insects, like sampling for other land animals, is tedious and expensive because they are highly aggregated in specialized feeding sites and large numbers of samples are needed for adequate coverage. We therefore sample flying insects, in the air where fine scale distributions are randomized and where many species occur in the same samples. In making a choice of taxon the physical size of the individuals is important because it is correlated with density per unit areas and this determines the sampling method and the cost (Taylor, 1973).

We have chosen to work mainly with the larger moths for detailed analysis because the taxon is large enough, with about 900 species, yet identification is straightforward. However, unlike aphids, which are common but difficult and therefore expensive to identify, and which we also sample for other purposes (see later), the aerial density of moths is low and an attractant, in this case light, is needed to concentrate the individuals so that a statistically adequate sample can be obtained. This introduces a behavioral component into the sampling method that must be accounted for (see later). Nevertheless, it has enabled us to devise a sampling procedure that is standard and simple and makes possible the large-scale coverage necessary to use Great Britain as the experimental arena (Taylor, 1974a).

The moths considered here are the nocturnal Macrolepidoptera (Heterocera), comprising mainly the super-families Noctuoidea and Geometroidea with smaller numbers of Hepialoidea, Cossoidea, Bombycoidea, Sphingoidea and Notodontoidea. These are the families already used by Williams (1953) for earlier investigations into species diversity and are those listed in South (1907), excluding the day-flying species. All nocturnal species are identified except those of the genus *Eupithecia*, and identification of the catches by volunteer operators of light traps are checked until proficiency is attained; difficult genera are confirmed at Rothamsted.

Another branch of the Rothamsted Insect Survey is based on the use of suction traps mainly for sampling aphids (Fig. 1).

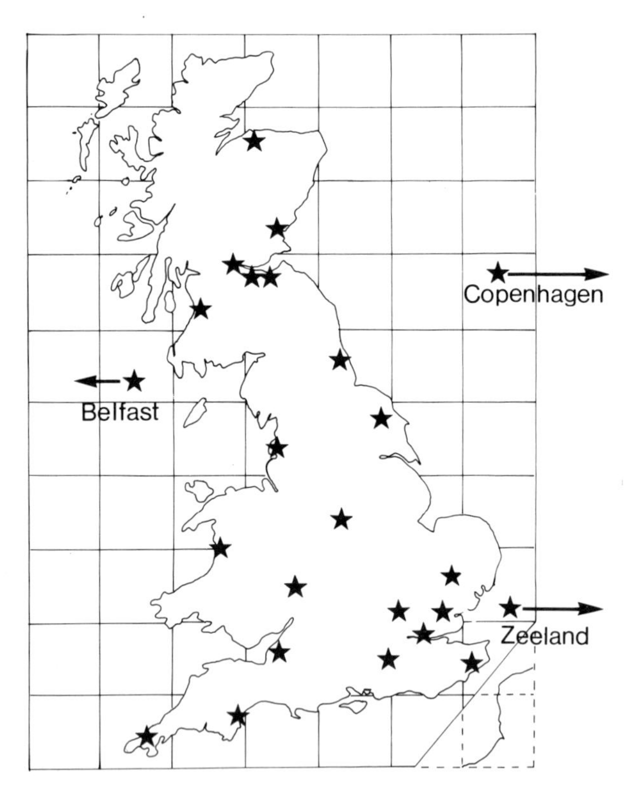

Fig. 1. Sampling network of suction traps in Great Britain.

This survey provides a more immediate service to agriculture and is exclusively professional (Taylor, 1974a). It is more highly standardized than the moth survey and technically more sophisticated but more costly to operate. However, it provides a useful criterion against which to consider the moth program. In addition to the aphids and moths caught in these two sampling systems, other insects are obtained and passed to other entomologists for study. The Trichoptera (Crichton, 1971, 1976), Neuroptera (Bowden, 1973a, 1976), and more recently the Coccinellidae and Syrphidae, have been counted from a wide range of sampling sites. Other groups have been identified mainly to provide information on distribution, e.g., Vespoidea and Apoidea from the Hymenoptera, and Tipulidae and Culicoides from the Diptera. Trap catches have also been used in studies to help chart the distribution of industrial melanism in insects (Lees, 1971), to investigate morphological variation in moth species and provide information on migration and flight activity on several species of economic importance (Taylor and French, 1973; Taylor, 1977).

B. The Sampling Method

The moths are sampled by a standard Rothamsted light trap (Williams, 1948) which has been used intermittently at Rothamsted since 1933 (Williams, 1939, 1940, 1951, 1953; French and Taylor, 1963; Taylor, 1968, 1973). The trap, which is 1.2 m above ground, is protected from rain by a roof and uses a 200 W tungsten lamp switched on from sunset to sunrise every night of the year except December 31 and February 29. The catch, which is collected in a killing jar, is removed and identified daily. The sample is small, averaging 10 moths per night, manageable yet statistically viable. This small catch also ensures that the effect on the population is minimal. A mark-recapture experiment on a population of *Tipula pagana* (Meigen) confined to a small lawn, 37 x 19 m, surrounded by woodland and houses, gave a trapping rate of 2.5% for this type of light trap (Alma, 1973). Also, the sample of moths taken daily for a decade from a small, 1.3 ha, woodland shows no tendency to decline in either number of individuals or species (Table I). This fulfills the sampling principle, that the population being investigated must not be affected by taking the samples, and also ensures that conservation principles are not disregarded.

Ideally, flying insects should be sampled nonselectively so as to relate the sample absolutely to the parent population. Selective traps, needed for large insects, depend for their catch on a behavioral response that is difficult to measure, and a program to standardize the Rothamsted trap is in progress. Mathematical models for sampling in wind (Taylor and Brown, 1972) and in shelter (Taylor and French, 1974) have been developed, and this Rothamsted trap, with a roof over the light, has been found to provide a more consistent sample than traps with the light exposed above. This is important because of the sampling artifacts produced in some light trap catches by the competitive effects of moonlight (Bowden, 1973b; Bowden and Church, 1973). The effects on the sample size expected from changing day length or twilight, associated with high latitudes, are also not marked with the Rothamsted trap.

Absolute standards, obtained by comparing light traps with suction traps, have not yet been published for separate species, although the effect of the sampling method on the measurement of diversity has been dealt with (Kempton and Taylor, 1974). Experiments to relate sample size to area are being analyzed and the conversion of catch to density is different for different species. This means that the resulting species structure of the population contains an unknown factor caused

TABLE I *Yearly Moth Totals of Number of Individuals (N_m), Number of Species (S) and Diversity ($\hat{\alpha}$) for the 10 Years from 1966 to 1975 in Geescroft Wilderness, Rothamsted*

	1966	1967	1968	1969	1970	1971	1972	1973	1974	1975
N_m	4,255	5,446	7,037	9,544	10,705	7,043	4,447	5,552	5,287	5,000
S	177	177	172	193	205	189	168	194	183	178
$\hat{\alpha}$	37.3	35.0	31.8	34.3	36.0	35.7	34.5	39.1	36.8	36.0

by the differential attraction to light. The total sample cannot, therefore, yet be expressed as absolute numbers. The case for regarding this as no detriment in comparative analysis has been argued by Williams (1960).

C. Volunteer and Professional Monitoring

The Rothamsted light trap survey was started in 1959 but the initial stages of establishing a national network developed very slowly because it depended on personal contacts rather than existing organizations (French and Taylor, 1963; Taylor, 1968). It was not until 1968 that enough sites were operating to make the first density distribution maps and by 1976 there were 172 traps, mostly operated by volunteers (Table II). Sites have been, and still are, chosen to be as diverse as possible and include urban sites whenever available, but we also aim eventually to provide an adequate geographical coverage of the whole country. An effort is made to fill in obvious gaps in the network which thus yields, in effect, a stratified sample. Scotland is still not adequately covered (Fig. 2).

In many branches of natural history the large number of interested and knowledgeable amateurs is an important asset and nearly all recording schemes rely heavily on this interest for their existence and success. This is true for the light trapping network of the Rothamsted Insect Survey Other dedicated amateur naturalists have collected valuable survey records in Britain. The Atlas of Breeding Birds (Sharrock, 1976) was based on the observations of more than 10,000 volunteers. The Atlas of the British Flora (Perring and Walters, 1962) involved over 3,000 volunteers for six years. Volunteers, working with little or no financial support, are well suited to broad-based, long-term, monitoring surveys of this kind, provided the demands made on their time are not unreasonable. Those particular schemes were less concerned with the actual biological processes of urbanization, perhaps because the time scales of population change in birds and plants are less suited to the kinds of measurement required. Our experience, like that of The Biological Records Centre at Monks Wood Experimental Station, The British Trust for Ornithology, British Lichen Society and The Botanical Society of the British Isles, has been that, while close contact with the volunteer is essential to maintain critical standards, the quality of the final data can be comparable to that from a wholly professional team. Indeed, the acute shortage of professional scientists trained in the appropriate taxonomy and observation means that without such voluntary participation few large-scale monitoring surveys would be possible.

TABLE II Categorization of Light Trap Operators in 1976

Category	*Number of sites*
Private individuals	*46*
Schools	*22*
Universities, colleges, museums	*22*
Field centers and nature reserves	*31*
Official organizations (forestry, agricultural advisory and similar organizations)	*51*
Total	*172*

Because we also operate the aphid survey, which is entirely professional, we can compare the relative advantages and hazards of the two systems. We conclude that it is equally important in both professional and voluntary surveying to begin slowly. However careful the planning, faults will become apparent with time and they need to be corrected before the system becomes too unwieldy to change without destroying the continuity. It is especially important to avoid rigid preconceptions about analysis, for, in long-term programs of this kind, analytical techniques and approaches will undoubtedly change, especially as new objectives develop. For these reasons, our approach is initially empirical, with computer modeling kept to a minimum. Only later, when solidly based on good data and well-tried statistical procedures, is a theoretical approach adopted, like the development of a model for the distributive process in population dynamics (Taylor and Taylor, 1977).

D. Statistics for Monitoring Population Change

It is of interest and concern to follow the changing fortunes of certain, named, species as land use changes and so the simple statictic, N_S, the number of individuals of a particular species in a year's sample is the primary data. However, to appreciate the quantitative change in the collective fauna at several sites with different species contents, species names must be omitted and some collective population function intro-

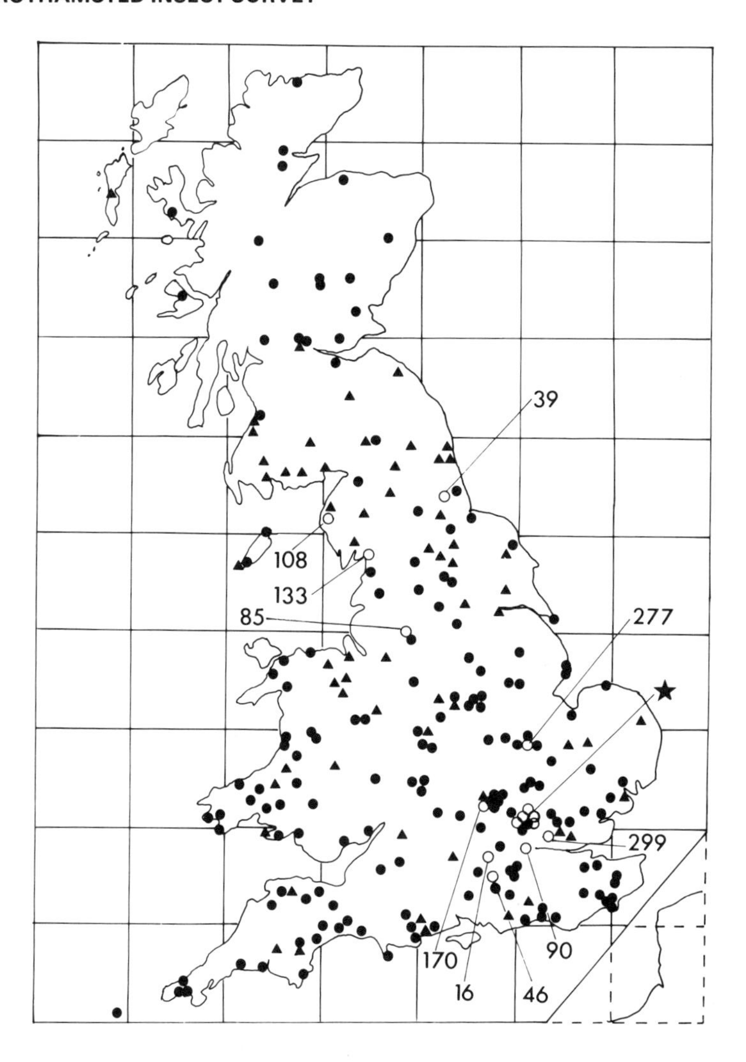

Fig. 2. Sampling network of light traps in Great Britain.
● *Sites with one or more complete year's sample.*
○ *Sites referred to in text and numbered.*
▲ *Recently started sites.*
★ *Sites around Rothamsted: 1, 22, 34, 251, 253.*

duced. This property, commonly called diversity, requires a descriptive statistic for the distribution of individuals in species, a function of N_m, the total number of individuals in the population, and S, the number of species. We shall take somewhat for granted the property and its measurement, but the ability to make sound quantitative use of a diversity statistic, like other statistics such as the mean or median, depends on having a stable yet moderately flexible single model that takes reasonable account of the frequency distribution without an obsessive insistance on goodness of fit (Kempton and Taylor, 1976).

The Rothamsted light trap was used to obtain the original samples that Fisher analyzed to develop the log-series distribution for the number of individuals per species (Fisher *et al.*, 1943). With the advantage of greatly improved data from samples collected over longer periods, simultaneously at many sites, we have now investigated in depth this frequency distribution and its derived statistic, α.

As the effects of urbanization become more extreme, species frequency distributions become more hollow (Fig. 3). The log-normal is then perhaps the best fitting descriptive model because of its flexibility (Kempton and Taylor, 1974; Kempton, 1975). However, detailed analysis of our very extensive data shows that the log-series model, with its parametric statistic, α, consistently provides the most repeatable, stable and sensible measure of the population species structure (Taylor *et al.*, 1976).

The nonparametric statistics derived from information theory and Simpson's index are too sensitive to the transient fluctuations in the numbers of the commonest species in a population and, as a consequence, their ability to discriminate between different populations is small (Table III) (Taylor, *et al.*, 1976). The statistic derived from the log-normal distribution, $c\ S^*/\sigma$, although a better discriminator than the nonparametric statistics, is more sensitive than α to both tails of the distribution, i.e., the number of very rare species and the number of individuals in the most common ones, and thus is slightly less effective at discriminating between sites than is α (Kempton and Taylor, 1976).

The observed log-series statistic $\hat{\alpha}$ is normally distributed in our samples (Fig. 4) and is used in the subsequent discussion as if it described the property, diversity, with which we are concerned. It provides a measure of species richness of the sample that is independent of sample size, so far as is possible, and is stable through time at the same site so long as the site remains undisturbed. If the commonest one or two species are unduly weighted in the population statistic used, their logarithmic fluctuations overweight the required measure of diversity. The log-series, which emphasizes the species of medium commoness, avoids this overweighting, so that the trend of the species complex is not lost in the violent reactions of a few abundant species to each step in the process of environmental change.

The productivity of any site is reflected in the total number of individuals of all species in a sample (N_m) and this in turn is affected by trap efficiency (Taylor and French, 1974);

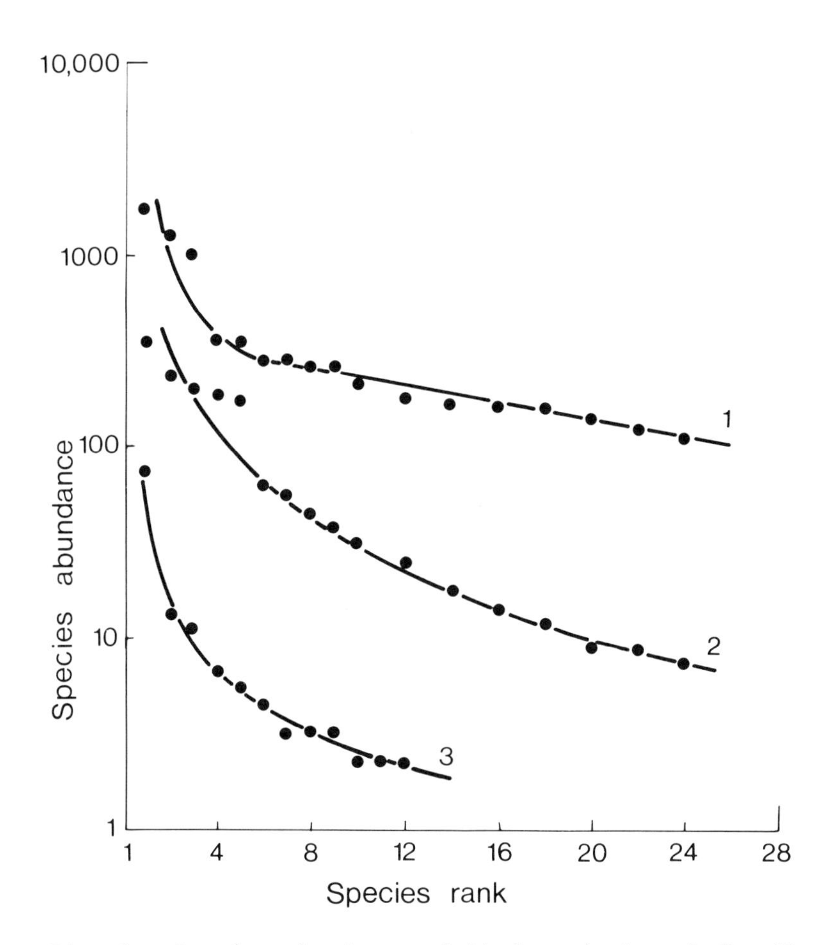

Fig. 3. Species abundance plotted against rank for the most abundant species at three sites in 1970. 1, Geescroft Wilderness (site 22); 2, Allotments (site 34); 3, Isleworth (site 90).

$\hat{\alpha}$, the measure of species richness, is independent of trap efficiency. Using the changes of these two statistics we shall attempt to express the effect of land use on the fauna. Although it has been widely used, the number of species (S) is a poor population statistic. Named species are considered separately.

Perhaps the least satisfactory aspect of land use analysis at present is the categorization of the land use itself. Our provisional assessment is based on a series of arbitrary scores for several crude categories which merely distinguish between

TABLE III Analysis of Variance of Diversity Statistics for Seven Replicate Samples of Moth Catches from Light Traps at 14 Sites in the Rothamsted Insect Survey

		Variance mean square		
Source of variation	*d.f.*	*M*	*I*	$\hat{\alpha}$
Between sites	*13*	*0.00174*	*0.977*	*1,037.5*
Within sites	*78*	*0.00014*	*0.018*	*6.4*
Between to within sites variance ratio		*12.1*	*54.0*	*163.1*

Ability of the statistics to discriminate between sites is indicated by the size of the variance ratio, between to within sites. The higher the ratio the better the discrimination.
$\hat{\alpha}$*, Diversity parameter of the log-series.*
M, Diversity statistic derived from Simpson's index.
I, Diversity statistic derived from the Shannon-Weaver information statistic.

the site area occupied by buildings, grass, arable land, gardens, woodland and hedgerows. The site index, ($\hat{\phi}$), is the sum of the products of the percentage area and the category score arrived at by iterative maximization of the $\hat{\alpha} \times \hat{\phi}$ regression (Fig. 5). So far this regression accounts for only 20% of the variance of $\hat{\alpha}$. However, the specific changes in sample with urbanization reveal more than appears from this figure of 20%, and the current stage in this analysis is the progressive refinement of the site categories.

III. URBANIZATION IN BRITAIN

Great Britain is a small, overcrowded (Taylor, 1970) island in a maritime, temperate, climate over most of which the climax vegetation is deciduous woodland and dominated by oak (*Quercus robur* L.), ash (*Fraxinus excelsior* L.) or beech (*Fagus sylvatica* L.) (Tansley, 1939). It has a highly efficient agriculture occupying 53% of the total land surface of 23.2 million ha, all that can be effectively farmed (Cooke, 1970). Most of the remaining land has lost its original tree cover and is now, at best, high wet, marginal grazing land for sheep or game

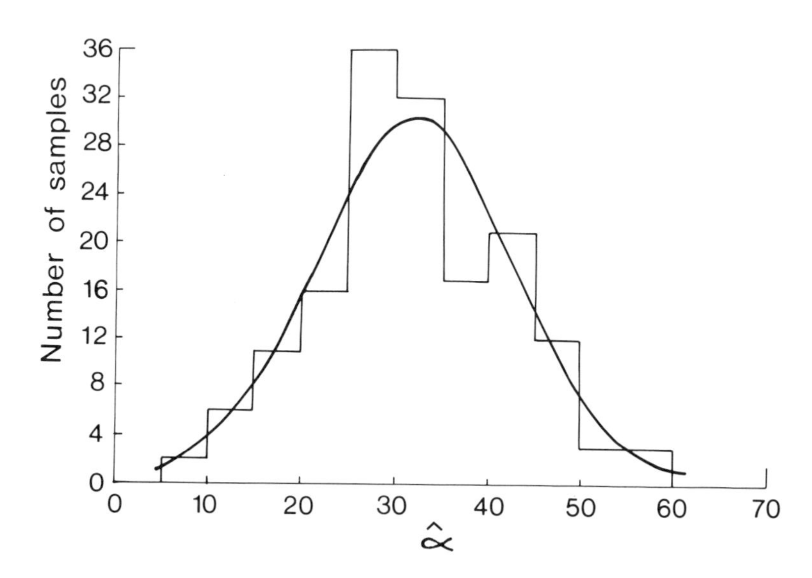

Fig. 4. Distribution of $\hat{\alpha}$ for 159 sites with fitted Normal curve (χ^2_8 = 8.99; mean = 32.1; S.D. = 10.4).

(Fig. 6). In the decade preceding 1970 nearly 161,943 ha of land were lost to agriculture (Cooke, 1970) mainly to grow timber and to house a population of 54 million people (Thompson, 1970) most of whom live in conurbations occupying about three percent of the land surface. Urbanization and the loss of land accelerate while farming intensifies on the remaining land.

The process of urbanization can be very rapid. Apparently stable woodland can be built over within a single season, and this catastrophic rate of change sometimes distracts attention from the more usual sequence of events. In the past there has been a gradual conversion of climax forest, by exploitation, into managed woodland and eventually to permanent grass or arable crops in small fields with hedgerows. Minor residential occupation has then developed into suburbia and eventually to full industrial urbanization. However rapid this process may have been in small areas, over the whole island it has taken several millenia. The resulting pattern is a scatter of urban centers (Fig. 6), each surrounded by a ring of suburbs merging into a background of agricultural land. This in turn encloses patches of managed woodland, and any original forest that still remains, and gives way to marginal grazing on the uplands. In Britain, the processes that lead to urbanization are so well advanced that the man-made landscape predominates, but it has been an evolutionary rather than a revolutionary process and

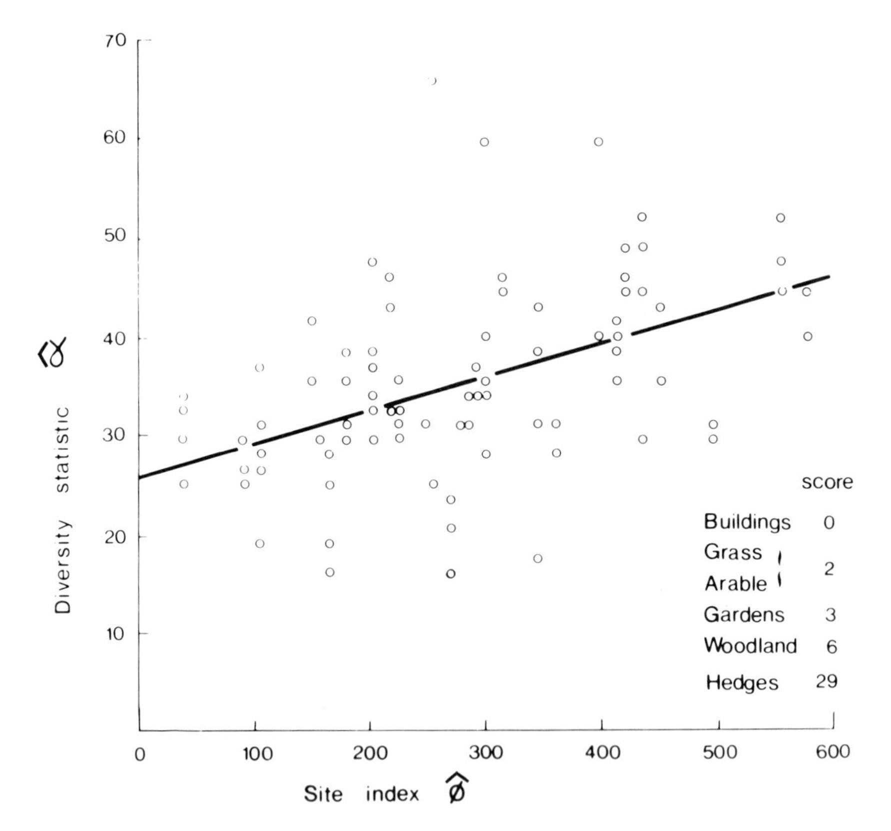

Fig. 5. Regression of moth diversity ($\hat{\alpha}$) on land use (ϕ) within a radius of 64.4 m of the sample point. Fitted regression is $\hat{\alpha} = 25.94 + 0.034\ \phi$. The index ϕ is a weighted area measure using the scores given in the figure.

the resulting biological mosaic has, in many instances, reached a quasi-stable condition on an ecological time scale.

A. The Regional Distribution of Moth Populations

Examination of records for the 68 butterflies (Rhopalocera) recognized as having occurred in Great Britain shows that the distribution of 14 species is known to change often, while a further 21 species are known to have changed their distribution over the last century or so. Few of these changes have ever been assigned to urbanization (Heath, 1974), for none of the species were confined to areas adjacent to towns. Changes in land use through drainage, plowing permanent grass, clearing

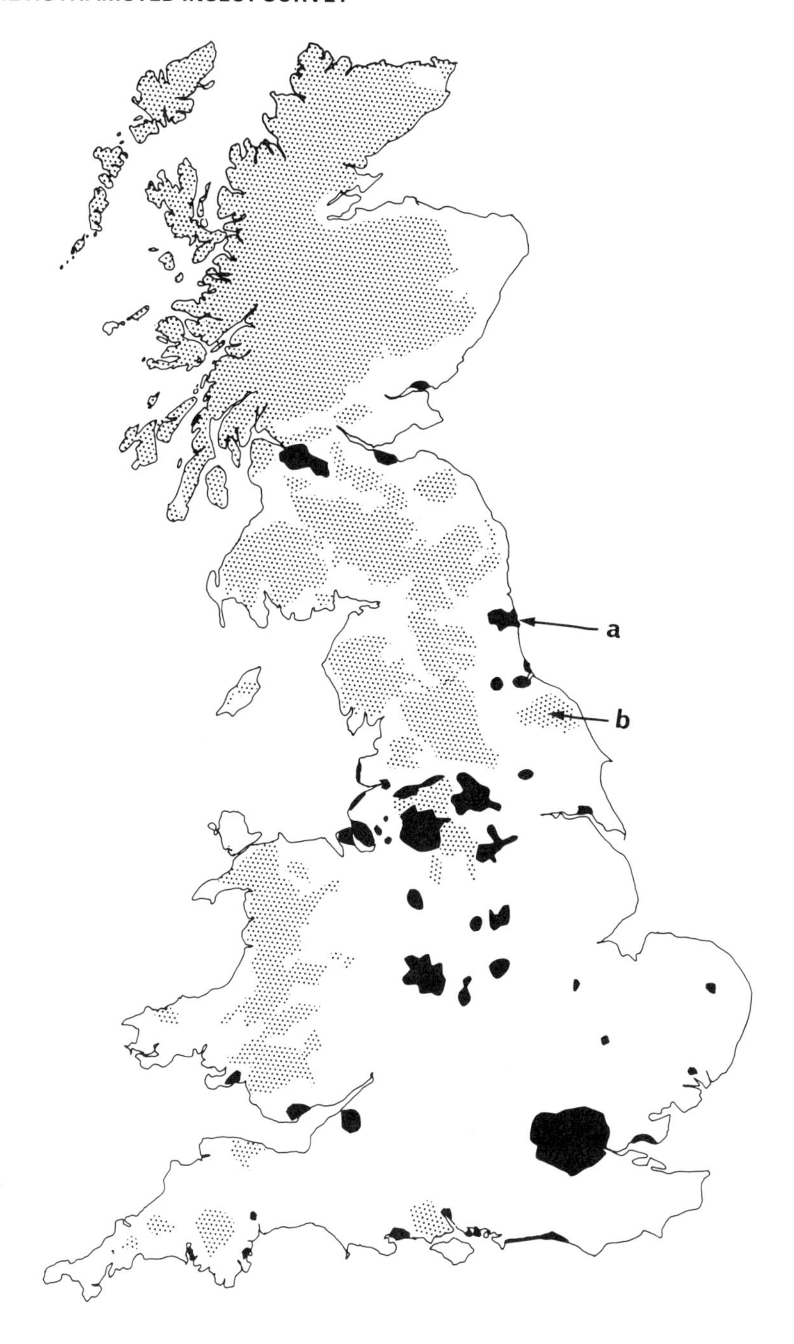

Fig. 6. Land use in Great Britain; (a) main urban areas, and (b) areas of upland pasture and rough grazing.

old woodland and scrub for farming or forestry, control of the rabbit by myxomatosis, insect collecting, and the weather, have all been held responsible at one time or another for loss of ground by one or more species. Sometimes there is good evidence; at other times species have increased their range with no known cause.

1. *Single Species Distribution*

To quantify these kinds of observations we have mapped the annual distribution of more than 100 species of moths (Heterocera) over six years (Fig. 7a) using records from the Rothamsted Insect Survey. Many of these records have been published annually since 1965 (e.g., Taylor and French, 1969, 1976). All the maps show density distribution patterns which change from year to year in some degree but which are characteristic for each species. As a consequence, when the maps are sequentially summed and a mean obtained (Fig. 7b), the patterns gradually merge and would eventually coalesce and be lost. In other words, all species are spatially fluid and differ only in their rate of spatial change (Taylor and Taylor, 1977).

These maps show that the location of areas of low density, or "holes" in the distribution for any one year, and their association with causative factors such as urbanization is complex. Holes appear in different places in different years. Even for crop pests, the pattern is related in no simple way to crop distribution. Urban areas are not easily identifiable on individual years' maps for individual species (Fig. 7a), not because they have no effect, but because their effect is lost against a background of other, more geographically mobile holes. These result from environmental limiting factors which may often be biological restraints, such as parasites and predators, that are as mobile as the moths we are mapping. The resulting normal fluctuation in density at a fixed point in space may be too great to separate the ephemera from the trend and, except with long runs of data, yield quite misleading results.

2. *Multispecies Distribution: Productivity and Diversity*

When samples are summed over species and time, to yield mean geographical distribution of the individuals in the whole multispecies population (N_m), urban areas (see Fig. 6) can be recognized (Fig. 8). However, the areas of low N_m are not as extensive nor as clearly defined as expected, for in disturbed situations adaptable species move in, if they are not already present, and their numbers increase to replace less adaptable species so that N_m falls very low only in extreme conditions. In contrast, the map for species diversity (Fig. 9), in which trap efficiency and sample size have been eliminated, shows

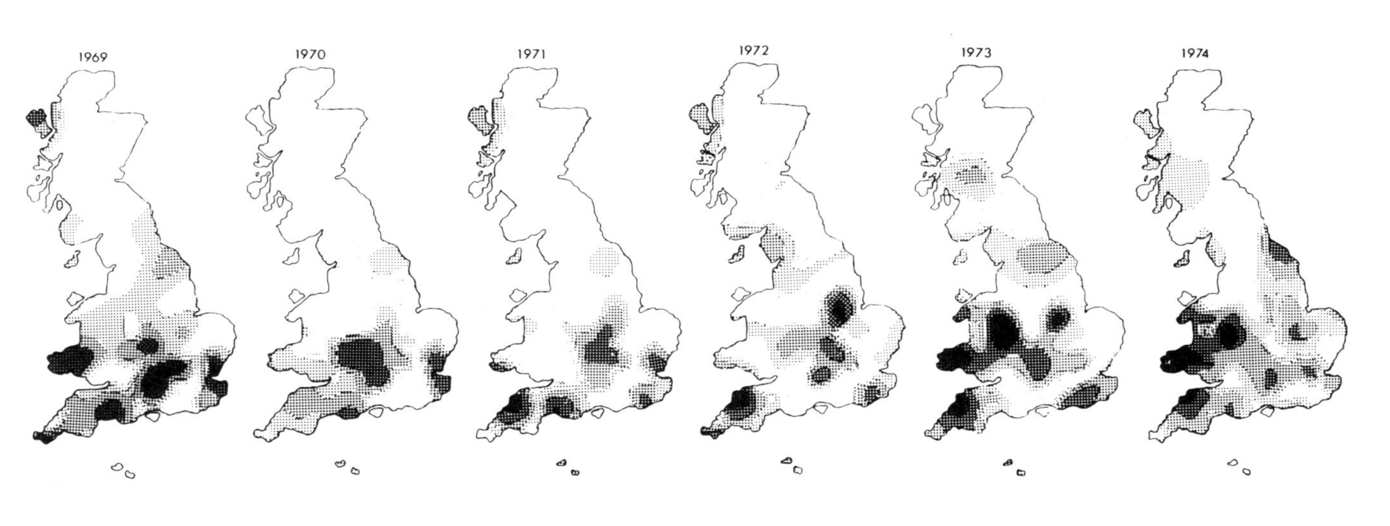

Fig. 7a. Abraxas grossulariata *(L.). Annual adult density distribution (1969-1974).*

Fig. 7b. Abraxas grossulariata *(L.). Geometric mean (1968-1974). Density layers 0, 1-2, 3-9, 10-31, 32-99, 100-315, 316-999, individuals per sample unit.*

Fig. 8. Macrolepidoptera. Mean number of individuals, (1968-1974). Density layers 199-412; 413-1,090; 1,091-3,230; 3,231-10,000; individuals per sample.

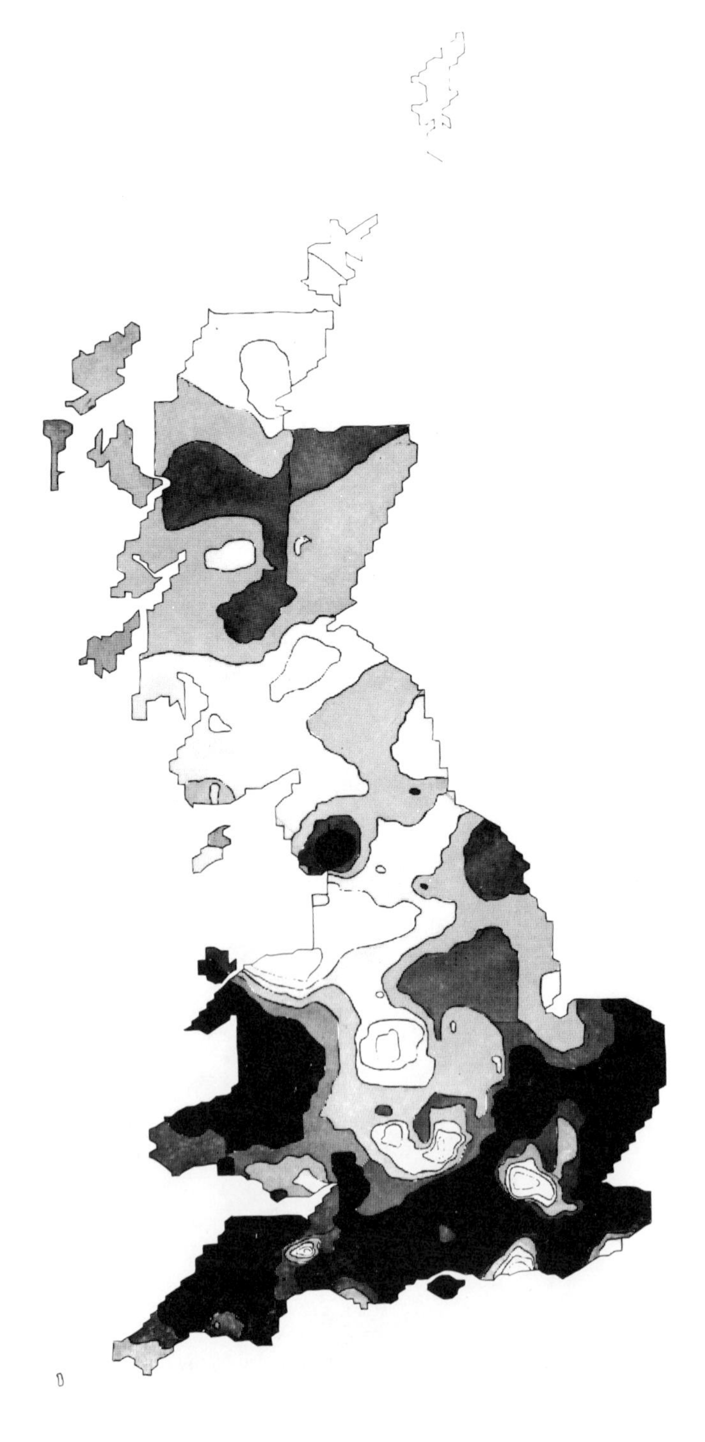

much more clearly the areas of urbanization although there are still some artifacts due to trap siting. The main characteristic, however, is the complexity of this $\hat{\alpha}$ pattern (Fig. 9) which, like the map for N_m, is produced from seven years' data. Unlike the map of N_m however, which approaches more closely to a uniform pattern as more years are added (Fig. 8), or the maps for separate species, which change annually and reach spatial stability only with loss of pattern (Fig. 7b), $\hat{\alpha}$ is more stable at each point in space and its complex spatial pattern shows no diffusion with succeeding years (c.f., Fig. 22; p. 234, Taylor, 1974a).

B. Selected Experimental Sites

The maps (Figs. 8 and 9) of mean numbers of individuals of all species (N_m) and of diversity ($\hat{\alpha}$) show that the urban fauna cannot be considered in isolation but must be related to the available evidence for a climax fauna. The problem in Great Britain is to find this climax fauna as a standard for comparison of the present degree of disturbance and to measure the rate of change when land use reverts and the flora and fauna readjust. The difficulty lies mainly in the fine scale of the mosaic of soil type and topography and the long and complex history of land use which prevents direct comparison of successive stages in development through the whole land use series. An experimental approach is generally impractical.

The most complete series of land use stages we have examined is in the area around Rothamsted. Ancient woodland provides some evidence for the climax end of the series and a London site for the urban end. A similar, but more scattered series of sites in Northern England is then given for comparison. Both series of sites, marked on Figure 2, are now examined in some detail.

1. *The Climax Background*

Rothamsted Experimental Station lies on the Chiltern plateau of clay-with-flints overlaying Cretaceous chalk, at about 130 m above sea level. The Chiltern climax woodland was probably oak-dominated (Brown, 1964) and the nearest surviving approach to original forest in this area is a small piece of mature woodland with century-old standing beech and oaks on the Ashridge estate (site No. 253; Fig. 10c), 15 km from Roth-

Fig. 9. Alpha diversity for moth (1968-1974). Contour intervals <15, 15-20, 20-25, 25-30, 30-35, 35-40, 40-45, 45-50, >50.

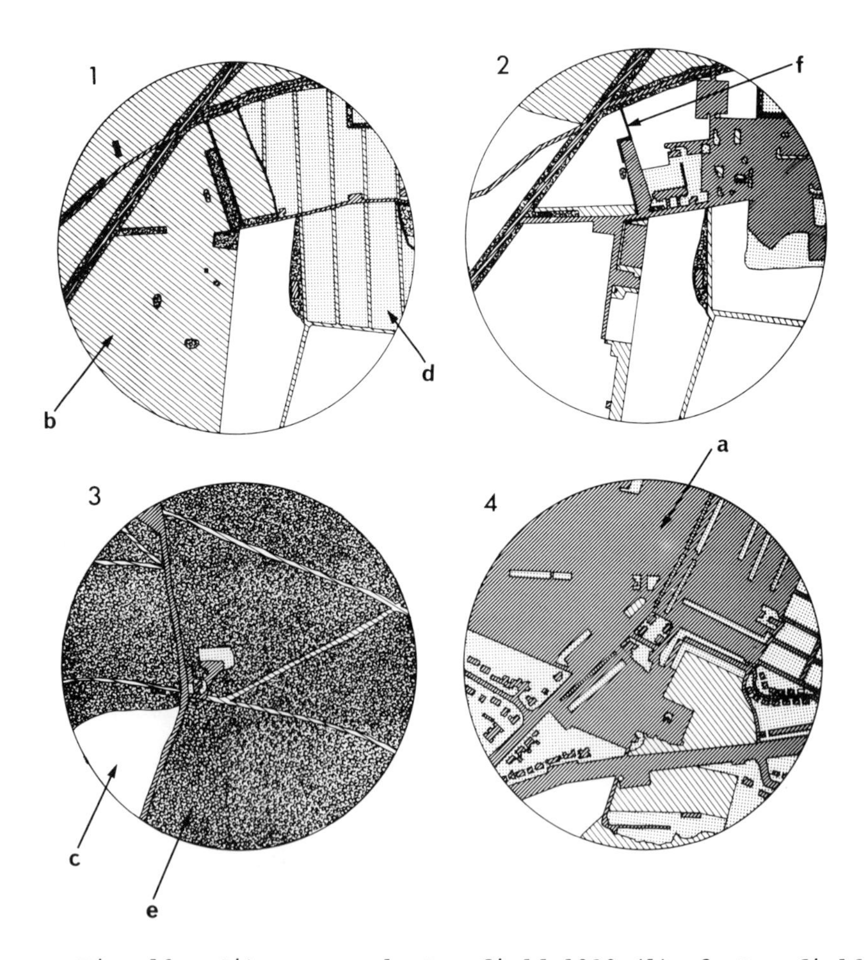

Fig. 10. Site maps: 1. Barnfield 1930 (1); 2. Barnfield 1970 (1); 3. Ashridge 1975 (253); 4. Chester-le-Street 1975 (39). a. buildings and paved roads; b. grassland; c. arable land; d. gardens and allotments; e. woodland; f. hedges.

amsted, at an altitude of 192 m but with the same exposure and on the same soil series. Two other, younger, clay woodlands, Howe Park Wood (site No. 170), 37 km to the north-west of Rothamsted at an altitude of 106 m, and 68 km away to the north-east at a height of between 7.5 to 39.5 m, the famous entomological collecting site of Ewingswode (site No. 277) in Monks Wood (Steele and Welch, 1973), provide comparisons.

London is on clay and gravel and the clay woodlands already mentioned are likely representations of the London climax. Ilford (site No. 299) in Essex, is a London suburb near to the Epping Forest, a similar old woodland to Ashridge, on clay at an altitude of 15 m.

On the Rothamsted estate, a small (1.3 ha) experimental woodland known as Geescroft Wilderness provides a vital link between agriculture and the original forest. This was a piece of agricultural land, possibly cultivated since the first centuries A.D., sown to clover in 1888 and then allowed to revert. The soil is still becoming more acid (Jenkinson, 1971), and samples made over the years (Lawes, 1895; Hall, 1905; Brenchley and Adam, 1915; Thurston, 1958) show the growth and establishment of a flora now approaching climax woodland. Geescroft Wilderness (site No. 22) has been sampled daily for moths since 1965 and provides the standard of stability for the other neighboring sites, including climax forest.

2. Agriculture and Urbanization in the South

Also on Rothamsted estate, 613 m from Geescroft, another site is on the western edge of an experiment in which root crops were grown every year for 126 years until 1969 in a field, Barnfield (site No. 1), that remains arable but is now encroached upon by tennis courts and pavilions, houses with lawns, and a laboratory. In addition, farm practice has become more mechanized and rough hedgerows removed or reduced. These changes towards suburbia are clearly reflected in the site maps (Fig. 10 [1] and [2]) and Fig. 1 in Taylor (1974b) and in the moth samples. The trap on Barnfield was operated for four years from 1933 to 1936, again from 1946 to 1949, and continuously since 1960.

Well-established, 40 year-old, suburban gardens in Harpenden, about one kilometer from Barnfield, are sampled at site 251. The garden is surrounded by other mature gardens and houses of about the same age in a twentieth century dormitory town of 26,000 people 40 km from the center of London.

Ilford is a satellite town to the north-east of London with a population of over 170,000 and a more urban center than Harpenden. The site No. 299, is in a younger garden, 15 to 20 years old, in more intensive housing and is visibly more urban.

The London site used here, No. 90, is at Isleworth, on a town college playing field with little vegetation other than grass nearby. This area is urban but not exclusively industrial.

3. A Northern Series

Other urban sites are given for comparison. Chester-le-Street, a small industrial town in the north-east of England (population 21,000), has a site (No. 39) in one of the few gardens in an area of old terrace housing (see Fig. 10d). At Prestwich (site No. 85) in urban Manchester, part of the heavily industrial conurbation of south Lancashire with a population of about four million, the site is a school six kilometers from the city center. The site at Egremont (No. 108) is also at a school on the edge of a small market town (population over 7,000), near the sea. As a northern rural site, Leighton Moss (No. 133) is a small orchard on limestone 58 km south-east of Egremont near a large area of reeds and willow scrub, deciduous woodland and grazed pasture.

4. *A Short-term Experiment*

Also on Rothamsted estate and adjacent to the laboratory buildings, is the site known as Allotments (No. 34) selected in 1966 when the area consisted of 2.8 ha of mature allotments, or vegetable gardens, founded about 1852 (Lawes, 1877) and cultivated continuously until 1967. In 1968 the gardens were cleared and plowed and the area sown to wheat and beans. In 1969 and 1970 the immediate surroundings of the trap were fallow with high weeds. In 1971 it was all buried under soil and debris from a building site and part was paved for parking cars. Since then there has been repeated growth of weeds, periodically suppressed by weed-killers, and the site has been cleared by bulldozers, and used for dumping. This site is remarkably informative about the effects of the early stages of urbanization on the fauna when considered in relation to the controls in Geescroft and elsewhere.

C. Samples from the Named Sites

1. *Variability in a Stable Environment*

We define stability empirically, the most stable site being that with minimum variability from year to year. Of the sites measured, Geescroft Wilderness is among the most stable (Kempton and Taylor, 1974); N_m has ranged from 4,255 to 10,705 over a period of 10 years, S from 168 to 205 and $\hat{\alpha}$ from 31.8 to 39.1 (Table I), with a mean of 35.4 and standard deviation of 2.0. This compares with a site at Stratfield Mortimer (No. 16) which has an $\hat{\alpha} \pm$ S.D. of 43.3 ± 2.4 and another at Alice Holt (site No. 46) with 60.8 ± 2.5, these being the three most stable long-term sites in the trapping network. With this known minimum variance in mind we can consider the statistics from the various sites.

2. The Urban Sequence

The mean diversities at Ashridge ($\hat{\alpha}$ = 38.0) Howe Park (36.0) and Ewingswode (40.1) fall very near the upper range at Geescroft Wilderness. Barnfield, the agricultural equivalent of Geescroft is slightly, but consistently, lower in diversity and considerably lower in productivity. The old suburban garden in Harpenden is very similar to Barnfield, marginally higher in both productivity and diversity. The younger garden at Ilford has a markedly lower productivity and diversity than Geescroft Wilderness while the urban site at Isleworth is much lower again so that the six year mean for Isleworth gives an $\hat{\alpha}$ of 13.2, with a standard deviation of 3.6, while N_m has ranged from 62 to 359 (Fig. 11).

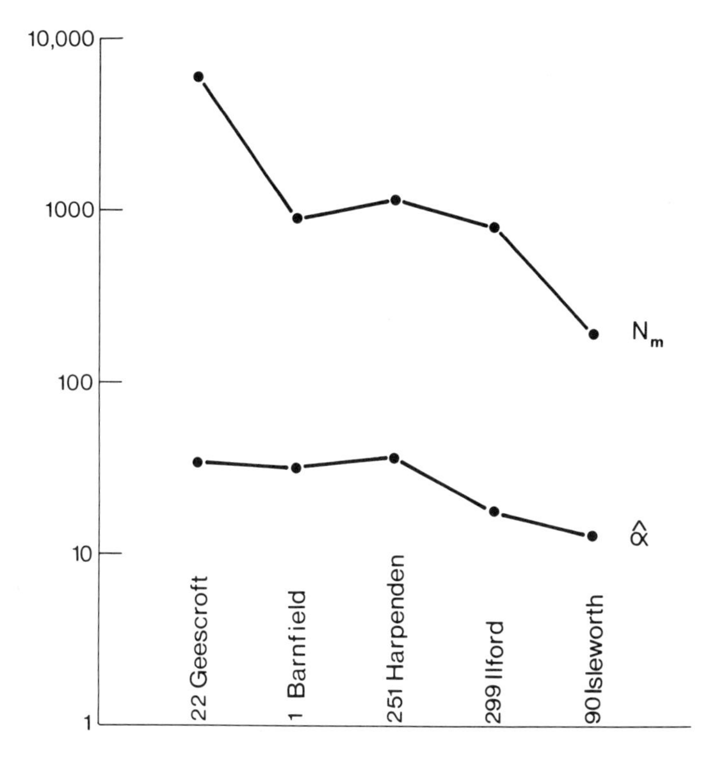

Fig. 11. N_m and $\hat{\alpha}$ for the southern woodland/urban series. Mean values.

Outside the London area the picture is less clear because the data are less complete and the comparisons less direct. At Chester-le-Street in the north-east of England and at Egremont and Prestwich in the north-west of England, N_m = 348, 390 and 200, respectively, and $\hat{\alpha}$ = 23.7, 20.5 and 18.6, respectively, falling between Isleworth and Ilford although Chester-le-Street consist of much older terraced houses in an urban setting in a smaller town. These low values are not just an effect of latitude or climate because the site at Silverdale (Leighton Moss), which lies geographically between the other three northern sites, has an N_m of 2,542 and an $\hat{\alpha}$ of 42.9 (Fig. 12), higher than Ashridge.

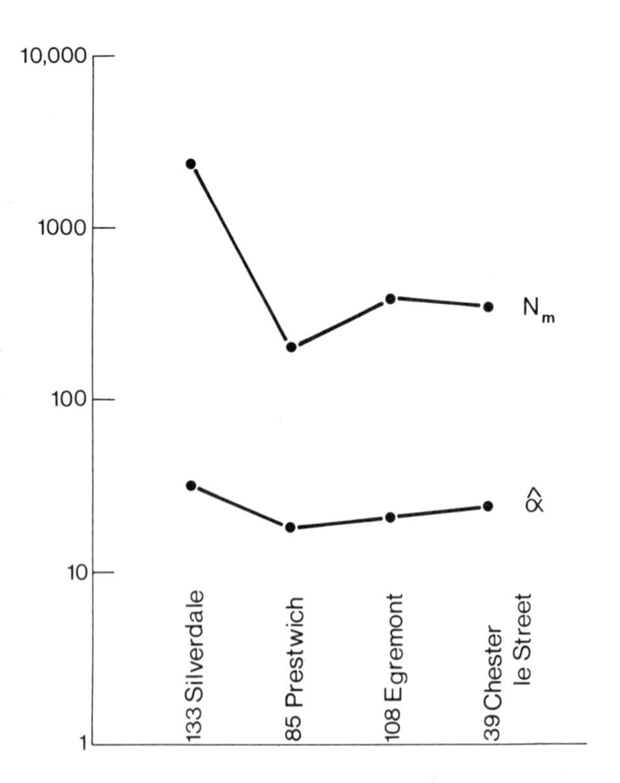

Fig. 12. N_m *and* $\hat{\alpha}$ *for the four northern sites. Mean values.*

3. The Allotment Experiment

As a result of disturbance it appears that productivity falls rapidly and later diversity, a process that can best be seen in the experiment on Allotments. In a series of cycles, this site was partly developed, allowed to revert, again developed and so on. The details of the successive "treatments" have not yet been published and it suffices here to show the outcome with some comments.

In 1966 the site was surrounded by mature gardens and the productivity and diversity were very similar to the nearby Barnfield trap (Fig. 13). When the gardens were abandoned the moth productivity increased but not diversity. This can be attributed to increased plant productivity, without regular cultivation, but with no profound change in the general composition of the flora. When the plot was cleared and plowed in 1968 the productivity decreased dramatically. However, if anything, diversity increased, possibly because a number of abundant, opportunist species were most severely affected by this treatment. The area was left fallow for the next two years and weed cover increased, the opportunist moth species recolonized and productivity again increased but diversity declined. There followed a series of disturbances producing large fluctuations in productivity and a steady decrease in diversity until 1972, when $\hat{\alpha}$ started to increase until it now has a value similar to that of the Barnfield site nearby, although the trend in productivity is towards a much lower level. This may be because, although disturbances have continued, there has not been a complete removal of weed cover in the last few years and this has enabled some of the perennial plants to become reestablished. As a result more moth species can breed in the vicinity of the trap, although the biological productivity of the area is still very low.

4. *Species Lists*

So far we have only considered changes in the population at a site, or differences between sites, in terms of total productivity, which is related to N_m, and a diversity parameter derived from the species frequency distribution. The problems associated with using named species for these purposes have been outlined earlier. However, it is possible to use a list of names to illustrate examples of these changes.

Table IV lists the 40 species caught most frequently by the Rothamsted Insect Survey light traps. These species have been ordered by their frequency of occurrence in the Ashridge trap as this is considered to be the nearest to the climax fauna. The three other southern woodland sites are then listed, for comparison, then the agricultural-suburban-urban series in the London area. Two years from the rapidly changing Allotments site are included, the first year when the area was still mature gardens (1966) and the other sample when the trap has reached its lowest productivity (1974). Finally,

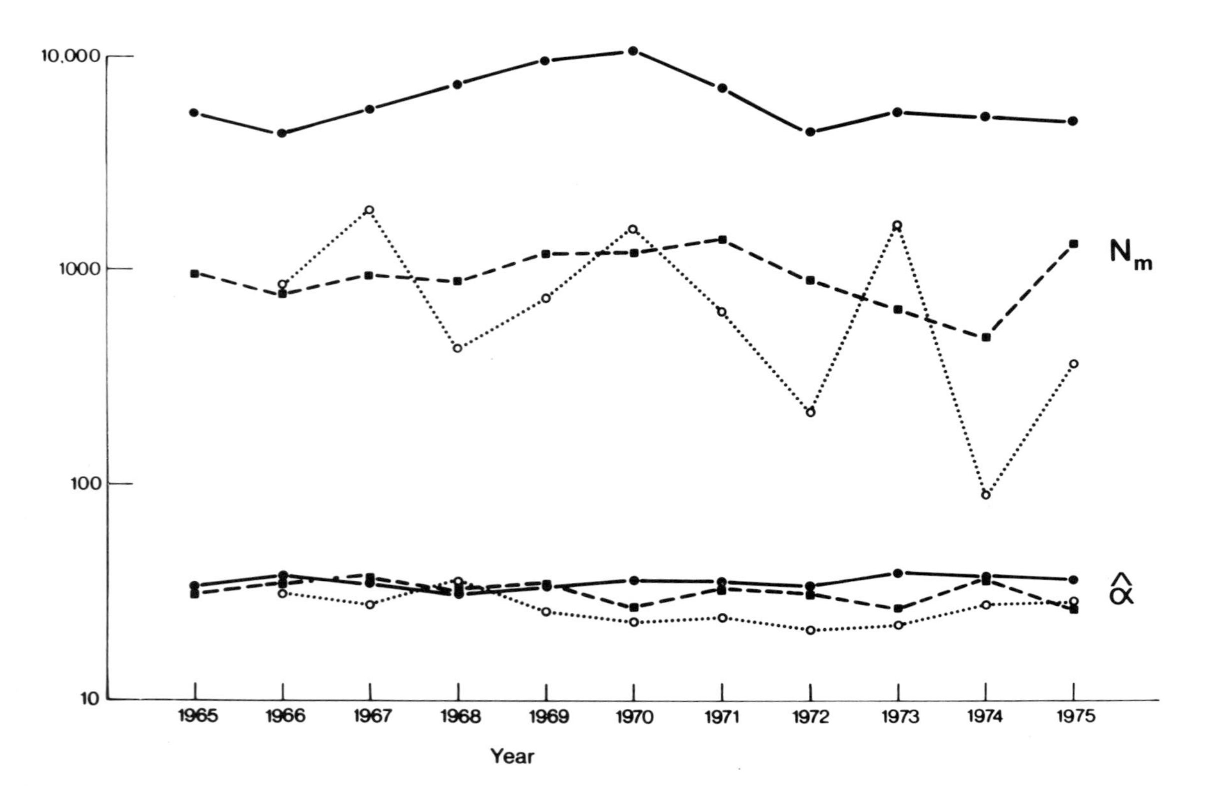

Fig. 13. N_m *and* $\hat{\alpha}$ *for the three traps at Rothamsted from 1965 to 1975.* ● *Geescroft Wilderness (22);* ■ *Barnfield (1);* ○ *Allotments (34).*

the three northern urban sites are listed at the end of the table. Care must be taken in the interpretation of this table for the reasons already outlined and also because different sites ran in different years and this introduces an added element of variability.

The most obvious feature is the general similarity in the abundance of the species in the four woodland sites and the sudden drop between Geescroft, the reverted woodland, and Barnfield, the typical agricultural site. This is particularly evident for the first 10 species on the list. The second group of sites with similar abundances includes Barnfield, Allotments 1966, Harpenden and possibly Ilford. The remainder have a very impoverished list of species. This visual impression coincides well with what has already been said about productivity at the various sites.

Large differences in species abundances between sites can sometimes be attributed to simple food plant distribution. Thus *Petrophora chlorosata* (Scopoli) is very abundant at Ashridge but almost absent elsewhere because it feeds on bracken *(Pteridium aquilinum* L.) which is uncommon at the other sites, urban or otherwise. The first two species in the list, *Hydriomena furcata* (Thunberg) and *Erannis defoliaria* (Clerck), show a distinctive cut off between Geescroft and Barnfield. These are both tree feeders and hence are casualties of one of the first processes of land development, the removal of trees. Other species appear to be able to maintain a reasonable population at any site, for example *Noctua pronuba* (L.), *Xestia xanthographa* (D. and S.), *Agrotis exclamationis* (L.) and *Caradrina morpheus* (Hufnagel). These all feed on a variety of low growing, mainly annual, plants and are probably good examples of opportunist species that can quickly colonize new areas as soon as suitable food plants appear and hence are sometimes pests. Only one species appears to be generally commoner in the urban sites than in the woodland ones, *Luperina testacea* (D. and S.). This species feeds on grass roots and is probably one of the few to thrive on the closely cropped grass of the suburban lawns and school playing fields near many of our urban sites.

IV. SUMMARY AND CONCLUSIONS

We have described the light-trapping network of the Rothamsted Insect Survey and explained its purpose and the way

TABLE IV Comparison of the 40 Commonest Species of Moth Caught in the Rothamsted Insect Survey with the Number of these Species Caught at 12 Sites. Site Number and Year are Indicated for each Site

Species	Ashridge 253 1975	Howe Park 170 1972	Ewingswode 277 1974	Geescroft 22 1973
Hydriomena furcata *(Thunberg)*	438	1249	189	55
Erannis defoliaria *(Clerck)*	362	50	34	38
Idaea biselata *(Hufnagel)*	264	353	930	275
Orthosia gothica *(L.)*	240	159	154	59
Operophtera brumata *(L.)*	238	137	105	133
Idaea aversata *(L.)*	186	65	216	80
Diarsia mendica *(Fabricius)*	185	19	199	247
Rusina ferruginea *(Esper)*	143	77	144	74
Xanthorhoe montanata *(D. and S.)*	142	113	60	78
Spilosoma luteum *(Hufnagel)*	138	49	54	53
Lomaspilis marginata *(L.)*	130	59	105	0
Petrophora chlorosata *(Scopoli)*	128	0	2	2
Opisthograptis luteolata *(L.)*	86	154	397	201
Xanthorhoe ferrugata *(Clerck)*	79	22	386	96
Noctua pronuba *(L.)*	68	158	19	115
Selenia dentaria *(Fabricius)*	66	185	453	44
Cerapteryx graminis *(L.)*	57	1	0	0
Hypena proboscidalis *(L.)*	61	227	90	88
Mythimna pallens *(L.)*	54	3	5	1
Xestia xanthographa *(D. and S.)*	53	171	33	183
Mesapamea secalis *(L.)*	52	113	3	155
Xanthorhoe spadicearia *(D. and S.)*	51	27	182	23
Spilosoma lubricipeda *(L.)*	40	25	39	5
Eulithis pyraliata *(D. and S.)*	31	128	10	50
Poecilocampa populi *(L.)*	26	36	190	5
Diarsia rubi *(Vieweg)*	26	52	32	26
Agrotis exclamationis *(L.)*	25	106	16	27
Ochropleura plecta *(L.)*	23	9	7	21
Autographa gamma *(L.)*	21	10	2	42
Mythimna impura *(Hübner)*	16	27	16	21
Xestia c-nigrum *(L.)*	14	157	5	28
Xanthorhoe fluctuata *(L.)*	12	3	13	97
Omphaloscelis lunosa *(Haworth)*	11	61	0	7
Eilema lurideola *(Zincken)*	10	5	21	3
Luperina testacea *(D. and S.)*	9	1	12	2
Apamea monoglypha *(Hufnagel)*	6	38	8	45
Hydraecia micacea *(Esper)*	4	17	5	8
Lycophotia porphyrea *(D. and S.)*	1	0	0	0
Abraxas grossulariata *(L.)*	1	232	89	38
Caradrina morpheus *(Hufnagel)*	0	4	35	19

TABLE IV (cont'd)

Barn-field	Allot-ments	Allot-ments	Harpen-den	Ilford	Isle-worth	Chester-le-Street	Prest-wich	Egre-mont
1	34	34	251	299	90	39	85	108
1966	1966	1974	1975	1975	1973	1973	1973	1973
2	0	0	2	0	1	1	1	1
0	0	0	0	0	0	0	0	0
1	1	0	7	0	0	0	0	0
17	17	1	68	5	0	2	0	4
1	0	0	6	0	4	0	0	0
8	11	4	44	8	0	2	3	7
6	6	0	8	0	0	0	8	5
12	12	1	21	0	0	1	0	0
15	38	0	12	1	0	7	0	0
8	18	1	3	0	0	1	3	4
2	1	0	0	0	0	1	1	0
0	0	0	1	0	0	0	0	0
23	6	1	19	8	1	1	4	0
41	27	5	0	0	0	0	1	2
10	9	0	7	22	1	42	21	38
4	4	0	8	1	0	0	0	1
0	0	1	0	0	0	0	0	0
4	4	1	10	0	0	1	0	4
13	11	5	52	24	1	19	0	1
72	57	1	12	62	4	2	12	7
9	9	1	58	44	0	14	3	33
4	12	1	1	2	0	0	0	0
10	18	1	0	1	0	1	0	1
0	6	0	4	0	0	0	0	0
4	0	1	0	0	0	0	0	0
28	55	9	2	0	0	12	0	18
31	9	5	24	32	1	11	24	16
8	5	1	2	1	0	0	1	3
14	38	0	54	53	1	65	1	0
30	43	2	5	13	0	29	8	4
25	23	1	5	4	1	0	3	1
14	12	4	27	97	2	31	8	13
1	0	0	2	5	3	0	0	0
3	5	2	2	0	0	0	0	0
20	8	3	9	27	16	13	13	35
3	4	0	0	3	0	18	8	22
4	19	4	0	0	0	9	0	9
0	0	0	0	0	0	0	0	0
7	7	2	19	0	0	0	1	4
9	25	3	35	13	0	14	17	4

it functions. The monitoring of moth populations throughout Great Britain for several years shows how patterns of distribution for single species change and make it difficult to recognize urban areas because of this. The fauna of a town is evidently not isolated and is probably constantly supplemented from the surrounding countryside unless the urban area is very large.

The property of diversity is more stable geographically, and urban areas show up more clearly on diversity maps than on maps of density.

Selected sites have then been used to illustrate population change during the process of urbanization, from climax background, through agriculture and suburban stages to full urban status. The first effect of land development is to cut down the moth population numerically. This does not necessarily immediately change its species structure. That is to say, although the total population, N_m, declines rapidly, diversity, $\hat{\alpha}$, changes more slowly. This can be seen in the sequence of samples from sites near Rothamsted Experimental Station on the Chiltern chalk-and-clay hills near London. The fauna of old established deciduous forest gives way, upon the advent of agriculture, to a loss of many species especially those associated with perennial herbs. This process is exacerbated with the further encroachment of concrete and eventually results in either an insect fauna that is as ephemeral as the flora, and thus includes all the pest species associated with ephemeral crop plants, or to a more stable fauna where houses with gardens gradually mature to become oases in the concrete desert.

If the pressure of development is relaxed, there is a rapid increase in population density but with it a large change in the diversity ($\hat{\alpha}$) because a few opportunist species quickly take advantage of the situation. Given time, the population recovers its structure, with a diversity not very different from the original, but at a lower population density than formerly.

We do not yet know how low N_m can go before $\hat{\alpha}$ breaks down but, as we have no very high diversities at very low densities, we presume there is a lower threshold of density for "healthy" diversity.

REFERENCES

Alma, P. J. 1973. A population study and light trap captures of *Tipula pagana* (Meigen) (Dipt., Tipulidae). *Entomol. Mon. Mag. 109*: 240-246.

Bowden, J. 1973a. Neuroptera in the Rothamsted insect survey, 1970-72. *Rep. Rothamsted Exp. Sta. 1972*, Part 2: 208-210.

Bowden, J. 1973b. The significance of moonlight in photoperiodic responses of insects. *Bull. Entomol. Res. 62*: 605-612.
Bowden, J. 1976. Neuroptera in the Rothamsted Insect Survey, 1973-75. *Rep. Rothamsted Exp. Sta. 1975,* Part 2: 126-128.
Bowden, J., and B. M. Church. 1973. The influence of moonlight on catches of insects in light-traps in Africa. Part II. The effect of moon phase on light-trap catches. *Bull. Entomol. Res. 63*: 129-142.
Brenchley, W. E., and H. Adam. 1915. Recolonisation of cultivated land allowed to revert to natural conditions. *J. Ecol. 3*: 193-210.
Brown, J. M. B. 1964. Forestry. *In* The soils and land use of the district around Aylesbury and Hemel Hempstead (B.W. Avery, ed.). H.M.S.O., London, pp. 191-198.
Cooke, G. W. 1970. The carrying capacity of the land in the year 2000. *In* The optimum population for Britain (L.R. Taylor, ed.). Academic Press, London and New York, pp. 15-42.
Crichton, M. I. 1971. A study of caddis flies (Trichoptera) of the family Limnephilidae, based on the Rothamsted Insect Survey, 1964-1968. *J. Zool. (London) 163*: 533-563.
Crichton, M. I. 1976. The interpretation of light trap catches of Trichoptera from the Rothamsted Insect Survey. *Proc. First Int. Symp. on Trichoptera, 1974.* Junk, The Hague, Netherlands.
Fisher, R. A., A. S. Corbet, and C. B. Williams. 1943. The relation between the number of species and the number of individuals in a random sample of an animal population. *J. Anim. Ecol. 12*: 42-58.
French, R. A., and L. R. Taylor. 1963. A survey of British moths. *Bull. Amat.Entomol. Soc. 22*: 81-83.
Hall, A. D. 1905. On the accumulation of fertility by land allowed to run wild. *J. Agric. Sci. (Camb.) 1*: 241-249.
Heath, J. 1974. A century of change in the Lepidoptera. *In* The changing flora and fauna of Britain (D.L. Hawksworth, ed.). Academic Press, London and New York, pp. 275-292.
Jenkinson, D. S. 1971. The accumulation of organic matter in soil left uncultivated. *Rep. Rothamsted Exp. Sta. 1970,* Part 2: 113-137.
Kempton, R. A. 1975. A generalised form of Fisher's logarithmic series. *Biometrika 62*: 20-38.
Kempton, R. A., and L. R. Taylor. 1974. Log-series and lognormal parameters as diversity discriminants for the Lepidoptera. *J. Anim. Ecol. 43*: 381-399.
Kempton, R. A., and L. R. Taylor. 1976. Models and statistics for species diversity. *Nature (London) 262*: 818-820.
Lawes, J. B. 1877. The Rothamsted Allotment Club. *J. Roy. Agric. Soc. 13*: 387-393.

Lawes, J. B. 1895. Upon some properties of soils which have grown a cereal crop and a leguminous crop for many years in succession. *Agric. Stud. Gaz. (Cirenc.) 7* (New ser.): 65-72.

Lees, D. R. 1971. The distributuon of melanism in the pale brindled beauty moth *Phigalia pedaria,* in Great Britain. *In* Ecological genetics and evolution (R. Creed, ed.). Blackwell, Oxford, pp. 152-174.

Perring, F. H., and S. M. Walters. 1962. Atlas of the British flora. Nelson, London and Edinburgh.

Sharrock, J. T. R. 1976. Atlas of breeding birds in Britain and Ireland. British Trust for Ornithology, Tring.

South, R. 1907. The moths of the British Isles. Series 1 and 2. (New edn. 1961). Frederick Warne, London.

Steele, R. C., and R. C. Welch. 1973. Monks Wood: a nature reserve record. The Nature Conservancy 1973, Monks Wood, Huntingdon, England.

Tansley, A. G. 1939. The British Islands and their vegetation. Cambridge University Press, Cambridge, England.

Taylor, L. R. 1968. The Rothamsted insect survey. *Nat. Sci. Sch. 6*: 2-9.

Taylor, L. R. 1970. Introduction. *In* The optimum population for Britain (L.R. Taylor, ed.). Academic Press, London and New York, xi-xiii.

Taylor, L. R. 1973. Monitor surveying for migrant insect pests. *Outlook on Agric. 7*: 109-116.

Taylor, L. R. 1974a. Monitoring change in the distribution and abundance of insects. Rep. Rothamsted Exp. Sta. 1973, *Part 2:* 202-239.

Taylor, L. R. 1974b. Insect migration, flight periodicity and the boundary layer. *J. Anim. Ecol. 43*: 225-238.

Taylor, L. R. 1977. Migration and the spatial dynamics of an aphid, *Myzus persicae*. *J. Anim. Ecol. 46*: (in press).

Taylor, L. R., and E. S. Brown. 1972. Effects of light-trap design and illumination on samples of moths in the Kenya highlands. *Bull. Entomol. Res. 62*: 91-112.

Taylor, L. R., and R. A. French. 1969. Rothamsted insect survey. *Rep. Rothamsted Exp. Sta. 1968,* Part 1: 207-214.

Taylor, L. R., and R. A. French. 1973. The Rothamsted insect survey. *Rep. Rothamsted Exp. Sta. 1972,* Part 1: 195-201.

Taylor, L. R., and R. A. French. 1974. Effects of light trap design and illumination on samples of moths in an English woodland. *Bull. Entomol. Res. 63*: 583-594.

Taylor, L. R., and R. A. French. 1976. The Rothamsted insect survey, seventh annual summary. *Rep. Rothamsted Exp. Sta. 1975,* Part 2: 97-127.

Taylor, L. R., R. A. Kempton, and I. P. Woiwod. 1976. Diversity statistics and the log-series model. *J. Anim. Ecol. 45*: 255-272.

Taylor, L. R., and R. A. J. Taylor. 1977. Aggregation, migration and population mechanics. *Nature (London) 265*: 415-421.

Thompson, J. H. 1970. The growth phenomena. *In* The optimum population for Britain (L.R. Taylor, ed.). Academic Press, London and New York, pp. 1-14.

Thurston, J. M. 1958. Geescroft Wilderness. *Rep. Rothamsted Exp. Sta. 1975*: 94.

Williams, C. B. 1939. An analysis of four years captures of insects in a light trap. Part 1. General survey; sex proportion; phenology; and time of flight. *Trans. Roy. Entomol. Soc. Lond. 89*: 79-132.

Williams, C. B. 1940. An analysis of four years captures of insects in a light trap. Part II. The effect of weather conditions on insect activity; and the estimation and forecasting of changes in the insect population. *Trans. Roy. Entomol. Soc. Lond. 90*: 227-306.

Williams, C. B. 1948. The Rothamsted light trap. *Proc. Roy. Entomol. Soc. Lond. A, 23*: 80-85.

Williams, C. B. 1951. Changes in insect populations in the field in relation to previous weather conditions. *Proc. Roy. Soc. Lond. B, 138*: 130-156.

Williams, C. B. 1953. The relative abundance of different species in a wild animal population. *J. Anim. Ecol. 22*: 14-31.

Williams, C. B. 1960. The range and pattern of insect abundance. *Am. Nat. 94*: 137-151.

EDUCATIONAL AND AESTHETIC VALUE OF INSECT-PLANT RELATIONSHIPS IN THE URBAN ENVIRONMENT

J. D. Shorthouse

Department of Biology
Laurentian University
Sudbury, Ontario, Canada

I. INTRODUCTION

Of the many pending concerns for mankind, as we enter the late nineteen seventies, among the most serious is the unprecedented rate of worldwide environmental degradation. Both the great population increase and the spectacular rise in that proportion inhabiting urban areas, have generated problems for the environment on a scale never before encountered. While environmentally related problems such as food shortages, the misuse of biocides, and depletion of the earth's resources may at present enjoy the limelight of public and scholarly attention, far too few people are aware neither of the alarming rate at which urban sprawl now occurs, nor have they stopped to consider its consequences.

By the year 2000, it is estimated that 85% of all Americans

ISBN 0-12-265250-9

will live in cities (Salter, 1974). In Canada, which surprising to most is already largely urban, the two greatly enlarged cities of Toronto and Montreal will comprise 30% of the entire Canadian population (McKeating, 1975). Urban sprawl will continue so rapidly, as is now occurring in northeastern U.S.A., that major cities will merge forming the so-called megalopolis (Salter, 1974). Such sprawl obliterates enormous tracts of natural and agricultural land at a rate estimated to be approximately 1,200,000 ha per year in the U.S.A. and over 120 ha per day in the state of California (Gilliam, 1972).

One consequence of such expansion is that it becomes increasingly difficult for people to identify with land in either a natural or seminatural state. Although daily contact with nature, in at least a minor form, is considered by many an inherent biological necessity (Treshow, 1976), projections of continued urbanization indicate that such experiences will become uncommon for a large sector of the population. Even more alarming is the revelation that once periodic contact with nature is lost, it becomes difficult to obtain public support for the preservation of natural ecosystems within or near cities (McHarg, 1964).

Although cities may be classic examples of biological instability and simplification due to human activity, most urban environments in North America and Europe are far from completely impoverished. The past retention of land for parks and the presence of open spaces and terrain unsuitable for construction has left sufficient habitats in most cities for the persistence of an amazing array of hardy flora and fauna (Kieran, 1959; Gill and Bonnett, 1973; Dagg, 1974; Dagg and Campbell, 1975; McKeating, 1975). Although studies of urban flora (Li, 1969; Lanphear, 1971; Schmid, 1975) and fauna (Dagg, 1970; Recher, 1972; Noyes and Progulske, 1974; Schlauch, 1976) with emphasis on their roles in the urban ecosystems are becoming more numerous, it is surprising that similar studies for insects are almost nonexistent. Although entomologists would unanimously agree that increasing the public's awareness of insects would be advantageous to all, little effort has been made to explain the roles and importance of insects in the urban environment.

It is unfortunate that most literature on urban insects is concerned with destructive species (Ebeling, 1975) since this gives the impression that all urban insects are pests. Although many people may be aware that pollination, for example, is mutually beneficial to both plants and insects, it is likely that visions of defoliated plants lead to a common assumption that phytophagous insects could be considered anything but educational or aesthetically pleasing. Consequently, this added responsibility of convincing people that the majority of insects are not only harmless to man, but indeed are valuable, affords the entomologist the opportunity of making a substantial contribution to the public's awareness of ecosystem complexities. My objective in the following discussion is therefore to examine some of the benefits derived from promoting the understanding and appreciation of urban insects.

II. EDUCATIONAL VALUE OF URBAN INSECTS

Insects are common inhabitants of urban ecosystems and provide a contact with nature for urban residents. Most cities have ample habitats acceptable to insects and little effort is required to encourage their presence (Pyle, 1974; Jackson, 1977). Urban wildlife biologists, in contrast, must often undertake habitat restoration before many species of birds and mammals can be established (Thomas *et al.*, 1973; Noyes and Progulske, 1974; Thomas and DeGraaf, 1975). Parks, institutional grounds and cemeteries, greenstrips, open spaces such as vacant lots, and private gardens all support insect populations.

Although the insect fauna of such sites may have only marginal resemblance to that once present, the array of insects is still surprisingly diverse. The eminent F. E. Lutz, for example, once collected 1,402 species of insects including 35 species of butterflies in a typical 23 x 61 m yard located near New York City. His fascinating book (Lutz, 1941) has become somewhat of a classic since it is one of the few examples of entomological literature aimed at explaining the natural history of common urban insects. In Great Britain 343 species of macrolepidoptera, or 14% of the total number of species found in the British Isles, have been found near the center of London (de Worms, 1965), while 122 of the 682 butterflies of North America have been recorded from the San Francisco Bay region of California (Tilden, 1965).

Gardens and parks perhaps provide the best opportunity for increasing the public's awareness of insects once the obstacle of man's territorial behavior over his home site is overcome. According to Howard (1974), man is usually so entrenched with the hypocracy of thinking he is competing with nature that he cannot bring himself to planting part of a garden for the fauna he has displaced. A more thorough understanding of urban insect ecology however, would likely illustrate that in all probability gardens with the greatest insect diversity are also the healthiest. It is generally accepted that species diversity in natural ecosystems contributes to stability (Odum, 1971), and similarly it is possible that gardens with a variety of insect species would have fewer pest populations increasing to damaging levels. Most gardeners are also surprised to learn that in some instances, moderate insect damage has been shown to increase plant growth (Harris, 1974). Such suggestions appear feasible, however, when it is realized that we are often advised to inflict damage by pruning branches or removing apical buds in order to increase the ultimate amount of plant biomass. Recent experiments by Dyer and Bokhari (1976) indicate that even grasshoppers may ultimately benefit their host plants by injecting growth promoting substances during the chewing process.

Entomologists often express concern that so few teachers are aware of the potential provided by insects for explaining basic principles of biology (Jantzie, 1972; Fischang, 1976; Tipton, 1976a, b). This lack of attention is unfortunate, for insects contribute some of the best examples for illustrating concepts such as diversity, adaptability, and species survival. Indeed, there is hardly a branch of biology that has not been enriched by knowledge obtained from insects.

Studying insects is also an ideal activity for family teaching units (Tipton, 1976b) that encourages parents to fulfull their roles as teachers. The empathy of children for living things along with the availability of insects near homes and schools provides an exciting teaching tool for both parents and teachers. Fortunately, a growing number of authors of children's books have recognized the potential of urban insects (Hutchins, 1966; Davis, 1971; Sadler, 1971; Simon, 1971; Hogner, 1974; Ordish, 1975), and there are also elementary instruction books for rearing insects (Borden and Herrin, 1972; Schneider, 1972; Clark, 1974; Roberts, 1974; Simon, 1975; Saurer, 1976). However, the full educational impact of this resource can be best realized when attempts are made to identify insect traits which illustrate interdependency of ecosystem components.

The diversity of feeding habits among phytophagous insects provides such opportunity. Many phytophagous insects such as aphids, bees, leaf beetles, and larvae of Lepidoptera are sufficiently familiar that most people acknowledge that contacts between insects and plants do exist. Few urbanites are aware, however, that the very existence of both natural and urban ecosystems is critically dependent upon intricate relationships between insects and plants. Approximately 50% of the North American species of insects are dependent upon flowering plants for their food and many species of plants support incredible insect loads. Corn, for example, is known to be host for about 200 species, apples 400, and oaks over 1,000 (Frost, 1959). Although some of these insects are notorious for the damage caused by their feeding, the majority feed without serious ill effects. Most urbanites are also unaware that the diversity of feeding habits has led to such specialized groups as leaf miners, borers, seed feeders, gall formers, and leaf rollers, besides those which feed externally causing noticeable defoliation.

Insect pollination provides one of the best examples of how various components of the urban ecosystem are interdependent, since most people are unaware of the extent to which we rely on insects for the propagation of garden plants such as carrots, tomatoes, and apples. It is common for pollination to be considered a simple process whereby an insect unconsciously brushes against the anthers as it collects nectar, becomes dusted with pollen, then carries it to the next flower. The range of modifications, however, both by the plant for attracting insects and by the insect for collecting nectar and pollen (Macior, 1971), provides almost limitless examples of adaptation. Many flowers are adapted to specialized feeders, such as carnations for butterflies, night bloomers like morning glory for moths, and flowers such as magnolia which avail themselves to beetles. All these plants, moreover, are commonly found in most urban areas.

Accepting the importance of pollinators such as butterflies and moths leads to another prevalent topic in environmental education - concern for the preservation of natural vegetation and ecosystems. Although it can be easily shown for example that butterflies and moths depend on flowers for adult food, it is important to stress that the abundance of these insects also depends on the presence of suitable host plants for their larvae (Rindge, 1965; Pyle, 1976a, b). These host plants, which are seldom the same species required by adults (Proctor and Yeo, 1973), are often absent from urban parks and gardens and are instead restricted to remaining seminatural open spaces.

Besides being the sole food source for many phytophagous insects, it is also useful for students to understand that flowers are an important source of energy for many parasitic and predacious insects. There are several studies in rural areas which indicate that higher levels of parasitism occur in the presence of flowers than in their absence (Leius, 1967; Syme, 1975). Since nectar and pollen are known to be essential for the maturation of ovaries in beneficial predators such as syrphids (Pollard, 1971), the results of experiments whereby various flowers are located near insect attacked vegetable gardens may well provide an interesting lesson in insect ecology.

There are many other phytophagous insects in the urban environment that exhibit remarkable adaptations, but two groups which have attracted a great deal of attention by both amateur and professional entomologists are the leaf miners and gall formers. The interest shown to these two groups is derived in part from their characteristic feeding patterns; the leaf miners which spend most of their life cycle between the two epidermal layers of a leaf and the gall formers which induce their host plants into surrounding them with thick layers of plant cells. Both groups are notorious for their highly specific feeding patterns and examples from either group also effectively illustrate the extent to which insects have filled available niches.

Leaf miners are found in four insect orders and form natural rather than a taxonomic group of species (Hering, 1951). They are well adapted to their restricted environment by having flattened bodies, wedge shaped heads, and reduced appendages. Leaf mines can also be conveniently preserved for the classroom, and the mines often reveal much of the insect's life history. Host specificity of the leaf miners also involves the serious student in plant taxonomy.

However, it is the phenomenon of gall formation that presents the student with the most fascinating of the insect-plant relationships. Galls have been known since ancient times (Hippocrates, 406 to 377 B.C., wrote on the medicinal properties of galls), but it was not until the late eighteenth century that the connection between galls and the insects found in them was discovered (Plumb, 1953). Galls are atypical growths commonly found on either the leaves, stems, or roots that are produced by specialized insects (Mani, 1964). This host response provides the insect with an almost unlimited supply of highly nutritious food and a shelter which affords the inhabitants protection from desiccation and predators.

Like the leaf miners, gall formers are a natural rather than a taxonomic group. They are found in at least eight insect orders, but the majority are restricted to the families Cecidoymiidae (Order Diptera) and Cynipidae (Order Hymenoptera). There are approximately 1,450 species of gall formers in North America (Felt, 1940), and many are found in the urban environment associated with either oaks, roses, willows, poplars, or composites. Eighty-seven different galls were once collected in the vicinity of New York City (Beutenmuller, 1904) and it is likely that even more could be found in other cities.

All galls, especially the more complex such as those formed by cecidomyiids and cynipids, have characteristic sizes and shapes. Many galls exhibit striking coloration while others are recognized by their spiny or hairy protrusions (Darlington, 1968; Hutchins, 1969). The adults of most gall forming insects are extremely difficult to separate taxonomically; however, they can usually be identified by association with their galls which are structurally distinct. The specific shape of an insect gall is dependent upon the genus of insect producing the gall rather than the plant on which the gall is produced.

Although a few gall insects are found on more than one host species, nearly all are specific to a single host genus. Cynipids, for example, have found optimal conditions on the oaks since about 85% of the known species are associated with this genus. Most of the remaining cynipids are associated with members of the Rosaceae and about 7% of these are restricted to the genus *Rosa*.

Galls initiated by cynipid wasps on urban oaks and roses are ideal for student study since they are easily handled and stored in the classroom. Those found on wild and domestic roses are perhaps the most suitable since they are easiest to collect and identify. There are about 30 species of rose gall wasps in North America, compared to over 450 species on oaks, and brief descriptions and illustrations of their galls are available (Felt, 1940; Weld, 1957, 1959).

Maturing galls on either the leaves or stems of roses can be easily opened with a blade to reveal the larvae, and good success can be obtained in emerging the adults once the galls have overwintered under natural conditions (Shorthouse, 1973a). One of the most common galls in urban areas is the spiny stem gall formed by *Diplolepis multispinosus* (Gillette) (Fig. 1). This gall is usually found in clusters with the recently initiated galls, which are soft and light reddish green, growing alongside old and dried galls of previous seasons. Large collections of these galls are most easily made during the winter months in northern climates, when it is most difficult to obtain living insects, since the galls become very conspicuous when snow covers the ground.

Fig. 1. Spiny stem galls of the cynipid Diplolepis multispinosus *(Gillette) found on domestic rose growing in an urban garden in Sudbury, Ontario. The gall to the left has been recently initiated while the gall to the right was initiated the previous season.*

Cynipid galls are commonly inhabited by numerous species of insects besides the gall formers. The localized accumulations of nutritive plant tissues often attract phytophagous insects and these in turn attract entomophagous species. One European gall is reported to have over 75 species of insects associated with it (Mani, 1964). The assemblage of these inhabitants, many of which have their life cycles restricted to specific galls, constitute small but distinct communities (Shorthouse, 1973a). These nongall forming insects also are emerged easily in the classroom providing students with the opportunity of examining aspects of insect community ecology (Shorthouse, 1973a, b). Collections of rose galls often contain so many parasites, predators, and inquilines (Shorthouse, 1973a) that they outnumber the population of gall formers (Fig. 2).

III. AESTHETIC VALUE OF URBAN INSECTS

The study of aesthetics is an important branch of philosophy that has arisen as a result of man's desire to explain such perceptions as beauty and pleasure. Philosophers have identified many types of aesthetic experiences (Beardsley, 1970), and it is generally accepted that regular aesthetic

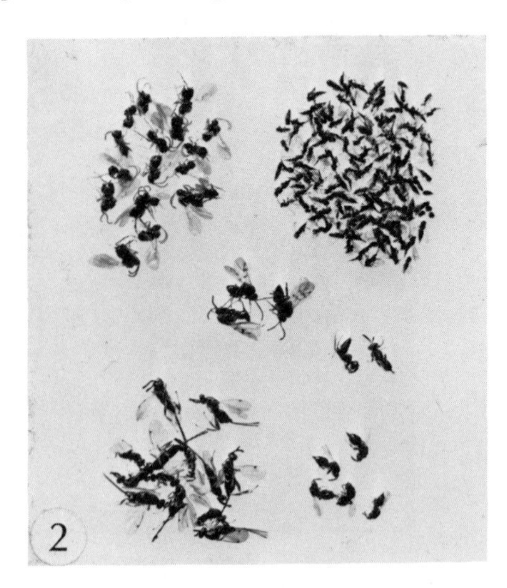

Fig. 2. An assemblage of insects emerged from a collection of cynipid galls found on domestic rose. The three insects at center are cynipid gall formers while those to the upper left are phytophagous species which feed on gall tissues. All others are parasites that feed on the larvae of both gall formers and phytophagous species.

experience is a basic human requirement (Smith and Smith, 1970). Although most of the concepts concerning aesthetic experiences have developed in association with the fine arts (Osborne, 1970), the same criteria for such experiences can be applied to many other fields. Most people, for example, are intrigued by the serene beauty in the natural world; however, few probably consider their enjoyment as being in the realm of aesthetic experiences. According to McHarg (1964), the response which nature can induce - tranquility, calm, introspection, openness to order and purpose is very similar to that evoked by works of art. Wildlife is also considered by many to have high aesthetic value (Geist, 1975; Treshow, 1976). Others have noted that encounters with wild creatures, in any habitat, strike deep chords of response thought to reflect the close ties man once had with the natural world (Shepard, 1967).

Although most philosophers would also agree that the beauties of nature can provide enjoyment, few have fully explored the potential of natural history for providing aesthetic experiences. Hepburn (1973), for example, reports that serious aesthetic concern with nature is today uncommon. Others have noted that although the natural world may never stimulate as many individuals as does art, proper education can show the student that the natural world is also rich in aesthetic value potential (Smith and Smith, 1970). According to philosophical theory, art provides aesthetic experience due to the appreciation one develops for the piece of work (Osborne, 1970). Similarly, we can learn to appreciate nature since appreciation also can be nourished and sustained by knowledge (Sadler, 1974; Geist, 1975).

Several authors have expressed concern that modern man is losing contact with the natural world (Evans, 1968; Gill and Bonnett, 1973; Allen, 1974). More alarming, however, is the suggestion that man may suffer psychologically if he does not regularly experience nature (Smith and Smith, 1970; Parlour and Roberts, 1974), and it is the people living in cities who undoubtedly find the greatest difficulty in maintaining a regular contact. Moreover, since these contacts are usually brief, it is important to identify and promote all aspects of the environment which might provide aesthetic enjoyment.

Public gardens and parks with either their natural or exotic vegetation are both the conscious and unconscious symbol that man is inextricably related to nature (Demaray, 1969) and have long provided enjoyment. It has been suggested that when our ancestors began to remove the natural cover of vegetation to build cities, the loss of nature compelled them to plant trees and to establish gardens and lawns for beauty and pleasure (Li, 1969). According to Shepard (1967) the garden is a formal recognition that there is art in the beauty of the forest edge and amenity in forest clearings. Others think there is psychological justification for parks and gardens based simply on the emotional needs which cause urbanites to seek open spaces and solitude (Gold, 1973). Since many species of birds and mammals have adapted well to such ecosystems and it has been established that people benefit from their presence, there are now efforts to promote urban wildlife management (Gill and Bonnett, 1973; Dagg, 1974; DeGraaf and Thomas, 1974; Geist, 1975; Thomas and DeGraaf, 1975). However, enjoyment derived from urban animals need not be restricted to birds and mammals. In the following discussion, I hope to illustrate that even encounters with various insects have the potential of providing aesthetic experience.

Several of the species of insects which have adapted well to urban parks and gardens have played important roles in the development of human culture, poetry, music and art (Schimitschek, 1968). From the Sung Dynasty of China (960 to 1278 A.D.) to the present, people have appreciated the songs of insects and have encouraged their presence near homes (Evans, 1968; Davis, 1971). The presence of pollinating insects such as bees and butterflies is considered by many to add an element of life and motion to the otherwise static beauty of a garden. It is reported that the late Sir Winston Churchill was so captivated by the beauty of butterflies that he had artificially reared specimens released on his property (Smith, 1975).

Insect verse is worldwide, spanning virtually the entire history of literate mankind, and much of this verse was composed due to the presence of insects associated with garden plants. Kevan (1974) has compiled a collection of over 500 verses associated exclusively with "grigs" or orthopterans. There are also numerous children's books such as Kreidolf's (1968) "Der Traumgarten" (The Dream Garden) and Aldridge's (1973) "The Butterfly Ball" which fancifully illustrate the lives of common garden insects. The recent appearance of books such as Sandved and Emsley's (1975) "Butterfly Magic" and Dalton's (1975) "Borne on the Wind" with their incredible photographs are not only a delight to the layman, naturalist and entomologist, but show without doubt that many insects also are of high aesthetic value.

Aesthetic experiences, however, can be obtained from the natural world by means other than perception of colors and sounds. For example, it has been observed that aesthetic experiences are also procured as one develops a realization or understanding of natural phenomena (Smith and Smith, 1970; Hepburn, 1973; Sadler, 1974; Geist, 1975). Lutz (1941) observed that we often minimize or even despise phenomena in the natural world of which we are ignorant, whereas we are enthusiastic about things that we understand. As an example, he suggested that people appreciate butterflies more than moths because there are so many species of moths that most people cannot easily learn to recognize any but the larger and more conspicuous. Mysterious and unfamiliar aspects of nature, according to Kolnai (1968), are responsible for arousing our interest, and it is this aroused state that contributes to a feeling of enjoyment. Kolnai also argues that interest grows as we continue to observe extraordinary matters and situations that set thinking in motion, and, as a result, aesthetic experiences intensify. It can be shown consequently that even an insect fauna has a greater potential for providing aesthetic experiences than most people realize, and perhaps the best examples to illustrate this are found in the realm of insect-plant relationships.

Insects attracted to garden flowers are familiar to everyone and most would agree that any pleasure derived from their presence is due to colorful bodies and wing patterns. However, the presence and activities of pollinating insects can be far more meaningful and provide more enjoyment if their significance in the urban ecosystem is understood (Pyle, 1976b). For example, monarch butterflies feeding at late season flowers become more fascinating when one associates their activities with the storage of fuel reserves for the long migration south.

Although bees and butterflies may be recognized by many urbanites as pollinators, other common insects such as syrphid flies are equally as important (Oldroyd, 1966). Syrphids may first attract attention simply because they are colorful insects (Fig. 3) and have the peculiar habit of hovering in mid-

Fig. 3. Syrphid fly feeding at a garden flower. These insects attract attention because of their coloration, flight activities, and our knowledge of their role in the garden ecosystem.

air above flowers. But to the entomologist and well informed gardener, syrphids are particularly interesting because of their role in the garden ecosystem. The adult flies live entirely on nectar and are important pollinators (Oldroyd, 1966). Their significance becomes evident when one realizes that many flowers could not reproduce without their presence. Syrphids become even more intriguing when one learns that their larvae are important predators of garden pests such as aphids. It is suggested, therefore, that even the casual observer serves to gain by the presence of such insects simply by obtaining a better understanding of their significance in the garden ecosystem.

Plant galls are another example of an insect-plant relationship that provides differing levels of aesthetic experience depending upon how one relates to their presence. To most casual observers, galls may simply be acknowledged as abnormal swellings on certain plants and warrant little further attention. Others may be further attracted to galls because of their striking coloration while still others have been sufficiently fascinated by the array of gall sizes and shapes that they have been incorporated into various artistic designs. Galls formed on goldenrod by the tephritid fly *Eurosta solidaginis* Fitch, (Family Tephritidae, Order Diptera), which is often found in abandoned city lots, are commonly used in table centerpieces (Morris, 1976) and even have been used in the construction of wall hangings (Fig. 4).

However, for those more intrigued by the natural world, the mysterious and unfamiliar presence of galls on an otherwise typical plant evokes a feeling of bewilderment and thus

Fig. 4. Insect galls of a tephritid fly used artistically in construction of a wall hanging.

provides an aesthetic experience at a different level. Perhaps the most rewarding experience occurs when the observer comes to realize that galls are initiated by specialized insects that have evolved the ability of tricking their hosts into furnishing food supply and a readymade shelter. Plant galls can therefore provide enjoyment and aesthetic experiences for both those observers who are simply attracted by the strange and mysterious shapes of nature and for those with more demanding curiosity who are rewarded with a feeling of awe and wonderment gained by at least partial understanding of nature's most intriguing insect-plant relationship.

IV. DISCUSSION

Since it is likely that the current rates of urbanization now occurring in North America and other regions of the world will continue, it is evident that fewer city residents will have the opportunity for contact with nature in rural areas. It is vital, therefore, that such opportunities be provided in the urban environment and that efforts be made to encourage the preservation of natural and seminatural areas near and within cities. It is encouraging that some influential urban designers now recognize the need for applying ecological principles in the development of urban landscapes (McHarg, 1969; Doxiadis, 1974); however, more citizens must be informed of these pressing needs.

It is also painfully evident that a growing number of urbanites must be both educated and convinced that the presence of nearby living things is of value to all society (Thomas and DeGraaf, 1975). Although we are fortunate that publicity given to environmental matters has alerted most citizens to the dangers of environmental degradation, steps must be taken to assure the public also develops positive attitudes. Convincing the public of the benefits accrued by such appreciation is an important and challenging task for biologists, and it is important to accept that much of this education can be accomplished only if natural and unnatural ecosystems are allowed to persist. Nature in the urban landscape is an underrated resource whose aesthetic and educational values have yet to be explored, yet there are unlimited opportunities for biologists in nearly all disciplines to assure that this attitude is changed.

Insects, since they are important components of the environment and yet so poorly understood by the majority of citizens, are particularly suitable for instituting appreciation of urban biota. Habits of urban insects provide endless examples of natural marvels and consequently, it is feasible that if entomologists were able to further promote awareness of insect fauna, such regard would soon extend to other aspects of the environment.

Entomologists should insure that educators are encouraged to include insect topics in their curricula. For this to occur, information on insects must be provided by entomologists which stirs the curiosity and stimulates interest. Promoting popular articles and pamphlets would be an ideal activity for many entomological societies, and there is urgent need for books on urban insects such as Hogue's (1974) "The Insects of the Los Angeles Basin."

Undertaking entomological research in urban areas is another way of drawing the public's attention to insects, besides revealing new information in neglected areas. At present, most entomological research is conducted outside urban areas, yet many intriguing projects, such as monitoring changes in insect populations as they adapt to reduced habitats, remain untouched. It is also imaginable that urbanization will necessitate a readjustment of entomological priorities which reflect the needs of metropolitan populations and that urban entomology will soon be recognized as a separate discipline.

We also can expect that continuing rises in the cost of living will promote an increase in urban food production and when this occurs, we can be sure that contacts with insects will also increase. Urban entomologists had better be ready to provide identification and control information if pesticides are to be used only when necessary. For this to occur the lay public must be introduced to concepts such as insect diversity, food chains, and an understanding that most insects are harmless. Entomophobia may also become more prevalent; however, the timely circulation of appropriate information would reduce its severity (Olkowski and Olkowski, 1976).

Developing more appreciation for aesthetic offerings of the natural world certainly can be accomplished in the urban setting. Although it may require changes in our current thinking, educators would be wise to promote aesthetics in biology and environmental studies, since it is conceivable that students able to acquire such experiences will subsequently become more sympathetic to environmental issues. Leisure time spent in the urban environment could also become more rewarding for many people once they discover the significance of living things that regularly occur in their backyards and neighborhood parks.

We might also accept the premise that both biologists and casual observers stand to gain from each other if their personal experiences and views of nature can be encouraged to surface. Perhaps more individuals would recognize that there is less distinction between scientific and artistic views of nature than we are led to believe. Again this might be done if scientists were encouraged to explain to the public some of the phenomena that they commonly witness and appreciate and to provide information which permits the casual observer to understand the significance of what he is seeing. Once this can be done, even insect associations such as the examples used in this paper can provide enjoyment for a far greater percentage of the population than exists at present.

ACKNOWLEDGMENTS

This paper was supported in part by the National Research Council of Canada, Grant Number A0230, and by the President's Research Fund of Laurentian University. I am indebted to Dr. G. M. Courtin of the Department of Biology and Dr. R. L. Nash of the Department of Philosophy, both of Laurentian University, for their comments and criticisms of the manuscript. I also thank Laurentian University photographer Mr. M. A. Derro for his help with the illustrations. The photograph of the wall hanging was kindly provided by G. W. Frankie of the University of California, Berkeley.

REFERENCES

Aldridge, A. 1973. The butterfly ball and the grasshopper's feast. Jonathan Cape Ltd., London.

Allen, D. L. 1974. Philosophical aspects of urban wildlife. *In* Wildlife in an urbanizing environment (J.H. Noyes and D.R. Progulske, eds.). Univ. Massachusetts Cooperative Extension Serv., Amherst, pp. 9-12.

Beardsley, M. C. 1970. Aesthetic theory and education theory. *In* Aesthetic concepts and education (R.A. Smith, ed.). Univ. Illinois Press, Urbana, pp. 3-20.

Beutenmuller, W. 1904. The insect galls of the vicinity of New York City. Guide Leaflet No. 16. American Museum of Natural History, New York.

Borden, J. H., and B. D. Herrin. 1972. Insects in the classroom. British Columbia Teachers' Federation, Vancouver.

Clark, J. T. 1974. Stick and leaf insects. B. Shurlock and Co., Winchester (Great Britain).

Dagg, A. I. 1970. Wildlife in an urban area. *Naturaliste canad. 97*: 201-212.

Dagg, A. I. 1974. Canadian wildlife and man. McClelland and Stewart, Toronto.

Dagg, A. I., and C. A. Campbell. 1975. Studies in urban nature. *Bull. Conserv. Coun. Ont. 22*: 10-14.

Dalton, S. 1975. Borne on the wind. E.P. Dutton & Co., New York.

Darlington, A. 1968. The pocket encyclopedia of plant galls. Blanford Press, London.

Davis, B. J. 1971. Musical insects. Lothrop, Lee and Shepard, New York.

DeGraaf, R. M., and J. W. Thomas. 1974. A strategy for wildlife research in urban areas. *In* Wildlife in an urbanizing environment (J.H. Noyes and D.R. Progulske, eds.). Univ. Massachusetts Cooperative Extension Serv., Amherst, pp. 53-56.

Demaray, H. D. (Ed.) 1969. Gardens and culture: eight studies in history and aesthetics. Eastern Press, Beirut.

de Worms, C. G. M. 1965. An analysis of the Macrolepidoptera recorded from the garden of Buckingham Palace. *Lond. Nat. 44*: 77-81.

Doxiadis, C. A. 1974. Marriage between nature and city. *Internat. Wildl. 4*: 4-11.

Dyer, M. I., and U. G. Bokhari. 1976. Plant-animal interactions: studies of the effects of grasshopper grazing on blue grama grass. *Ecology 57*: 762-772.

Ebeling, W. 1975. Urban entomology. Univ. Calif. Div. Agric. Sci., Berkeley, California.

Evans, H. E. 1968. Life on a little-known planet. E.P. Dutton, New York.

Felt, E. P. 1940. Plant galls and gall makers. Comstock Publishing Co., New York.

Fischang, W. J. 1976. Another wasted resource. *Am. Biol. Teach. 38*: 204.

Frost, S. W. 1959. Insect life and insect natural history. Second Edition. Dover Publications, New York.

Geist, V. 1975. Wildlife and people in an urban environment - the biology of cohabitation. *In* Wildlife in urban Canada (D. Euler, F. Gilbert, and G. McKeating, eds.). Univ. Guelph, Guelph (Ontario), pp. 36-47.

Gill, D., and P. Bonnett. 1973. Nature in the urban landscape: a study of city ecosystems. York Press, Baltimore.

Gilliam, H. 1972. For better or for worse: the ecology of an urban area. Chronicle Books, San Francisco.

Gold, S. M. 1973. Urban recreation planning. Lea and Febiger, Philadelphia.

Harris, P. 1974. A possible explanation of plant yield increases following insect damage. *Agro-Ecosystems 1*: 219-225.

Hepburn, R. W. 1973. Aesthetic appreciation of nature. *In* Contemporary aesthetics (M. Lipman, ed.). Allyn and Bacon, Boston, pp. 340-354.

Hering, E. M. 1951. Biology of the leaf miners. W. Junk, Gravenhage (Netherlands).

Hogner, D. C. 1974. Good bugs and bad bugs in your garden. Thomas Y. Crowell Co., New York.

Hogue, C. L. 1974. The insects of the Los Angeles Basin. Natural History Museum of Los Angeles County. Science Ser. 27.
Howard, W. E. 1974. Why wildlife in an urban society? *In* Wildlife in an urbanizing environment (J.H. Noyes and D.R. Progulske, eds.). Univ. Massachusetts Cooperative Extension Serv., Amherst, pp. 13-18.
Hutchins, R. E. 1966. Insects. Prentice-Hall, New Jersey.
Hutchins, R. E. 1969. Galls and gall insects. Dodd, Mead and Company, New York.
Jackson, B. S. 1977. How to start a butterfly garden. *Nature Canad. 6*: 10-14.
Jantzie, D. N. 1972. University education and high school biology. *In* Entomology and Education. *Quaest. Entomol. Suppl. 8*: 1-3.
Kevan, D. K. McE. 1974. Land of the grasshoppers. Lyman Entomological Museum and Research laboratory. Memoir No. 2, Macdonald College of McGill Univ., Montreal.
Kieran, J. 1959. A natural history of New York City. Houghton Mifflin, Boston.
Kolnai, A. 1968. On the concept of the interesting. *In* Aesthetics in the modern world (H. Osborne, ed.). Weybright and Talley, New York, pp. 166-187.
Kreidolf, E. 1968. Der Traumgarten. Rotapfel-Verlag, Zürich.
Lanphear, F. O. 1971. Urban vegetation: values and stresses. *Hortscience 6*: 332-334.
Leius, K. 1967. Influence of wild flowers on parasitism of tent caterpillar and codling moth. *Canad. Entomol. 99*: 444-446.
Li, H. I. 1969. Urban botany: need for a new science. *BioScience 19*: 882-883.
Lutz, F. E. 1941. A lot of insects. Entomology in a suburban garden. G.P. Putnam's Sons, New York.
Macior, L. W. 1971. Co-evolution of plants and animals - systematic insights from plant-insect interactions. *Taxon 20*: 17-28.
Mani, M. S. 1964. Ecology of plant galls. Junk, The Hague.
McHarg, I. 1964. The place of nature in the city of man. *Ann. Am. Acad. Pol. & Soc. Sci. 352*: 1-12.
McHarg, I. 1969. Design with nature. Natural History Press, New York.
McKeating, G. B. (Ed.) 1975. Nature and urban man. Canadian Nature Federation. Special Publication No. 4, Ottawa.
Morris, J. R. 1976. Personal communication. Dept. of Biology, Laurentian University, Sudbury, Ontario.
Noyes, J. H., and D. R. Progulske. 1974. Wildlife in an urbanizing environment. Univ. Massachusetts Cooperative Extension Serv., Amherst.

Odum, E. P. 1971. Fundamentals of ecology. Third Edition. W.B. Saunders, Philadelphia.
Oldroyd, H. 1966. The natural history of flies. W.W. Norton & Company, New York.
Olkowski, H., and W. Olkowski. 1976. Entomophobia in the urban ecosystem, some observations and suggestions. *Bull. Entomol. Soc. Am. 22*: 313-317.
Ordish, G. 1975. The year of the butterfly. Charles Scribner's Sons, New York.
Osborne, H. 1970. The art of appreciation. Oxford Univ. Press, London.
Parlour, J. W., and E. N. R. Roberts. 1974. Conservation of urban open space. *In* Conservation in Canada (J.S. Maini and A. Carlisle, eds.). Government of Canada Publication, No. F047-1340-1, Ottawa, pp. 343-371.
Plumb, G. H. 1953. The formation and development of the Norway spruce gall caused by *Adelges abietis* L. *Conn. Agric. Exp. Sta. Bull. 566*: 1-77.
Pollard, E. 1971. Hedges *VI*. Habitat diversity and crop pests: a study of *Brevicoryne brassicae* and its syrphid predators. *J. Appl. Ecol. 8*: 751-780.
Proctor, M., and P. Yeo. 1973. The pollination of flowers. Collins, London.
Pyle, R. M. 1974. Watching Washington butterflies. Seattle Audubon Society, Seattle.
Pyle, R. M. 1976a. Conservation of Lepidoptera in the United States. *Biol. Conserv. 9*: 55-75.
Pyle, R. M. 1976b. The scientific management of butterfly and moth populations; a new thrust of wildlife conservation. *Discovery 11*: 68-77.
Recher, H. F. 1972. The vertebrate fauna of Sydney. *In* The city as a life system (H.A. Nix, ed.). Proc. Ecol. Soc. Aust. 7, Canberra, pp. 79-87.
Rindge, F. H. 1965. The importance of collecting - now. *J. Lepid. Soc. 19*: 193-195.
Roberts, H. R. 1974. You can make an insect zoo. Childrens Press, Chicago.
Sadler, D. 1971. Studying insects. McGraw-Hill Co. of Canada, Toronto.
Sadler, D. 1974. In appreciation ... *Ontario Naturalist 14*: 5-7.
Salter, P. S. 1974. Toward an ecology of the urban environment. *In* The environmental challenge (W.H. Johnson and W.C. Steere, eds.). Holt, Rinehart and Winston, New York, pp. 238-263.
Sandved, K. B., and M. G. Emsley. 1975. Butterfly magic. Viking Press, New York.
Saurer, R. J. 1976. Rearing insects in the classroom. *Am. Biol. Teach. 38*: 216-221.

Schimitschek, E. 1968. Insekten als Nahrung, in Brauchtum, Kult und Kultur. Handbuch der Zoologie. IV Band: Anthropoda - 2. Hälfte: Insecta, Berlin.

Schlauch, F. C. 1976. City snakes, suburban salamanders. *Nat. Hist. 85*: 46-53.

Schmid, J. A. 1975. Urban vegetation: A review and Chicago case study. Research Paper No. 161. Univ. Chicago Press, Chicago.

Schneider, G. 1972. Conservation teaching in the cities. *In* Environmental education: A sourcebook (C.J. Troost and H. Altman, eds.). John Wiley and Sons, New York.

Shepard, P. 1967. Man in the landscape: A historical view of the aesthetics of nature. Alfred A. Knopf, New York.

Shorthouse, J. D. 1973a. The insect community associated with rose galls of *Diplolepis polita* (Cynipidae, Hymenoptera). *Quaest. Entomol. 9*: 55-98.

Shorthouse, J. D. 1973b. Common insect galls of Saskatchewan. *Blue Jay 31*: 15-20.

Simon H. 1971. Our six-legged friends and allies: Ecology in your backyard. Vanguard Press, New York.

Simon, S. 1975. Pets in a jar. Viking Press, New York.

Smith, A. U. 1975. Attracting butterflies to the garden. *Hort. 52*: 34-35.

Smith, R. A., and C. M. Smith. 1970. Aesthetic and environmental education. *J. Aesthetic Educ. 4*: 125-140.

Syme, P. D. 1975. The effects of flowers on the longevity and fecundity of two native parasites of the European pine shoot moth in Ontario. *Environ. Entomol. 4*: 337-346.

Thomas, J. W., R. O. Brush, and R. M. DeGraaf. 1973. Invite wildlife to your backyard. *Natn. Wildl. 11*: 5-16.

Thomas, J. W., and R. M. DeGraaf. 1975. Wildlife habitats in the city. *In* Wildlife in urban Canada (D. Euler, F. Gilbert and G. McKeating, eds.). Univ. Guelph, Guelph (Ontario), pp. 48-68.

Tilden, J. W. 1965. Butterflies of the San Francisco Bay region. Univ. Calif. Press, Berkeley.

Tipton, V. J. 1976a. Insects: A success story. *Am. Biol. Teach. 38*: 205-207.

Tipton, V. J. 1976b. Changing scene. *Insect World Digest 3*: 3-4.

Treshow, M. 1976. The human environment. McGraw-Hill, New York.

Weld, L. H. 1957. Cynipid galls of the Pacific slope. Privately printed. Ann Arbor, Michigan.

Weld, L. H. 1959. Cynipid galls of Eastern U.S. Privately printed. Ann Arbor, Michigan.

ECOLOGY OF MEDICALLY IMPORTANT ARTHROPODS IN URBAN ENVIRONMENTS

Bernard C. Nelson

Vector and Waste Management Section
California Department of Health
Berkeley, California

I. INTRODUCTION

Medically-important arthropods comprise a vast array of taxonomically and ecologically diverse groups. Perusal of the literature of medical entomology discloses that most orders of insects and arachnids contain some members that are considered arthropods of medical importance, because either they 1) bite; 2) sting; 3) feed on blood or other tissues; 4) are parasites; 5) are vectors and intermediate hosts of parasites and pathogens; 6) induce allergies; 7) are nuisances; or 8) invoke entomophobia. The scope of medical entomology is broad and in-

ISBN 0-12-265250-9

cludes all arthropods of medical, veterinary, and wildlife importance. This is reasonable as certain of these arthropods affect humans as well as domestic and wild animals. Of greatest medical importance are those vector species that transmit infectious diseases, primarily zoonoses, from domestic and wild animals to humans. This chapter is limited primarily to those arthropods that are clearly of medical importance, both vector and non-vector species, in urban environments.

The focus of this work centers on ideas and concepts basic to investigations into the ecology of arthropods of medical importance in urban environments and on the development of new ideas and concepts. The purpose is to crystallize thinking and to stimulate further research in the emerging field of urban entomology. An effort has been made to place the status of medically-important arthropods in perspective with the field of urban entomology.

A discussion is presented that illustrates ecological factors and requirements that have allowed certain medically-important arthropods to occur and persist in urban environments. Two questions are indicative of the parameters of this section of the paper: 1) What medically-important arthropods have succeeded in adapting to the urban environment? 2) What conditions are created through urbanization which allow for successful establishment of certain medically-important arthropods or which limit the occurrence and persistence of others that are less adapted to urban environments? Major species that have adapted to the urban setting are discussed, whereas only representative members of the less-adapted species that enter urban settings are considered. The latter illustrate the relationship of urbanization to species that occur in that setting under only favorable conditions, that occur only incidentally, and that use certain urban settings transitionally during urbanization.

Certain groups of medically-important arthropods and topics relating to them are discussed elsewhere in this book (e.g., cockroaches, wasps, and entomophobia) and will not be repeated here. Of those arthropods discussed, greater emphasis will be given to vectors of infectious diseases and haematophagous arthropods as a body of literature exists that attempts to integrate the occurrence and persistence of these arthropods and their associated disease cycles into ecological and evolutionary perspectives. Two groups of largely non-vector or non-haematophagous arthropods will also be discussed: synanthropic flies and house dust mites.

II. OBSTACLES IN ANALYZING THE ECOLOGY OF MEDICALLY-IMPORTANT ARTHROPODS IN URBAN ENVIRONMENTS

Any discussion of medically-important arthropods in reference to urban environments is made difficult and complicated by the distribution patterns of the arthropods themselves and by certain traditional and practical approaches to studies of public health-related problems. The foremost obstacle is that no medically-important arthropod has an exclusively urban distribution. All occur in urban, rural, and wild settings to a greater or lesser degree. For example, populations of an indigenous species, such as *Aedes aegypti* (Linnaeus) in Africa, may be found in all three settings, whereas populations of introduced species of medical importance generally occur in domestic urban and rural settings in close association with mankind, as does *A. aegypti* in the Americas, Asia, and Oceania. Even then, after introduction into Queensland, Australia, a population of *A. aegypti* has returned to a semi-wild setting (Mattingly, 1957).

Because of this general distribution pattern, more studies (Kleevens, 1971) and more comprehensive autecological studies (e.g., Trpis and Hausermann, 1975) of medically-important arthropods have been made in rural settings than in urban ones. Rural study sites are generally preferred, because they are small and tend to have natural boundaries whereby the investigator is more easily able to determine, measure, and control variables that occur and to initiate and maintain effective public relations. In urban settings there are greater chances than in rural situations for traps and study areas to be disturbed by people and pets. Sufficient numbers of studies have been made in both urban and rural settings of species that are adapted to urban environments (e.g., *A. aegypti* and *Culex pipiens* complex) to demonstrate the general applicability of results of some rural studies to urban situations.

Because of the role of arthropods in transmission of agents of diseases affecting human health, a major research emphasis has been placed on epidemiological studies related to the arthropods involved. Paramount to these studies is the need to assess the sites of exposure or attack by medically-important arthropods. A classification has been developed that reflects habits and habitats of these arthropods in relation to humans, their dwellings, and their areas of activities, and has great epidemiological significance, far greater than the classification of habitats as urban and rural. Thus, use of the concepts of domestic, peridomestic, and wild (feral is sometimes substituted for wild) habitats and habits of these arthropods has predominated in both epidemiological and ecological studies of medically-important arthropods.

Trpis and Hausermann (1975) have defined domestic habitat as a man-made shelter or dwelling; peridomestic habitat as that area immediately surrounding but outside man-made shelters; and feral as the natural ecosystem, which evolved independent of man and still remains more or less independent from man and his dwellings. The domestic habit of an arthropod was defined by Lewis (1973) as "the habit of remaining within a man-made shelter throughout the whole or a definite part of the gonotrophic cycle." Domestic arthropods primarily breed, feed, and rest indoors. Peridomestic arthropods primarily breed, feed, and rest outdoors but within the immediate areas of man-made shelters; occasionally they may feed and rest indoors. Wild arthropods dwell in natural habitats where they breed, feed, and rest; occasionally they enter the peridomestic habitat where the two habitats are approximated (Trpis and Hausermann, 1975).

Lewis (1973) noted that "domestic" is an indefinite term, perhaps because domesticity in a species or population is a matter of degree. Nevertheless, Lewis emphasized that the domestic habit of medically-important arthropods is important in relation to disease, sampling, and control. The worldwide malaria eradication program of the World Health Organization has been based largely on the concept of control of anopheline mosquitoes in domestic habitats (Mattingly, 1962). Lewis found these terms useful in reference to species of phlebotomines (sand flies). Trpis and Hausermann (1975) demonstrated that there was a genetic basis, reflected taxonomically and behaviorally, for domestic, peridomestic, and wild populations of *A. aegypti*. Most urban species of medically-important arthropods are domestic and peridomestic species, but not all domestic and peridomestic species occur in urban environments. Thus, use of this classification of domesticity is a major stumbling block in accurately assessing which medically-important arthropods can be incorporated into the field of urban entomology.

The final obstacle faced in an analysis of medically-important arthropods in urban environments is the lack of a universally accepted definition of urban. Because there is a continuum of inhabited sites from the family farm or primitive village to the giant metropolis, division between urban and rural is arbitrary. Surtees (1971) defined urbanization as, "the replacement of a natural ecosystem by a dense focus created by man and containing man as the dominant species and environmentally organized for his survival." The dense focus is the urban site, but there is no agreement as to what is dense. The Population Division, United Nations Bureau of Social Affairs (1969), gives two criteria to distinguish urban from rural: one is quantitative in terms of populations, using 20,000 inhabitants as the arbitrary minimum figure for urban centers and the other is qualitative, using characteristics of the economy and mode of living. Urban areas are usually associated with commerce and industry, whereas rural areas are usually associated with agriculture. Traditionally, medical entomologists, epidemiologists, and sociologists have used the figure of 100,000 inhabitants as the dividing line between urban and rural (Davis, 1969; Kleevens, 1971; Graham and Gratz, 1975). To further complicate the issue, census bureaus of the United States and India use population levels of 2,000 and 5,000, respectively, as the lower limits of the urban category (Murphey, 1969).

Acceptance of the United Nations' figure of 20,000 seems reasonable, but is not helpful when confronting the vast literature on medical entomology. Although many authors refer to their study sites as urban or rural, many other authors use the domestic classification of habitat and fail to designate the urban or rural status of their study sites. In general, without a description of the qualitative and quantitative characteristics of the latter study sites, it is often difficult to determine the applicability of a study to urban entomology. In the future, authors are encouraged to designate their study sites in terms of urban and rural as well as domestic, peridomestic, and wild. On a practical basis, it is easier to analyze medically-important arthropods in relation to the process of urbanization (as defined by Surtees, 1971) which is dynamic ecologically and evolutionarily. This approach precludes acceptance of any arbitrary population figure and is used herin.

III. CONCEPTS UNDERLYING THE ECOLOGY AND EVOLUTION OF MEDICALLY-IMPORTANT ARTHROPODS

The need for a general theoretical framework for understanding the ecology and evolution of medically-important arthropods and the diseases that they transmit was stressed by Audy (1958). The theoretical framework that best meets this end is the doctrine of natural nidality of transmissible diseases of Pavlovsky (1966) or, as it is sometimes referred to, the landscape theory of epidemiology in reference to arthropod-borne diseases or the theory of natural foci of diseases. Pavlovsky first formalized his doctrine in 1939, although concepts underlying this doctrine were recognized earlier by a number of workers (Audy, 1958). Nevertheless, Pavlovsky must be given credit inasmuch as the formalization of this doctrine has led to greater understanding of arthropod-borne diseases by reducing the bewildering complexity of these diseases to a common denominator (Olsen, 1970) and by directing research to unravel more easily this complexity (Audy, 1958).

A natural nidus, or nest of the disease (focus is generally substituted for nidus in the western literature), is "that portion of territory of a definite geographic character, on which there has evolved a definite interspecies set of relationships between the disease agent, the animal donors, and the recipients of the agent and its vectors among factors of the external environment favoring or at least not hindering the circulation of the agent" (Pavlovsky, 1955, in Galuzo, 1968). Focus of the disease is maintained in nature by the fact that the disease agent is a member of a historically formed biocenose (community) belonging to a definite natural geographic landscape. The disease agent circulates uninterruptedly in nature, passing from one species of animal to another, from donor animals through arthropod vectors to susceptible animal recipients. Natural foci of diseases arose along with biocenoses and were formed in the process of the evolution of geographic landscape, independent of the life and activity of man. The relationship between structure of the natural focus and the biocenose of a definite geographic landscape determines length of existence of the focus. Natural foci may become extinct when changes occur in the geographic landscape, as these changes produce conditions under which the existence of the biocenose (the biological basis of a focus) becomes impossible. Succession to the biocenose in which the natural focus exists may be brought about in a nature by the participation of man as well as without his help (Galuzo, 1968). Pavlovsky (1966) recognized two kinds of natural foci: 1) *autochthonous*, arising in the process of evolutionary formation of biocenoses without even the tangential participation of man; and 2) *anthropurgic*, or man-made, arising in nature as the result of some human activity.

In his review of concepts and examples of the theory of localization of arthropod-borne diseases, Audy (1958) also regarded the components associated with the focus (pathogen, vectors, reservoir hosts, and recipient hosts) as a biocenose, i.e., a species-network or community. Whereas Pavlovsky developed a distinct terminology associated with his doctrine of epidemiology, Audy defined, clarified, and discussed the ecological phenomena associated with the focus and biocenose in the familiar ecological terms and concepts of Elton (1927, 1966). Thus, Audy's discussion and examples of the ecology and evolution of disease foci centered on concepts of niche, habitat, community, ecosystem, edge-effects and the like, which are more familiar to biologists who may not be versed in epidemiological or medical terminology. Audy, as true of Pavlovsky and other workers, adhered to a holistic outlook and regarded an ecosystem or biocenose as a kind of super-organism, which has some new quality of its own. Audy contended that natural selection may act on a biocenose as a whole rather than separately on each component. Audy's paper should be read in its entirety as there are more data and philosophical concepts therein that can be mentioned or summarized here. Nevertheless, it is emphasized that Audy discussed man's role in the development of new foci through domestication and urbanization. In this regard, he implied that the occurrence and persistence of both vector and non-vector species of arthropods in natural or man-made habitats can be more easily understood in the framework of the concepts of nidality.

A. Applicability of the Concept of Nidality to Medically-important Arthropods in Urban Environments

The theoretical framework, which is based on ecological and evolutionary principles, underlying the concept of nidality seems clearly applicable to analysis and greater understanding of the ecology of medically-important arthropods in urban environments. According to the definition of urbanization by Surtees (1971), urbanization is a dynamic ecological process, replacing natural ecosystems by destroying and modifying original landscapes and communities and creating new ones. Urbanization, therefore, may be viewed in an evolutionary sense as one of the selective forces that determines extent of occurrence and persistence of medically-important arthropods in urban environments. A struggle for survival ensues between components of the biocenose in autochthonous foci and urbanization. Owing to the severity of certain pathogens that some vectors transmit, the autochthonous focus may prevail over urbanization. For example, onchocerciasis, caused by a filarial worm transmitted by the black fly, *Simulium damnosum* Theobald, is reported to be the most important single deterrent to large-scale development of the fertile river valley of the Volta River basin in West Africa (World Health Organization, 1976). Dorozynski (1976) has stated that 100 million km^2 of fertile land in Africa have been abandoned to the tsetse, *Glossina* spp., and trypanosomiasis (sleeping sickness). Therefore, arthropod vectors have actually prevented the development of urbanization.

Usually urbanization succeeds through destruction and modification of landscapes and eradication of one or more components of the biocenoses associated with the foci. The gradual disappearance of malaria in the United States historically may be associated with the urbanization of rural districts and not simply with eradication of *Anopheles quadrimaculatus* Say breeding sites (Watson, 1949). Also, the enzootic plague focus in the San Bruno Mountains area of San Mateo County, California (Kartman and Hudson, 1971), has diminished in size owing to the continuous encroachment of urbanization into grassland and chaparral habitats that support the reservoir hosts, *Microtus californicus* (Peale) and *Peromyscus maniculatus* (Wagner). The struggle between proponents of open space (saving this focus is not their objective) and those favoring development that is in progress will almost certainly determine continuity or extinction of this focus. Apparently when urbanization succeeds, rapidity of the demise of autochthonous foci is determined by speed and extent of urbanization. In contrast, under certain conditions of war and natural disasters, an autochthonous focus may be reestablished in urban habitats that begin to revert to the original ecosystem. Traub and Wisseman (1974) reported that foci of scrub typhus, a rickettsial disease transmitted by chigger mites (Trombiculidae), became established in suburbs of cities in Asia during World War II in previously well-kept gardens, on golf courses, and on race tracks. These landscapes reverted to secondary or transitional vegetation necessary as habitat for rodent reservoirs and mite vectors.

B. Concept of Fringe Habitat

The activities of man in the process of urbanization produce a mosaic pattern of habitats (=landscapes) (i.e., business centers, industrial centers, urban residential centers, suburbs, slums, open space, parks; and, associated with many large metropolises, "islands" of farms, ranches, orchards, wildlife reserves, and wilderness areas that have become surrounded by the encroachment of urban sprawl) along with a multiplicity of associated fringe habitats. These potentially allow for the formation of new biocenoses and foci, each with its complement of medically-important arthropods that have become adapted or were preadapted to conditions found in each habitat. These man-made conditions provide food, water, and harborage for the arthropods or the hosts of these arthropods. Although urbanization may produce an increase in certain natural conditons that are necessary for survival of the arthropod or host components of the biocenose, it more often produces artificial and simulated conditions that allow for development of unusually dense populations. Urbanization selects those arthropods that will succeed in urban environments by offering substitutes for their natural requirements.

The concept of fringe habitat is important in understanding the occurrence and persistence of many medically-important arthropods in natural as well as urban environments. The term denotes a special biotope found along borders where two kinds of ecological formation or vegetation meet, producing a third habitat, which has an abundance of food and shelter to provide for the localized concentration of organisms (Traub and Wisseman, 1974). At the junctures of these biotopes, special "edge-effects" (Audy, 1958) influence distribution and behavior of the organisms living there. As a result, certain species (vectors and hosts) may occur there in unusually large numbers, and this may intensify disease transmission. Audy further stated that humans during settlement create a patchwork of vegetation ("mosaic vegetation") which produces a multiplicity of fringe habitats, inviting occupancy by medically-important arthropods and for the formation of anthropurgic foci. Use of the concept of fringe habitat better illustrates how medically-important arthropods become associated with urban environments when autochthonous or rural anthropurgic foci and urbanized habitats are proximate and, during a period of transition that occurs, when urbanization invades areas of autochthonous foci.

The concept of fringe habitat is particularly enlightening concerning the major trends that have taken place in urban growth in both developed and developing countries during the post-World War II period. Besides the greatly increased population growth, there has been a phenomenal growth in rural-to-urban migration in the developing countries of Africa, Asia, and Latin America (Abu-Lughod, 1969; Morse, 1969; Ragheb, 1969; Turner, 1969; Graves and Graves, 1974; Bruce-Chwatt, 1975). Turner (1969) reported that an estimated 300 million people moved to cities in Africa, Asia, and Latin America during the 1960's. Bruce-Chwatt (1975) gave a nine percent annual increase in population growth of African cities. This migration has been so rapid that social services offered in urban centers have been overwhelmed. The result has been the development of areas of squalor occupied by a large segment of the migrant population. These areas are often marginal to the cities and have been called "septic fringes" (Bruce-Chwatt, 1975). In contrast, in developed countries, there has been a migration from inner cities to suburbs during the post-World War II period. These suburbs are called herein the "affluent fringe." Although the origin of the two kinds of fringes and the socioeconomic statures of the inhabitants are different, these fringes have produced conditions conducive to the occurrence and persistence of medically-important arthropods.

1. The Septic Fringe

"Septic fringes" have been created by migrants forced out of their rural settings by overpopulation and limited resources (Graves and Graves, 1974). They have come to urban centers as this is the only hope for survival and for bettering themselves (Turner, 1969). The tremendous influx of migrants has created immense housing and public health problems as cities have had insufficient physical resources to absorb their growing populations (Morse, 1969). Because housing is scarce, rural migrants quickly establish rural-type dwellings that are devoid of any sanitation or modern facilities (Ragheb, 1969), creating the shanty towns, the "marginal" settlements of Morse (1969) or the "autonomous" settlements of Turner (1969). These shanty towns are built on active or abandoned refuse dumps, along the banks of canals and rivers, along roads and railway tracks, on public or private land that has been invaded, or on private unimproved lands (Morse, 1969). Water is often in short supply, at least seasonally, is often obtained from community hydrants and wells, and is stored in containers for drinking and domestic use. These water sources are subject to pollution from infiltration of ground water and surface runoff. Treatment of water for contamination is inadequate or nonexistent. Along with this kind of water supply one usually finds primitive sewage and waste disposal practices. Owing to a lack of trained manpower and physicians, health facilities are unable to cope with resulting public health problems.

These conditions have created foci for a variety of infectious diseases including many of arthropod-origin. Inadequate housing, necessities for storage of water, inadequate and polluted water supply, and lack of sewage and waste disposal create substrates for such vectors as *Aedes aegypti* (Bruce-Chwatt, 1975), *Culex pipiens* Linnaeus (Subra, 1975), *Anopheles* spp. (Bruce-Chwatt, 1975), *Phlebotomus papataci* Scopoli (Lewis, 1974), *Triatoma barberi* Usinger (Zarate, 1977), rodent hosts (Gratz and Arata, 1975; Barnes, 1975), and synanthropic flies (Nnochiri, 1968). Many migrants come from rural areas that are endemic or hyperendemic for diseases such as yellow fever, dengue, other arboviruses, malaria, filariasis, leishmaniasis, and Chagas' disease, and they may harbor a concurrent infection during migration. Graves and Graves (1974) have established that certain segments of the migrant population return to their rural villages for regular visits. Thus, these patterns of migration and visits from hyperendemic areas allow for introduction and reintroduction of pathogens into septic fringes to complete and maintain these anthropurgic foci. Foci associated with septic fringes offer sources of infections and infestations, not only to inhabitants of septic fringes, but to inhabitants of the inner city as well.

2. The Affluent Fringe

The "affluent fringe" has developed by migration of middle to upper socioeconomic elements of the population from the older, often decaying, inner cities to the suburbs. Houses here are modern and have all the facilities for adequate water supply and waste disposal. Two kinds of suburbs are characteristic of the affluent fringe. In the first, which are largely built in natural ecosystems or on lands formerly used for agriculture and later abandoned and allowed to revert to a "natural" state, owners tend to maintain as much of the natural setting as is possible (Weber, 1977). In the second, most homes are built on land recently used for orchards and agriculture. These homes were designed to include spacious lots with patios, pools, and lush landscaping; landscaped vegetation matures in approximately 10 to 18 years (Ecke, 1964; Brooks, 1966) in California. The first kind of suburb may be associated with autochthonous foci, owing to invasion of natural ecosystems, whereas the second is more often associated with anthropurgic foci. However, where the second kind lies proximate to natural ecosystems, the precise designation of the origin of any foci present may be difficult.

Evidence of foci in the affluent fringe may be seen in the contemporary picture that has emerged in the eastern United States with large outbreaks of Rocky Mountain spotted fever (caused by *Rickettsia rickettsi*) expanded into previously abandoned farmlands and woodlands (Easton *et al.*, 1977). A similar situation exists for the occurrence of boutonneuse fever (caused by *Rickettsia conori*) in South Africa (Hoogstraal, 1972). Many affluent South Africans live in suburbs of Johannesburg in houses with lush gardens and spend time on weekends in the veld, a natural focus of boutonneuse fever. As a result of the introduction of ticks and the rickettsial agent into the suburbs, an urban focus has developed in these lush gardened suburbs where ticks are able to survive using rodents that occupy these garden sites as hosts for their immature stages and dogs for the adult stages. Humans have acquired boutonneuse fever from these garden sites.

In Southern California, a changing ecological picture for murine typhus (caused by *Rickettsia mooseri*) has taken place in the affluent suburbs in the foothills in the eastern greater Los Angeles area (Adams *et al.*, 1970). Here opossums (*Didelphis marsupialis* Linnaeus) and the cat flea (*Ctenocephalides felis* [Bouche]) have become associated with human cases. This transmission cycle differs from the classic rat-flea cycle implicating *Rattus* spp. and *Xenopsylla cheopis* (Rothschild) in rural and urban habitats.

Sylvatic plague (caused by *Yersinia pestis*) has been the source for human cases and intense rodent epizootics in suburbs and at the interface of suburbs and natural ecosystems. Weber (1977) has pointed out that more and more populations of Albuquerque and Santa Fe, New Mexico, have moved into suburbs in the Sandia Mountains and foothills of the Sangre de Cristo Mountains, respectively. Both areas are known sylvatic plague foci. These suburbs are of the autochthonous kind in which the "natural" setting is maintained so sylvatic hosts and their fleas coexist. Epizootics of sylvatic plague have also occurred in California in 1975 to 1977 at the interface of suburbs and natural ecosystems (California Department of Health records) and in Denver, Colorado, in 1970 in an urban park

(Hudson *et al.*, 1971).

Many examples of other medically-important arthropods of urban environments could be given. *Aedes sierrensis* (Ludlow), a tree-hole mosquito found in natural woodland areas, has become a suburban-urban nuisance in many communities surrounding San Francisco Bay, California (Mathis, 1962; Brannan, 1964), as more people have moved into wooded areas. Oaks are preserved to grace the landscape, and natural sources for *A. sierrensis* remain intact in suburban areas. This mosquito is but one example of the growing list of medically-important arthropods that were previously associated with natural, rural, or recreational habitats (Merritt and Newson, Chapt. 6, this volume), but now are more frequently found in suburban and urban habitats. This phenomenon is also seen with increased urbanization of the recreational areas, e.g., the Lake Tahoe Basin of California and Nevada. Here, snow mosquitoes (*Aedes* spp.) have become an urban problem. Similarly, artificial lakes, sewage lagoons, and reservoirs near suburban homes have become sources for chironomid midges (Magy, 1968; Frommer and Rauch, 1971; Newson, 1976; Merritt and Newson, Chapt. 6, this volume).

IV. ARTHROPODS OF MEDICAL IMPORTANCE IN URBAN ENVIRONMENTS

This preliminary discussion has focused on concepts that concern our understanding of the ecology of medically-important arthropods in urban environments. Examples have been given to illustrate these concepts. The following section attempts to reveal ecological factors and requirements that have allowed certain groups of arthropods to occur and persist in urban environments. Only selected examples can be covered. The role of some arthropods in disease cycles have been simplified in some instances. It is understood that ecological and epidemiological relationships are complex and that variations frequently exist. The picture given here is a generalized one for each arthropod discussed.

A. Culicidae-Mosquitoes

The epitome of the urban medically-important arthropod has to be *Aedes aegypti*, which has been wildly incriminated in the transmission of diseases of man in domestic and urban environments. Because this species is one of the easiest mosquitoes to be maintained in the laboratory (MacDonald, 1967), more may be known about its biology, ecology, genetics, and physiology than is known about any other species of medically-important arthropods.

The taxonomic status of the populations of *A. aegypti* was

established by Mattingly (1957). *Aedes aegypti* exhibits a continuous spectrum of color variation, from pale to dark forms, and there is good evidence for associating these color differences with important behavioral differences (Mattingly, 1967a). The dark subspecies, *formosus*, is found only in moist regions of sub-Saharan Africa where it prefers natural breeding places (tree holes) and shows a marked reluctance to bite humans. It is considered to be the ancestral form and sub-Saharan Africa the ancestral home of this species. The nominate subspecies, *aegypti*, is brownish, intermediate between black and pale forms, and is found in domestic and peridomestic regions throughout tropical, subtropical, and warmer temperate regions of the world where it readily bites man. The third subspecies, *queenslandensis*, is pale and is found in mixed populations with the nominate subspecies in all parts of the range in domestic situations. McClelland (1971) found that in Africa the color spectrum exhibited clines instead of subspecies.

Trpis and Hauserman (1975) provided evidence of a genetic basis for the behavioral traits associated with the domestic, peridomestic, and wild habits of three populations of *A. aegypti* in East Africa. These authors collected large numbers of larvae from each habitat type and reared them to adults in segregated cages. Two thousand specimens of each population were marked with a different colored fluorescent powder and released in a peridomestic setting. Of the 307 recaptures on man inside houses, 83% were domestic, 15.5% were peridomestic, and 1.5% were feral. Although peridomestic and feral specimens were found resting on grass and bushes and peridomestic specimens were found on outside walls of houses, no domestic specimens were found resting outside houses on walls or on vegetation. The authors concluded that ability to enter houses is genetically fixed in the domestic population, partly in the peridomestic population, and is nearly absent from the feral population. These behavioral traits were associated with the degree of coloration indicating that both behavioral and morphological traits had a genetic basis.

MacDonald (1967) summarized host feeding preferences of populations of *A. aegypti* in Africa. Precipitin tests of bloodmeals showed that the vast majority of specimens found resting indoors had fed on man, whereas those that rested outdoors had fed on man and a variety of domestic and wild vertebrates.

Crovello and Hacker (1972) and Hacker *et al.* (1977) demonstrated differences in life table characteristics between feral and urban strains, with the urban strains (the paler forms) having a larger intrinsic rate of increase (r_m) and a greater net reproductive rate (R_o) than the feral (dark) strains. They concluded that differences in R_o and r_m between subspecies suggest that the risk of an individual surviving to reproduction is higher in the urban environment than in the feral one. Apparently, the feral habitat is less variable than that in urban areas with respect to several meteorological parameters; therefore, there would be selective pressure for strains in urban areas to have larger reproductive potentials.

These experimental data present a strong rationale for the distribution and density of *A. aegypti* in urban areas. Soper (1967) believed that this species owes its virtual worldwide distribution to its adaptation to human habitation, to breeding in artificial water containers, and to traveling with man. Of special importance is the extended viability of its eggs and in particular their ability to survive desiccation. Cheong (1967) reiterated that the most important single factor influencing the distribution of *A. aegypti* has been humans, through their habits with regard to water storage and unwanted containers. Cheong (1967), Rao (1967) and Surtees (1967) listed the wide variety of artificial water containers, which substitute for the ancestral tree hole, and the conditions necessary for larval development in these containers. Surtees (1967) concluded that transport and urbanization of new areas are major causes of the spread and increased prevalence of this species. However, type of urbanization is important. The need for water storage and increase in artificial containers enhance the prevalence of *A. aegypti*, whereas expansion of closed water pipelines is believed to be the major cause for elimination of *A. aegypti* in the Mediterranean region (Holstein, 1967).

The importance of *A. aegypti* as a vector in urban environments is well illustrated through its role in yellow fever transmission (Gillette, 1971). In tropical Africa, yellow fever virus is transmitted among monkeys, in which the disease is relatively benign, by the canopy-dwelling mosquito, *Aedes africanus* (Theobald). This forest cycle, which is a four-factor one, involves the virus, *A. africanus*, the monkey reservoir, and rarely humans. During deforestation for development of villages or plantations, monkeys infected with yellow fever virus come down from tree canopies to feed in plantations. Here, *Aedes simpsoni* (Theobald) (a leaf-axil breeder dwelling readily in the plantations, because of the increase in larval development sites afforded by bananas and other vegetation) feeds on infected monkeys and transmits the virus in turn to humans working in plantations or living in nearby rural villages. This rural cycle is also a four-factor one. However, when infected workers or villagers enter the city, *A. aegypti* will transmit the virus from human-to-human. This is now a three-factor cycle, and the reservoir host is no longer available or necessary to maintain the urban chain of infection. In Latin America, *Haemagogus spegazzinii* Brethes is a vector in both the forest and jungle cycles, and man again introduces the virus into the urban environment, where *A. aegypti* maintains the urban cycle. Urban cycles of dengue and chickungunya viruses are also transmitted by *A. aegypti* (Gilotra *et al.*, 1967; Rudnick, 1967). Because of the urban nature of *A. aegypti*, Rudnick (1967) believes that new epidemics of diseases transmitted by *A. aegypti* are possible whenever a different arbovirus that produces a significant viremia is introduced into urban areas harboring this mosquito.

Another species, *Aedes albopictus* (Skuse) is also an important mosquito in Asia where it is indigenous. This species apparently uses plant cavities as its normal larval developmental site (Horsfall, 1955). Although *A. albopictus* is primarily an outdoor, sylvatic species, it readily invades the domestic environment where it utilizes the same kind of artificial containers as *A. aegypti* (Gilotra *et al.*, 1967). As a result, *A. albopictus* has become a pest in urban and suburban areas and in rural and woodland habitats (Gilotra *et al.*, 1967; Mattingly, 1967a; Rao, 1967; Rudnick, 1967), and it is an important urban vector of dengue viruses.

Apparently, the phenomenon of competitive displacement has been occurring in Southest Asia between *A. albopictus* and *A. aegypti* (Gilotra *et al.*, 1967). Because *A. aegypti* is a better competitor in most urban habitats, it is favored in urban habitats. *Aedes albopictus* is favored in outdoor habitats in suburban and rural areas; however, in small urban gardens there is a state of equilibrium where the densities of the two species are approximately equal.

Culex pipiens probably ranks second to *A. aegypti* in importance to humans, particularly in urban environments. The species is a nuisance wherever it occurs, and it serves as a vector of Bancroftian filariasis and several arboviruses. Systematics of the *pipiens* complex has been extremely confusing; however, Barr (1967), Mattingly (1967b), and Spielman (1967) agree that the species should be treated as members of a single polytypic strain. The large amount of variability, hybridization, genetic incompatibility, and strain differences concerning autogeny and anautogeny preclude taxonomic stability.

In spite of the taxonomic confusion, it is agreed that the subspecies *Culex pipiens quinquefasciatus* Say (=*C. p. fatigans* Wiedemann of European workers) is the major vector form in urban environments throughout the world. This subspecies is characteristically found in water heavily charged with organic matter, either in ground pools or in various kinds of containers (Barr, 1967). It freely enters houses and readily bites humans. It has thrived in man-made habitats resulting from urbanization and industrialization (Samarawickrema, 1967). This is particularly true in developing countries in the septic fringe environments. In Africa and Madagascar, the prevalence of *C. p. quinquefasciatus* and filariasis have increased owing to the increase in populations of urban centers, which in turn creates an increase in polluted water through inadequate water supplies, storage of water, and inadequate waste disposal (Subra, 1975). Abdulcader (1967), Samarawickrema (1967), and Singh (1967) each lists the various containers (drains, latrines, septic tanks, catch-pits, cesspools, ditches, bucket latrines, etc.) that provide the aquatic sources for this mosquito in Asia. Abdulcader also noted that cement-lined catch-pits used to receive wastes from bucket latrines became ideal sites for this mosquito. In California, larvae of *quinquefasciatus* occur in cesspools, septic tanks, dairy drains, sewage treatment ponds, winery and cannery wastes, industrial wastes, drains, street gutters, catch basins, and all types of man-made containers. This is often the most abundant mosquito in urban areas (Spiller, 1968). Gomez *et al.* (1977) demonstrated that *C. p. quinquefasciatus* was an excellent "r strategist", which has allowed its great success in urban environments. This opportunistic attribute can be illustrated in the United States where this mosquito is a vector of the urban cycle of St. Louis encephalitis (SLE). Most epidemics develop in urban or suburban sites, with *Culex* spp. being the principal vectors. How the virus is introduced and sustained to become endemic is uncertain (Reeves, 1972). Increases in *quinquefasciatus* populations associated with SLE epidemics result from the following events: a period of drought following heavy early rains, flooding, or sparse rains, with a subsequent period of high sustained temperatures. This results in pool formation and concentration of water polluted with human sewage or other organic wastes, particularly in pools at ends of culverts and street drain catch-basins, all suitable mosquito breeding sites (Brody and Browning, 1960; Kokernot *et al.*, 1967; Mack *et al.*, 1967; Sudia *et al.*, 1967; Kokernot *et al.*, 1969; Luby *et al.*, 1969).

Of the malaria-transmitting anophelines, only *Anopheles stephensi* Liston has become adapted to the urban environment. Besides its domestic feeding and resting habits, *A. stephensi* breeds readily in wells, cisterns, roof gutters, fountain basins, garden tanks, tubs, discarded tins, and receptacles of all kinds in cities of India and the Middle East (Covell, 1949). According to Watson (1949), other anopheline species most often emanate from aquatic sources in rural districts; urbanization appears to be partially responsible for a decrease in the prevalence of these species. Nevertheless, adult anophelines and malaria appear to exist in urban centers. The specific reasons for the occurrence of one or another of the species of *Anopheles* in cities are not always given - in part attributable to use of the concept of domesticity by malariologists (see discussion of domesticity above). Some general features are apparent, however, which determine the presence of anophelines in urban settings. Flooding and heavy rainfall have created proper breeding sites for *A. atroparvus* Van Thiel, *A. messae* Falleroni, *A. sacharovi* Favre, and *A. superpictus* Grassi in the cities of Europe (Hackett, 1949). As cities may be located near breeding sites of *Anopheles* spp., the flight range of many species, the migration patterns of species such as *A. freeborni* Aitken, and wind dispersal have accounted for the urban occurrence of anophelines (Russell *et al.*, 1946; Hackett, 1949; Garrett-Jones, 1962; Spiller, 1968; Snow and Wilkes, 1977, among others). *Anopheles gambiae* Giles species A and B populations may become dense locally as this species is a puddle-breeder and increases greatly when rainfall exceeds 12.7 cm a month (Haddow, 1942). Gillette (1971) reported rapid larval development (a minimum of two days) for this species. Thus, heavy rainfall and rapid development account for much of the urban distribution of this species throughout sub-Saharan Africa. Subra *et al.* (1975) also have found *A. gambiae* developing in reservoirs and underground cisterns in Madagascar. Although *A. claviger* (Meigen) in the Middle East and *A. culicifacies* Giles on the Indian subcontinent normally breed in natural waters, *A. claviger* during hot weather will breed in underground water cisterns (Hackett, 1949) and *A. culicifacies* will use burrow pits and other man-made pools and ponds (Russell *et al.*, 1946). Anophelines through various behavioral habits and under favorable conditions tend to occupy urban environments.

In similar ways, various species of *Aedes*, *Culex*, *Culiseta*, and *Psorophora* also occur in urban habitats, especially when the normal aquatic sources exist proximate to urban centers. Many salt marsh and flood water *Aedes* (*A. cantator*, [Coquillett], *A. dorsalis* [Meigen], *A. sollicitans* [Walker], *A. squamiger* [Coguillett], *A. taeniorhynchus* [Wiedemann], and *A. vexans* Meigen) have flight ranges varying from 6.4 to 64 km, bringing these species easily into urban habitats (Herms and Gray, 1944). *Culex tarsalis*, vector of western equine encephalitis virus, and *C. tritaeniorhynchus*, vector of Japanese encephalitis virus, are rural mosquitoes associated with agriculture and irrigation; however, during favorable conditions (flooding and heavy rainfall) and heavy population pressure, they enter urban habitats where they use both natural and man-made structures that hold water (Herms and Gray, 1944; Pratt *et al.*, 1959; Scherer *et al.*, 1959; Spiller, 1968). In this manner these viruses, usually associated with rural foci, are brought into urban environments.

B. Ceratopogonidae and Simuliidae

Species of the families Ceratopogonidae (biting midges,

punkies, and no-see-ums) and Simuliidae (black flies and black gnats) are problems to humans in wild and recreational habitats (Merritt and Newson, Chapt. 6, this volume); however, in numerous cases urban development has encroached on the natural sources of these flies (Nielsen, 1963; Hall, 1972). The feature of extensive flight ranges (up to 200 km for *Simulium damnosum* [Thompson, 1976]) by some of these species bring them into cities that are proximate to the sources of these flies. Species in these families have not become adapted to urban habitats, and the process of urbanization usually destroys sources of these flies.

C. Phlebotomidae - Sand Flies

Sand flies (*Phlebotomus, Sergentomyia, Lutzomyia,* and their relatives) are vectors of the agents causing leishmaniases, bartonellosis, and sand fly fever. Epidemiology of leishmaniases clearly demonstrates the evolution of urban anthropurgic foci in the Old World. In the New World where some leishmaniases probably were introduced, no urban cycles have yet evolved.

In the Old World there are two types of leishmaniases: viceral and cutaneous (Lysenko, 1971). Four types of foci (if the East African form of the disease is disregarded; disagreement exists over the epidemiology of this form): 1) Enzootic natural (autochthonous) foci with jackals incriminated as a reservoir. Humans are vulnerable when they invade these foci. 2) Rural endemic foci where dogs enter the jackal (fox)-sand fly-dog-sand fly-man cycle. Lysenko believes this to be an evolutionary transitional phase from a zoonosis to an anthroponosis. 3) Urban endemic (anthropurgic) foci occur when dogs are the main reservoir, and man becomes a part of the cycle. 4) Endemic (anthropurgic) foci of Indian kala azar is the final stage in this evolutionary sequence where only man is involved. The four-factor cycle has become a three-factor cycle. Cutaneous leishmaniasis occurs in two forms: 1) The wet, rural, or zoonotic form found associated with infections from the normal rodent hosts. Man acquires this kind of infection when he invades new areas or creates conditions conducive for rodent buildup near settlements (Petrischeva, 1971). 2) The dry, urban, or anthroponotic form found in ancient cities involving a man-sand fly-man cycle or a dog-sand fly-man cycle.

In the New World both visceral and cutaneous forms also occur under various names (Garnham, 1971). The cutaneous and a mucocutaneous form are associated with natural foci in forested areas with rodents acting as natural reservoirs. Man obtains infections when he enters the forested zones for occupation or settlement (Garnham, 1971; Herrer and Christiansen,

1976). Visceral leishmaniasis occurs in rural, nonforested areas, small towns, and settlements. Fox and dogs (perhaps also other wild canids and sylvatic rodents) maintain this zoonosis. The genus *Lutzomyia*, species of which vector leishmaniasis in the New World, is basically sylvatic; however, there is apparently some domesticity occurring in the epidemiology of the visceral form. Garnham (1971) believes there is evidence that the disease is spreading in Latin America. Currently, urbanization appears to be detrimental to the cycle, but perhaps the degree of domesticity that has appeared may indicate an early evolutionary phase toward an urban cycle.

Old World phlebotomines are associated with lairs of carnivores and burrows of rodents (Petrischeva, 1971). Here they obtain shelter and can breed and overwinter. The requirements for sand flies appear to be a stable microclimate with humidity nearly at a constant humid level. Carnivores and rodents supply the necessary organic matter in nitrogenous wastes on which larvae feed (Horsfall, 1962; Lewis, 1971, 1973). Humans enter into these natural foci where rodents (*Rhombomys, Meriones,* and *Psammonys*) dwell (Petrischeva, 1971; Nadim and Rostami, 1974; Ashford *et al.*, 1976), and the domestic forms (*Phlebotomus papatasi, P. sergenti* Parrot, *P. argentipes* Annandale and Brunetti, *P. chinensis* Newstead, *P. perniciosus* Newstead, and *P. longicuspis* Nitzulescu, among others) may enter into houses and feed on humans. Certain populations of these species have foregone the rodent burrow and now breed in rich, loose soil, cracks in stone walls, dirt floors of houses, edges of refuse piles, brick linings of pit privies, and cracks in caves. Use of caves is thought to have played an evolutionary step toward synanthropism or domesticity in these species (Lewis, 1973; Williams, 1976). Thus, the predilection of certain species of sand flies for use of houses and associated structures as areas for shelter and development and use of humans and dogs for blood meals have contributed to the development of established urban cycles of diseases transmitted by these species.

D. Synanthropic Flies

Synanthropic flies, which include representatives of calypterate Diptera (mostly Anthomyiidae, Calliphoridae, Muscidae, and Sarcophagidae), are those that have come to coexist with man over an extended period of time (Povolný, 1971). From their status as members of primary, natural biocenoses (communities), these flies have adapted to secondary, cultural biocenoses formed through man's activities and influence. In many instances, they have become better adapted to the new environments than they were in their original environments (Povolný, 1971). Synanthropy has resulted in response to waste products associated with man's use of his environment, for these wastes meet the trophic requirements of larvae and adults of these flies. Various waste material are of high organic content, moist, and warm enough for larval development. Three of five categories of synanthropes of Povolný are pertinent to the discussion of these flies in urban settings. Among these categories are flies that are mechanical and biological vectors of disease organisms, contaminators of food products, and persistent nuisances to mankind. Eusynanthropes are flies that become almost wholly dependent upon the activities of man for their trophic requirements and other biotic necessities. *Musca domestica* Linnaeus, *Phaenicia sericata* (Meigen), and *Fannia canicularis* (Linnaeus) are examples of eusynanthropes. Hemisynanthropes are flies that are generally independent of man's activities, but increase greatly whenever they enter the environments influenced by man. This group includes *Calliphora* spp., *Hylemya* spp., *Lucilia* spp., *Muscina* spp., *Phormia* spp., and *Sarcophaga* spp. Symbovines are those linked to man through their predilection for larval development in feces of domestic animals. *Musca autumnalis* De Geer, *M. sorbens* Wiedemann, *M. vetustissima* Walker, and *Stomoxys calcitrans* (Linnaeus) are common examples of this category. However, in Africa even *Glossina* spp. are known to enter cities (Iwuala and Onyeka, 1977).

It is ironic that synanthropic flies are products of the septic and affluent fringes in urban environments. In the former, they are associated with primitive sewage and waste management of both human and domestic animal wastes. In the latter, improper composting methods, improperly covered garbage receptacles (Schoof *et al.*, 1954; Ecke and Linsdale, 1967), and dog droppings (Schoof *et al.*, 1954; Wilton, 1963; Poorbaugh and Linsdale, 1971; Legner *et al.*, 1974) generate most of the flies developing herein. Furthermore, urban encroachment into agricultural areas has intensified fly problems. Modern housing tracts and suburbs have virtually surrounded old, well established feed lot, dairy, poultry, orchard, and cannery operations in many counties of California that have undergone a rapid change from agricultural to urban status since World War II. Fly production from these sites, normally associated with rural habitats, now occurs adjacent to or surrounded by affluent urban homes, schools, and business establishments. With increased waste production associated with the trend of expanding urbanization, fly problems will undoubtedly intensify as modern cities are confronted with the economics of waste disposal.

E. Fleas

Plague, another arthropod-borne disease that well illustrates the concept of nidality, is a four-factor disease in its enzootic state, involving the bacterium, relatively resistant wild rodent hosts, their fleas, and rarely humans. Home of plague is in relatively small autochthonous foci usually found in steppes, high plateaus, and mountainous areas (Pollitzer, 1954, 1960; Olsen, 1970). Man rarely contacts plague in these enzootic foci; no human case is known to be associated with the San Bruno Mountain focus in California. In the epizootic state, sylvatic plague is still a four-factor disease with susceptible wild rodents and their fleas replacing resistant rodents and their fleas. These rodents include tree and ground squirrels, chipmunks, marmots, gerbils, wood rats, among others. Man contacts plague when he invades areas where epizootics are in progress. Urban, or domestic, rodent plague develops from either the enzootic or, more often, the epizootic, sylvatic state, owing to contact between the wild and urban rodents at the interfaces of natural or rural habitats and urban habitats. Domestic rodents (primarily *Rattus rattus* [Linnaeus], *R. norvegicus* [Berkenhout], *R. exulans* [Peale]) and their fleas (primarily *Xenopsylla cheopis*, *X. astia* Rothschild, *X. brasiliensis* [Baker] and *Nosopsyllus fasciatus* [Bosc]) replace the sylvan rodents and their fleas (Pollizter, 1960; Velimirovic, 1972). In certain rural areas a three-factor system may occur, with rodents dropping out and *Pulex irritans* Linnaeus providing human-to-human transmission (Pollitzer, 1960; Velimirovic, 1972). A matter of major importance in plague epidemiology is the development of the pneumonic form, a two-factor cycle establishing human-to-human transmission via aerosol, or droplet, infection. In urban situations, the pneumonic cycle may give rise to the epidemics that occurred during the Middle Ages (McNeill, 1976).

Urban plague has occurred in the past and remains a potential threat because humans have created harborage for domestic rodents and their fleas in homes, buildings, sewer systems, and in the type of landscaping used around homes and buildings (Pollitzer, 1954; Ecke, 1964; Brooks, 1966; Dutson, 1973, 1974; Barnes, 1975; Gratz and Arata, 1975). Furthermore, extension of plague over great distances are made possible by transportation of domestic rodents and their fleas in caravans and ships (Traub and Wisseman, 1974).

Xenopsylla cheopis, the oriental rat flea, has been widely responsible for outbreaks of urban plague and murine typhus. This species mainly infests domestic rodents living in buildings (Pollitzer, 1954). Isaacson (1975) pointed out that distribution of fleas is dependent on numerous factors, with the important ones being atmospheric temperature and humidity.

Xenopsylla cheopis needs warm, relatively dry microclimates for adult and larval survival (Sharif, 1948); therefore, it is found in or under buildings (Haas *et al*., 1972) and in granaries and warehouses (Pollitzer, 1954). Larval food is also important (Sharif, 1948), and this flea successfully breeds in areas such as granaries, where cereals and foodstuffs are processed and stored. Both the domestic rodent and *X. cheopis* find excellent harborage in urban buildings and in transportation vehicles between urban centers.

The human flea, *Pulex irritans* Linnaeus, apparently had been associated with humans from the time they dwelt in caves (Jordan, 1948). Although this species was once prevalent in urban communities, its present distribution is associated more often with slum-type dwellings (Hudson *et al*., 1960). According to Keh (1977), this species was once common in many well-kept homes in San Francisco. Linoleum and wood floors were mopped frequently, allowing proper humidity for larval survival in cracks and corners. With the advent of the vacuum cleaner at a price that most people could afford, the human flea has become rare or absent in most homes (Busvine, 1951). This flea has been replaced in the home by the cat flea, *Ctenocephalides felis*, which has increased in numbers along with the mounting populations of cats and dogs as house pets. The high reproductive potential of this flea (over 800 eggs laid by each female), the great number of hosts, and the favorable microclimate (warm temperatures and relatively low humidities) afforded by modern homes have allowed it to become an extreme nuisance in urban environments throughout the world (Lunsford, 1949; Pollitzer, 1954; Hudson *et al*., 1960).

F. Ticks

Human cases of Rocky Mountain spotted fever (RMSF) have been increasing in number in eastern United States since the early 1960's (Burgdorfer, 1975), and a considerable number of cases have been contracted in suburban and urban environments (Hattwick *et al*., 1973, 1976). Smadel (1959) predicted this increase as suburbanization brought more people into ecological habitats harboring the vectors and reservoirs of RMSF. Although this trend has occurred in some eastern and southern states, there has been a decrease in cases associated with suburbanization in other states in the same general geographic region (Hattwick *et al*., 1976). Rothenberg and Sonenshine (1970) postulated that this decrease was associated with a more rapid degree of urbanization, and they predict that as the outward expansion of cities occurs, the current endemic areas for RMSF will fall. Therefore, the trend in human cases in the eastern United States fits the pattern predicted by Pavlovsky's theory: an increase in incidence with early invasion into autochthonous foci followed by a decrease after modification of the foci by human interference, namely urbanization.

Increase in human cases is associated largely with *Dermacentor variabilis* (Say), the American dog tick. Ecological conditions that have allowed for an increase in density of this tick have been cited above under discussion of the affluent fringe. McEnroe (1974) postulated that the increase in Massachusetts, as well as the recent range extension into the eastern part of the state, has occurred as a result of increased density of dogs associated with urbanization. This explanation is probably accurate for the more southeastern states also. The predilection of *D. variabilis* for domestic dogs increases the chances for human contact with these ticks.

Concurrent with the increase of RMSF in eastern North America, there have been suburban and urban cases of tick paralysis associated with *D. variabilis*. Usually, tick paralysis is acquired when persons enter wild or rural habitats (Gregson, 1973). In Australia, *Ixodes holocyclus* Neumann is responsible for most cases of tick paralysis in wild and suburban areas. In the greater Sydney area, this tick occurs in the suburbs (Roberts, 1970) because much natural vegetation is interspersed with suburbanized and urbanized areas, affording habitat for the bandicoot (*Parameles* sp.), the primary host for *I. holocyclus*.

Rhipicephalus sanguineus (Latreille), because it can complete its life cycle on domestic dogs, has increased in urban situations (Keh, 1964). From Africa this tick has been introduced into Europe, Asia, Australia, and the Americas where it is associated with dogs in homes and kennels (Haarlov, 1971; Hoogstraal, 1972). According to Sweatman (1967), this tick is dependent on relatively high temperatures, so in more temperate countries it is associated with dogs in houses. In warm climates, *R. sanguineaus* has become indentified with the epidemiology of RMSF in Mississippi (Sexton *et al.*, 1976) and the domestic cycle of boutonneuse fever in the Mediterranean area of Europe (Camicas, 1975).

Many species of ticks are now occurring proximate to urban-suburban situations, owing to the spread of cities. In California, *Ixodes pacificus* Cooley and Kohls and *Dermacentor occidentalis* Marx are frequently encountered at the edge of urban centers (Arthur and Snow, 1968; California Department of Health records), where the chaparral biome is maintained near or within the city limits.

G. Mites

As emphasized repeatedly here, many medically-important

arthropods enter houses and buildings to find food, shelter, and breeding sites, and these structures and their inhabitants have supplied requirements for survival. In like manner, mammalian and avian hosts of various ectoparasites (insects and mites) have come to use houses, churches, warehouses, grain elevators, schools, other buildings, and transportation structures (bridges, tressels, stations, and the like) for home sites. Certain features of these structures such as crevices, ledges, roofs, walls, attics, and chimneys resemble natural habitats, such as caves, cliffs, and hollow trees, sufficiently that some birds and mammals use these artificial sites almost exclusively. As a result hosts have brought with them the ectoparasites that dwell in their nests. When these hosts are killed through planned control efforts, die of natural causes, or leave the sites for a variety of reasons, the vagile ectoparasites (usually mites) may leave the nests when hungry and seek another host. In buildings the host is usually man. Bites of these mites may result in intense pruritus and dermatitis and, in the case of *Allodermanyssus sanguineus* (Hirst) from *Mus musculus* Linnaeus nests, in a rickettsial disease called rickettsialpox. The following mites or their close relatives attack humans on a cosmopolitan basis: *Ornithonyssus bacoti* (Hirst), (nests of mice and rats); *O. bursa* (Berlese), *O. sylviarium* (Canestrini and Fanzago), and *Dermanyssus gallinae* (De Geer), (nests of house sparrows, rock doves, starlings, and other domestic nesting birds); and *Chiroptonyssus robustipes* (Ewing) (roosts of bats) (Baker *et al.*, 1956; Hetherington *et al.*, 1971; Keh, 1975a). Man's increasingly close association with cats and dogs has also been responsible for an increase in human cases of canine scabies (Smith and Claypoole, 1967) and pruritus and dermatitis from *Cheyletiella yasguri* Smiley and *C. blakei* Smiley, mites of dogs and cats, respectively (Keh, 1973, 1975b; Brownswijk and Kreek, 1976).

The epidemiology of scabies (caused by *Sarcoptes scabiei* De Geer) resembles that of lice in many respects. War and natural calamities crowd people together and reduce facilities for bathing and cleaning clothes, resulting in increased incidence of both scabies and lice. Scabies is also associated with families or groups that live in intimate contact (Mellanby, 1972). From these focal groups scabies may be spread rapidly to other persons that come into direct contact with an infested member. Direct contact is the only known method of spread. Because this mite is very small (females average approximately 200 μm), an infestation is not easly recognized. Furthermore, these mites burrow into the skin and usually symptoms of intense itching do not develop for one to two months. During this time two to four generations of mites (generation time is 14 days) have developed with the potential for further transmission. In this manner, scabies may rapidly spread throughout a community as was reported in Dakar, Senegal, by Marchand *et al.* (1975).

House dust mites are medically-important arthropods that are currently undergoing intensive study. Although the relationship between house dust mites and house dust allergy was proposed by Dekker in 1928, it was not until 1964 (Voorhorst *et al.*, 1964) that a convincing incrimination of these mites appeared (Wharton, 1976). Since 1964, approximately 500 papers have been written on these mites. It is now agreed that house dust mites contribute a significant antigenic factor to house dust for people who suffer from house dust allergy, asthma, and rhinitis (Brandt and Arlian, 1976). Although the literature on taxonomy, ecology, and control of these mites and on medical aspects of house dust allergy is vast, most data have been gathered together in the recent review of Wharton (1976). The review forms the basis of this discussion.

Although many species from various families of mites are found in houses, hotels, motels, hospitals, dormitories, camps, rest homes, and transportation carriers, the most prevalent family is Pyroglyphidae. In this family several species are associated with house dust, and three species (*Dermatophagoides pteronyssinus* [Trouessart], *D. farinae* Hughes, and *Euroglyphus maynei* [Cooreman]), which are cosmopolitan in houses, are thought to be the main sources of house dust mite allergy. These species can exploit nests and stored food products and probably had their origin as nidicoles of bird and mammal nests. They develop well at room temperatures and need a humidity above 70%, but not high humidities. House dust mites seem to thrive best on high protein diets that contain fats of human origin. It is, therefore, not surprising that they are universally associated with beds, mattresses, and other bedding. In this environment the proper humidity and food from human skin scales are supplied for survival and propagation throughout the year in houses in a variety of macroclimates, both temperate and tropical. On floors, a seasonal peak in numbers occurs in late summer and early fall when favorable temperatures and humidities occur. Thus, the inner home environment supplies needed features for the occurrence and persistence of these mites.

Mumcuoglu (1976) has postulated that these same mites living in quarters of various domestic pets (dogs, cats, birds, hamsters, and guinea pigs) may be responsible for the seemingly allergic reaction to pets by many patients. The number of these mites obtained from beddings of pets support his hypothesis. Based on observations of many investigators, house dust mites play an extremely important role in urban communities and may determine the health and well being of many persons on a worldwide scale.

H. Bugs

Bed bugs (family Cimicidae) are another group of medically-important arthropods that have adapted to houses and other buildings used by humans. *Cimex lectularius* Linnaeus, the human bed bug, has a wide geographical distribution. According to Usinger (1966), the human bed bug probably originated with bats in caves; *C. lectularius* has been collected rarely in caves, but has been found associated with bats in a cave in Afghanistan. Probably because of the long association with humans living in such places, strains of this bed bug developed a predilection for human blood. Owing to the similarity of caves and certain kinds of houses, this bug was preadapted to buildings when humans left caves. These bugs prefer darkness inside walls and ceilings, rough surfaces for movement and oviposition, and cracks and crevices for shelter. Houses maintain temperatures and humidities similar to cave environments. The tropical human bed bug, *Cimex hemipterus* (Fabricius), was also originally a parasite of bats as was the third species frequently associated with man, *Leptocimex boueti* (Brumpt).

Although the actual course of events that led these species into adaptation to human habitat cannot be reconstructed with certainty, the probable course can be ascertained in general, especially through human contacts with the bat bug, *Cimex pilosellus* (Horvath), and the swallow bug, *Oeciacus vicarius* Horvath. Both bats and cliff swallows commonly use dead space of walls and roofs and eaves for roosting and nesting, respectively. In the absence of their normal hosts, these bugs seek out and feed upon humans; however, neither species can survive and reproduce on human blood. Apparently, this factor precludes their ability to become established on humans.

The conenose bugs (family Reduviidae, subfamily Triatominae) also illustrate gradual adaptation to human habitations. Most species occur in the Western Hemisphere. *Triatoma protracta* (Uhler), associated with the wood rat, *Neotoma fuscipes* Baird, may enter homes in suburbs owing to attraction to lights and in some instances to the invasion of homes by this wood rat. Because of the recent tendency to build homes in chaparral areas on the edges of cities, there has been an increase in the incidence of hypersensitivity to conenose bug bites and possible exposure to Chagas' disease caused by *Trypanosoma cruzi* (Ebeling, 1975). Fortunately, *T. protracta* appears to be a poor vector of this pathogen.

Usinger *et al.* (1966) sumarized the epidemiology of Chagas' disease in terms of the peridomesticity and domesticity of the vectors (e.g., *Triatoma* spp., *Rhodnius prolixus* Stal, and *Panstrongylus megistus* [Burmeister]) and the domestic and semidomestic hosts which are reservoirs of the trypanosome. These include bats, cats, dogs, opossums, armadillos, squirrels, and several species of rats and mice. The focus of infection is the primitive dwelling with poor social, economic, and sanitary conditions, which allow habitat for the vector and/or reservoir host species. Most studies on the transmission of Chagas' disease have been carried out in rural situations. Furthermore, it was noted that improvement in housing would preclude entry of the vector and hosts. Zarate (1977) found that in urban situations, occurring in what is termed herein the septic fringe of Oaxaca, Mexico, *Triatoma barberi* commonly is found in the various shanty homes.

I. Lice

The human louse, *Pediculus human* Linnaeus, consisting of two unstable environmental subspecies, *P. h. humanus* (the body louse) and *P. h. capitis* De Geer (the head louse) are cosmopolitan in distribution (Clay, 1973). During times of war and natural disasters devastating epidemics of epidemic typhus (caused by *Rickettsia prowazeki*), trench fever (caused by *R. quintana*), and louse-borne relapsing fever (caused by *Borellia recurrentis*) have been transmitted by these lice. The association of lice with man has been extremely long, resulting in strict host specificity. Therefore, transmission of lice to man involves direct contact and use of louse-infested clothing and implements. Buxton (1946) reported that crowding is important in the spread of lice among perons. Lice tend to infest most members of a family, workers on ships, soldiers housed in barracks, and prisoners in jails - persons who work and live together in close association. Slonka (1975) concluded that large families or groups living in intimate contact are the source of persistent pediculosis foci and provide the focus from which pediculosis may recrudesce in a community. From these family groups lice have rapidly spread to contacts of members of these units, particularly among students in schools. The pubic, or crab louse, *Pthirus pubis* Linnaeus, is spread generally, but not exclusively, through intimate sexual contacts. Permissive, sexually active infested persons form the foci for infestation in communities.

V. CONCLUSIONS

It can be concluded from the above discussion that the concept of nidality forms a sound theoretical basis for the study of medically-important arthropods in urban environments. This approach views arthropods as one component of the community of organisms. The urban environment produces substrates that allow development of the man-made community, or biocenose.

Substrates are those factors that supply various biotic and abiotic requirements for members of the community. Use of these urban substrates for feeding, oviposition, development, and shelter by arthropods has resulted from either preadaptation or selection by the population of arthropods that enter urban habitats. Undoubtedly, the processes of adaptation or preadaptation to urban environments evolved through domesticity to man's habitats and practices (his dwellings, and associated structures, pets and other domestic animals, methods of waste disposal, and methods of use, storage, and disposal of water).

Certain medically-important arthropods are highly successful in domestic and urban settings as the substrates produced by man meet all requirements for survival. These relatively few species apparently have been associated with mankind for long periods of time and are successful in stable domestic and urban settings that retain primitive housing and primitive water and waste management. In general, these species are cosmopolitan in distribution. They could be reduced or eliminated by improved housing and proper water and waste management practices, but are retained owing to traditional practices and to economic considerations. Therefore, the most successful urban arthropods of medical importance represent greater problems in developing countries, in particular within the septic fringe.

In modern developed countries housing, water and waste management practices generally supply only a portion of the requirements of a medically-important arthropod found therein. Therefore, arthropods found in these settings are those that are more independent of the substrates offered by man. These arthropods have only relatively recently become associated with man in urban settings. They are generally from the natural ecosystems surrounding an urban center, and their distributions are more localized and limited. The trend seems to favor increased contacts with arthropods currently regarded as rural and recreational pests (Merritt and Newson, Chapt. 6, this volume). It is among these arthropods that the present and future problems in modern urban environments lie as encroachment and influence of urbanization extends further into rural and recreational areas.

ACKNOWLEDGMENTS

I thank the following persons for their criticism and review of this manuscript, for translating certain papers, and for helpful discussions and suggestions concerning the subject: J. R. Anderson, D. P. Furman, B. Keh, R. S. Lane, R. W. Merritt, J. R. Walker, C. F. Williams, C. A. Wolf, and L. G. Zarate.

REFERENCES

Abdulcader, M. H. M. 1967. The significance of the *Culex pipiens fatigans* Wiedemann problem in Ceylon. *Bull. World Health Organ. 37*: 245-249.

Abu-Lughod, J. 1969. Migrant adjustment to city life: The Egyptian case. *In* The city in developing countries: Readings on urbanism and urbanization (G.Breese, ed.). Prentice-Hall, Inc., Englewood, New Jersey, pp. 376-388.

Adams, W. H., R. W. Emmons, and J. E. Brooks. 1970. The changing ecology of murine (endemic) typhus in southern California. *Am. J. Trop. Med. Hyg. 19*: 311-318.

Arthur, D. R., and K. R. Snow. 1968. *Ixodes pacificus* Cooley and Kohls, 1943: Its life-history and occurrence. *Parasitol. 58*: 893-906.

Ashford, R. W., M. L. Chance, F. Ebert, L. F. Schnur, A. K. Bushwereb, and S. M. Drebi. 1976. Cutaneous leishmaniasis in the Libyan Arab Republic: Distribution of the disease and identity of the parasite. *Ann. Trop. Med. Parasitol. 70*: 401-409.

Audy, J. R. 1958. The localization of disease with special reference to the zoonoses. *Trans. Roy. Soc. Trop. Med. Hyg. 52*: 308-328.

Baker, E. W., T. M. Evans, D. J. Gould, W. B. Hull, and H. L. Keegan. 1956. A manual of parasitic mites of medical or economic importance. Nat. Pest Contr. Assoc., Inc., New York.

Barnes, A. M. 1975. Problems of rodent control in rural tropical areas. *Bull. World Health Organ. 52*: 669-676.

Barr, A. R. 1967. Occurrence and distribution of the *Culex pipiens* complex. *Bull. World Health Organ. 37*: 293-296.

Brandt, R. L., and L. G. Arlian. 1976. Mortality of house dust mites, *Dermatophagoides farinae* and *D. pteronyssinus*, exposed to dehydrating conditions or selected pesticides. *J. Med. Entomol. 13*: 327-331.

Brannan, T. 1964. Control of tree-hole mosquitoes in Alameda County. *Proc. Calif. Mosq. Contr. Assoc. 32*: 53-54.

Brody, J. A., and G. Browning. 1960. An epidemic of St. Louis encephalitis in Cameron County, Texas, in 1957. *Am. J. Trop. Med. Hyg. 9*: 436-443.

Bronswijk, J. E. M. H., and E. J. de Kreek. 1976. *Cheyletiella* (Acari: Cheyletiellidae) of dog, cat and domesticated rabbit, a review. *J. Med. Entomol. 13*: 315-327.

Brooks, J. E. 1966. Roof rats in residential areas - The ecology of invasion. *Calif. Vector Views 13*: 69-74.

Bruce-Chwatt, L. J. 1975. Endemic diseases, demography and socioeconomic development of tropical Africa. *Canad. J. Pub. Health 66*: 31-37.

Burgdorfer, W. 1975. A review of Rocky Mountain spotted fever

(tick-borne typhus), its agent, and its tick vectors in the United States. *J. Med. Entomol. 12*: 269-278.
Buxton, P. A. 1946. The louse, 2nd edition. Williams and Wilkins Co., Baltimore, Maryland.
Camicas, J. L. 1975. Conceptions actuelles sur l'épidémiologie de la fièvre boutonneuse dans la région Ethiopienne et la sous-région Europêene Méditerranéenne. *Cah. O.R.S. T.O.M., ser. Entomol. med. Parasitol. 13*: 229-232.
Cheong, W. H. 1967. Preferred *Aedes aegypti* larval habitats in urban areas. *Bull. World Health Organ. 36*: 586-589.
Clay, T. 1973. Phthiraptera (Lice). *In* Insects and other arthropods of medical importance (K.G.V. Smith, ed.). British Museum (Natural History), London, pp. 395-397.
Covell, G. 1949. Malaria incidence in the Far East. *In* Malariology (M.F. Boyd, ed.). W.B. Saunders Co., Philadelphia and London, 2: 810-819.
Crovello, T. J., and C. S. Hacker. 1972. Evolutionary strategies in life table characteristics among feral and urban strains of *Aedes aegypti* (L.). *Evolution 26*: 185-196.
Davis, K. 1969. The urbanization of the human population. *In* The city in developing countries: Readings on urbanism and urbanization (G. Breese, ed.). Prentice-Hall, Inc., Englewood, New Jersey, pp. 5-20.
Dorozynski, A. 1976. The attack on tropical disease. *Nature 262*: 85-87.
Dutson, V. J. 1973. Use of the Himalayan Blackberry, *Rubus discolor*, by the roof rat, *Rattus rattus*, in California. *Calif. Vector Views 20*: 59-68.
Dutson, V. J. 1974. The association of the roof rat (*Rattus rattus*) with the Himalayan Blackberry (*Rubus discolor*) and Algerian Ivy (*Hedera canariensis*) in California. Proc. Seventh Vert. Pest Contr. Conf., pp. 41-48.
Easton, E. R., J. E. Keirans, R. A. Gresbrink, and C. M. Clifford. 1977. The distribution in Oregon of *Ixodes pacificus, Dermacentor andersoni,* and *Dermacentor occidentalis* with a note on *Dermacentor variabilis* (Acarina: Ixodidae). *J. Med. Entomol. 13*: 501-506.
Ebeling, W. 1975. Urban entomology. Div. Agric. Sci. Univ. Calif., Berkeley, California.
Ecke, D. H. 1964. Roof rat populations in Santa Clara County, California. Proc. Second Vert. Pest Contr. Conf., pp. 99-107.
Ecke, D. H., and D. D. Linsdale. 1967. Fly and economic evaluation of urban refuse systems. Part I. Control of green blow flies (*Phaenicia*) by improved methods of residential refuse storage and collection. *Calif. Vector Views 14*: 19-27.
Elton, C. 1927. Animal ecology. MacMillan Co., London.
Elton, C. 1966. The pattern of animal communities. Methuen

and Company, Ltd., London.

Frommer, S. I., and R. A. Rauch. 1971. Pupal duration, adult emergence, and oviposition periods for the midge *Dicrotendipes californicus* (Johannsen) (Diptera: Chironomidae). *Calif. Vector Views 18*: 33-38.

Galuzo, I. G. 1968. Twenty years of natural-nidal disease studies. *In* Natural nidality of diseases and questions of parasitology (N.D. Levine, ed.). Univ. Illinois Press, Urbana, Chicago, and London, pp. 9-16.

Garnham, P. C. C. 1971. American leishmaniasis. *Bull. World Health Organ. 44*: 521-527.

Garrett-Jones, C. 1962. The possibility of active long-distance migrations by *Anopheles pharoensis* Theobald. *Bull. World Health Organ. 27*: 299-302.

Gillett, J. D. 1971. Mosquitoes. Weidenfeld and Nicholson, London.

Gilotra, S. K., L. E. Rozeboom, and N. C. Bhattacharya. 1967. Observations on the possible competitive displacement between populations of *Aedes aegypti* Linnaeus and *Aedes albopictus* Skuse in Calcutta. *Bull. World Health Organ. 37*: 437-446.

Gomez, C., J. E. Rabinovich, and C. E. Machado-Allison. 1977. Population analysis of *Culex pipiens fatigans* Wied. (Diptera: Culicidae) under laboratory conditions. *J. Med. Entomol. 13*: 453-463.

Graham, J., and N. G. Gratz. 1975. Urban vector control services in the developing countries. *PANS 21*: 365-379.

Gratz, N. G., and A. A. Arata. 1975. Problems associated with the control of rodents in tropical Africa. *Bull. World Health Organ. 52*: 697-706.

Graves, N. B., and T. D. Graves. 1974. Adaptive strategies in urban migration. *Ann. Rev. Anthropol. 3*: 117-151.

Gregson, J. D. 1973. Tick paralysis: An appraisal of natural and experimental data. Canad. Dept. Agric. Monogr. No. 9.

Haarlov, N. 1971. The introduction into Denmark of the kennel tick (*Rhipicephalus sanguineus* [Latr. 1806]) with remarks on its reactions to different humidities. Proc. Third Internat. Congr. Acarol., Prague, pp. 467-472.

Haas, G. E., N. Wilson, and P. Q. Tomich. 1972. Ectoparasites of the Hawaiian Islands. I. Siphonaptera. *Contrib. Am. Entomol. Inst. 8*: 1-76.

Hacker, C. S., W. Ling, B. P. Hsi, and T. J. Crovello. 1977. An application of mathematical modelling to the study of reproductive adaptations in the yellow fever mosquito, *Aedes aegypti*. *J. Med. Entomol. 13*: 485-492.

Hackett, L. W. 1949. Conspectus of malaria incidence in northern Europe, the Mediterranean region and the Near East. *In* Malariology (M.F. Boyd, ed.). W.B. Saunders Co., Philadelphia and London, 2: 788-799.

Haddow, A. J. 1942. The mosquito fauna and climate of native huts at Kisumu, Kenya. *Bull. Entomol. Res. 33*: 91-142.

Hall, F. 1972. Observations on black flies of the genus *Simulium* in Los Angeles County, California. *Calif. Vector Views 19*: 53-58.

Hattwick, M. A. W., A. H. Peters, M. B. Gregg, and B. F. Hanson. 1973. Surveillance of Rocky Mountain spotted fever. *J. Am. Med. Assoc. 225*: 1338-1343.

Hattwick, M. A. W., R. J. O'Brien, and B. F. Hanson. 1976. Rocky Mountain spotted fever: Epidemiology of an increasing problem. *Ann. Internat. Med. 84*: 732-739.

Hermes, W. B., and H. F. Gray. 1944. Mosquito control. 2nd edition. The Commonwealth Fund, New York.

Herrer, A., and H. A. Christensen. 1976. Epidemiological patterns of cutaneous leishmaniasis in Panama. I. Epidemics among small groups of settlers. *Ann. Trop. Med. Parasitol. 70*: 59-71.

Hetherington, G. W., W. R. Holder, and E. B. Smith. 1971. Rat mite dermatitis. *J. Am. Med. Assoc. 215*: 1499-1500.

Holstein, M. 1967. Dynamics of *Aedes aegypti* distribution, density and seasonal prevalence in the Mediterranean area. *Bull. World Health Organ. 36*: 541-543.

Horsfall, W. R. 1955. Mosquitoes: Their bionomics and relation to disease. Ronald Press, New York.

Horsfall, W. R. 1962. Medical entomology: Arthropods and human disease. Ronald Press, New York.

Hoogstraal, H. 1972. The influence of human activity on tick distribution, density, and diseases. *Wiad. Parazytol. 18*: 501-511.

Hudson, B. W., B. F. Feingold, and L. Kartman. 1960. Allergy to flea bites. II. Investigations of flea bite sensitivity in humans. *Exper. Parasitol. 9*: 264-270.

Hudson, B. W., M. I. Goldenberg, J. D. McCluskie, H. E. Larson, C. D. McGuire, A. M. Barnes, and J. D. Poland. 1971. Serological and bacteriological investigations of an outbreak of plague in an urban tree squirrel population. *Am. J. Trop. Med. Hyg. 20*: 255-263.

Isaacson, M. 1975. The ecology of *Praomys (Mastomys) natalensis* in southern Africa. *Bull. World Health Organ. 52*: 629-636.

Iwuala, M. O. E., and J. O. A. Onyeka. 1977. The types and distribution patterns of domestic flies in Nsukka, East Central State, Nigeria. *Environ. Entomol. 6*: 43-49.

Jordan, K. 1948. Suctoria. Fleas. *In* A handbook for the identification of insects of medical importance (J. Smart, ed.). Trustees of the British Museum, London, pp. 211-245.

Kartman, L., and B. W. Hudson. 1971. The effect of flea control on *Yersinia (Pasteurella) pestis* antibody rates in the California vole, *Microtus californicus*, and its epizootiological implications. *Bull. World Health Organ. 45*: 295-301.

Keh, B. 1964. The brown dog tick, *Rhipicephalus sanguineus*, in California. *Calif. Vector Views 11*: 27-31.

Keh, B. 1973. Dermatitis in man traced to dog infested with *Cheyletiella yasguri* Smiley (Acari: Cheyletiellidae) in California. *Calif. Vector Views 20*: 77-79.

Keh, B. 1975a. Dermatitis caused by the bat mite, *Chiroptonyssus robustipes* Ewing, in California. *J. Med. Entomol. 11*: 498.

Keh, B. 1975b. Intense pruritis in man and concurrent infestation of *Cheyletiella bakeri* Smiley (Acari: Cheyletiellidae) on cats in a home in California. *Calif. Vector Views 22*: 2-4.

Keh, B. 1977. Personal communication. California Department of Health, Berkeley, California.

Kleevens, J. W. L. 1971. Housing, urbanization and health in developing (tropical) countries. Trans. Roy. Soc. Trop. Med. Hyg., Supplement, pp. 60-72.

Kokernot, R. H., J. Hayes, N. J. Rose, and T. H. Work. 1967. St. Louis encephalitis in McLeansboro, Illinois, 1964. *J. Med. Entomol. 4*: 255-260.

Kokernot, R. H., J. Hayes, R. L. Will, C. H. Tempelis, D. H. M. Chan, and B. Radivojevic. 1969. Arbovirus studies in the Ohio-Mississippi basin, 1964-1967. II. St. Louis Encephalitis Virus. *Am. J. Trop. Med. Hyg. 18*: 750-761.

Legner, E. F., B. B. Sugarman, Y. Hyo-sok, and H. Lum. 1974. Biological and integrated control of the bush fly, *Musca sorbens* Wiedemann, and other filth breeding Diptera in Kwajalein Atoll, Marshland Islands. *Bull. Soc. Vector Ecol. 1*: 1-14.

Lewis, D. J. 1971. Phlebotomid sandflies. *Bull. World Health Organ. 44*: 535-551.

Lewis, D. J. 1973. Phlebotomidae and Psychodidae (sand-flies and moth-flies). *In* Insects and other arthropods of medical importance (K.G.V. Smith, ed.). British Museum (Natural History), London, pp. 155-179.

Lewis, D. J. 1974. The biology of Phlebotomidae in relation to leishmaniasis. *Ann. Rev. Entomol. 19*: 363-384.

Luby, J. P., S. W. Sulkin, and J. P. Sanford. 1969. The epidemiology of St. Louis encephalitis: A review. *Ann. Rev. Med. 20*: 329-349.

Lunsford, C. J. 1949. Flea problem in California. *Arch. Dermatol. Syphilol. 60*: 1184-1202.

Lysenko, A. J. 1971. Distribution of leishmaniasis in the Old World. *Bull. World Health Organ. 44*: 515-520.

MacDonald, W. W. 1967. Host feeding preferences. *Bull. World Health Organ. 36*: 597-599.

Mack, T. M., B. F. Brown, W. D. Sudia, J. C. Todd, H. Maxfield, and P. H. Coleman. 1967. Investigation of an epidemic of St. Louis encephalitis in Danville, Kentucky, 1964. *J. Med. Entomol. 4*: 70-76.

Magy, H. I. 1968. Vector and nuisance problems emanating from man-made recreational lakes. *Proc. Calif. Mosq. Contr. Assoc. 36*: 36-37.

Marchand, J. P., Renault-Steens, G. Baquillon, and B. N'Diaye. 1975. La gale. A propos d'une épidémie actuelle au Sénégal et de ses complications. *Bull. Soc. Med. d'Afr. Noire Langue Francaise 20*: 74-82.

Mathis, H. L. 1962. Mosquito control and the urban sprawl in Marin County. *Proc. Calif. Mosq. Contr. Assoc. 30*: 50-51.

Mattingly, P. F. 1957. Genetical aspects of the *Aedes aegypti* problem. I. Taxonomy and bionomics. *Ann. Trop. Med. Parasitol. 51*: 392-408.

Mattingly, P. F. 1962. Mosquito behaviour in relation to disease eradication programmes. Ann. Rev. Entomol. 7: 419-436.

Mattingly, P. F. 1967a. Taxonomy of *Aedes aegypti* and related species. *Bull. World Health Organ. 36*: 552-554.

Mattingly, P. F. 1967b. The systematics of the *Culex pipiens* complex. *Bull. World Health Organ. 37*: 257-261.

McClelland, G. A. H. 1971. Variation in scale patterns of the abdominal tergum of *Aedes aegypti* (L.). World Health Organ./VBC/71.271.

McEnroe, W. D. 1974. Spread and variations of populations in eastern Massachusetts of the American dog tick. Univ. Massachusetts Agric. Exp. Sta. Res. Bull. No. 594.

McNeill, W. H. 1976. Plagues and peoples. Anchor Press and Doubleday, Garden City, N.Y.

Mellanby, K. 1972. Scabies. E.W. Classey Ltd., Middlesex, England.

Morse, R. M. 1969. Recent research on Latin American urbanization: A selective survey with commentary. *In* The city in developing countries: Readings on urbanism and urbanization (G. Breese, ed.). Prentice-Hall, Inc., Englewood, New Jersey, pp. 474-506.

Mumcuoglu, Y. 1976. House dust mites in Switzerland. I. Distribution and taxonomy. *J. Med. Entomol. 13*: 361-373.

Murphey, R. 1969. Urbanization in Asia. *In* The city in developing countries: Readings on urbanism and urbanization (G. Breese, ed.). Prentice-Hall, Inc., Englewood, New Jersey, pp. 58-67.

Nadim, A., and G. S. Rostami. 1974. Epidemiology of cutaneous leishmaniasis in Kabul, Afghanistan. *Bull. World Health Organ. 51*: 45-49.

Newson, D. H. 1976. Arthropod problems in recreational areas. Ann. Rev. Entomol. 22: 333-354.

Nielsen, B. O. 1963. The biting midges of Lyngby Aamose (Culicoides: Ceratopogonidae). *Natura Jutlandica 10*: 5-46.
Nnochiri, E. 1968. Parasitic disease and urbanization in a developing community. Oxford Univ. Press, London.
Olsen, P. F. 1970. Sylvatic (wild rodent) plague. *In* Infectious diseases of wild mammals (J.W. Davis, L.H. Karstad, and D.O. Trainer, eds.). Iowa State Univ. Press, Ames, Iowa, pp. 200-213.
Pavlovsky, E. N. 1966. Natural nidality of transmissible diseases with special reference to the landscape epidemiology of zooanthroponoses. Univ. Illinois Press, Urbana and London.
Petrischeva, P. A. 1971. The natural focality of leishmaniasis in USSR. *Bull. World Health Organ. 44*: 567-576.
Pollitzer, R. 1954. Plague. World Health Organ. Monogr. Ser. 22, Geneva, Switzerland.
Pollitzer, R. 1960. A review of recent literature on plague. *Bull. World Health Organ. 23*: 313-400.
Poorbaugh, J. H., and D. D. Linsdale. 1971. Flies emerging from dog feces in California. *Calif. Vector Views 18*: 51-56.
Population Division United Nations Bureau of Social Affairs. 1969. World urbanization trends, 1920-1960. *In* The city in developing countries: Readings on urbanism and urbanization (G. Breese, ed.). Prentice-Hall, Inc., Englewood, New Jersey, pp. 21-45.
Povolný, D. 1971. Synanthropy. *In* Flies and disease (B. Greenberg, ed.). Princeton Univ. Press, Princeton, New Jersey, pp. 17-54.
Pratt, H. D., R. C. Barnes, and K. S. Littig. 1959. Mosquitoes of public health importance. CDC Training Guide. U.S. Public Health Serv., Atlanta, Georgia.
Ragheb, I. 1969. Patterns of urban growth in the Middle East. *In* The city in developing countries: Readings on urbanism and urbanization (G. Breese, ed.). Prentice-Hall, Inc., Englewood, New Jersey, pp. 104-126.
Rao, T. R. 1967. Distribution, density and seasonal prevalence of *Aedes aegypti* in the Indian subcontinent and South-east Asia. *Bull. World Health Organ. 36*: 547-551.
Reeves, W. C. 1972. Recrudescence of arthropod-borne diseases in the Americas. Symposium on Vector Control and the Recrudescence of Vector-borne Diseases. Pan Am. Health Organ. Publ. 238.
Roberts, F. H. S. 1970. Australian ticks. CSIRO, Melbourne, Australia.
Rothenberg, R., and D. E. Sonenshine. 1970. Rocky Mountain spotted fever in Virginia: Clinical and epidemiologic features. *J. Med. Entomol.* 7: 663-669.

Rudnick, A. 1967. *Aedes aegypti* and haemorrhagic fever. *Bull. World Health Organ. 36*: 528-532.

Russell, P. F., L. S. West, and R. D. Manwell. 1946. Practical malariology. W.B. Saunders Co., Philadelphia, Pennsylvania.

Samarawickrema, W. A. 1967. A study of the age-composition of natural populations of *Culex pipiens fatigans* Wiedemann in relation to the transmission of filariasis due to *Wuchereria bancrofti* (Cobbold) in Ceylon. *Bull. World Health Organ. 37*: 117-137.

Scherer, W. F., E. L. Buescher, M. B. Flemings, A. Noguchi, and J. Scanlon. 1959. Ecological studies of Japanese encephalitis virus in Japan. III. Mosquito factors. Zootropism and vertical flight of *Culex tritaeniorhynchus* with observations on variations in collections from animal-baited traps in different habitats. *Am. J. Trop. Med. Hyg. 8*: 665-677.

Schoof, H. F., G. A. Mail, and E. P. Savage. 1954. Fly production sources in urban communities. *J. Econ. Entomol. 47*: 245-253.

Sexton, D. J., W. Burgdorfer, L. Thomas, and B. R. Norment. 1976. Rocky Mountain spotted fever in Mississippi: Survey for spotted fever antibodies in dogs and for spotted fever group rickettsiae in dog ticks. *Am. J. Epidemiol. 103*: 192-197.

Sharif, M. 1948. Nutritional requirements of flea larvae, and their bearing on the specific distribution and host preferences of the three Indian species of *Xenopsylla* (Siphonaptera). *Parasitol. 38*: 253-262.

Singh, D. 1967. The *Culex pipiens fatigans* problem in Southeast Asia with special reference to urbanization. *Bull. World Health Organ. 37*: 239-243.

Slonka, G. F. 1975. Epidemiology of pediculosis capitis. *Bull. Soc. Vector Ecol. 2*: 16-19.

Smadel, J. E. 1959. Status of the rickettsioses in the United States. *Ann. Intern. Med. 51*: 421-435.

Smith, E. B., and T. F. Claypoole. 1967. Canine scabies in dogs and in humans. *J. Am. Med. Assoc. 199*: 59-64.

Snow, W. F., and T. J. Wilkes. 1977. Age composition and vertical distribution of mosquito population in the Gambia, West Africa. *J. Med. Entomol. 13*: 507-513.

Spielman, A. 1967. Population structure in the *Culex pipiens* complex of mosquitoes. *Bull. World Health Organ. 37*: 271-276.

Spiller, D. 1968. Mosquito problems in California's Central Valley. Div. Agric. Sci. Univ. Calif., Berkeley, California.

Subra, R. 1975. Urbanisation et Filariose de Bancroft en Afrique et à Madagascar. *Cah. O.R.S.T.O.M., ser. Entomol.*

med. Parasitol. 13: 193-203.

Sudia, W. D., E. Fowinkle, and P. H. Coleman. 1967. St. Louis encephalitis in Memphis, Tennessee. *J. Med. Entomol. 4*: 77-79.

Surtees, G. 1967. The distribution, density and seasonal prevalence of *Aedes aegypti* in West Africa. *Bull. World Health Organ. 36*: 539-540.

Surtees, G. 1971. Urbanization and the epidemiology of mosquito-borne disease. *Abstr. Hyg. 46*: 121-134.

Sweatman, K. 1967. Physical and biological factors affecting the longevity and oviposition of engorged *Rhipicephalus sanguineus* female ticks. *J. Parasitol. 53*: 432-445.

Thompson, B. H. 1976. Studies on the flight range and dispersal of *Simulium damnosum* (Diptera: Simuliidae) in the rain-forest of Cameroon. *Ann. Trop. Med. Parasitol. 70*: 343-354.

Traub, R., and C. L. Wisseman, Jr. 1974. The ecology of chigger-borne rickettsiosis (scrub typhus). *J. Med. Entomol. 11*: 237-303.

Trpis, M., and W. Hausermann. 1975. Demonstration of differential domesticity of *Aedes aegypti* (L.) (Diptera, Culicidae) in Africa by mark-release-recapture. *Bull. Entomol. Res. 65*: 199-208.

Turner, J. F. 1969. Uncontrolled urban settlement: Problems and policies. *In* The city in developing countries: Readings on urbanism and urbanization (G. Breese, ed.). Prentice-Hall, Inc., Englewood, New Jersey, pp. 507-535.

Usinger, R. L. 1966. Monograph of Cimicidae (Hemiptera-Heteroptera). The Thomas Say Foundation, College Park, Maryland.

Usinger, R. L., P. Wygodzinsky, and R. E. Ryckman. 1966. The biosystematics of Triatominae. Ann. Rev. Entomol. 11: 309-330.

Velimirovic, B. 1972. Plague in South-east Asia. *Trans. Roy. Soc. Trop. Med. Hyg. 66*: 479-504.

Voorhorst, R., M. I. A. Spieksma-Boezeman, and F. T. M. Spieksma. 1964. Is a mite (*Dermatophagoides* sp.) the producer of the house-dust allergen? *Allerg. Asthma. 10*: 329-334.

Watson, R. B. 1949. Location and mosquito-proofing of dwellings. *In* Malariology (M.F. Boyd, ed.). W.B. Saunders Co., Philadelphia and London, pp. 1184-1202.

Weber, N. 1977. Plague in New Mexico. Health and Social Services Department, New Mexico.

Wharton, G. W. 1976. House dust mites. *J. Med. Entomol. 12*: 577-621.

Williams, P. 1976. The phlebotomine sandflies (Diptera, Psychodidae) of caves in Belize, Central America. *Bull. Entomol. Res. 65*: 601-614.

Wilton, D. P. 1963. Dog excrement as a factor in community fly problems. *Proc. Hawaiian Entomol. Soc. 18*: 311-317.
World Health Organization. 1976. Epidemiology of onchocerciasis. World Health Organ. Tech. Rep. Ser. 597.
Zarate, L. G. 1977. Personal communication. Division of Entomology and Parasitology, University of California, Berkeley, California.

ECOLOGY AND MANAGEMENT OF ARTHROPOD POPULATIONS IN RECREATIONAL LANDS[1]

Richard W. Merritt
H. D. Newson

Department of Entomology
Michigan State University
East Lansing, Michigan

I. INTRODUCTION

During the past decade an increased awareness and importance associated with pestiferous arthropods occurring in recreational lands has been recognized. A number of human factors can be attributed to this current trend. Increased population growth, mobility, leisure time and affluence have opened new recreational vistas to millions of people who had limited

---

[1] *Michigan Agricultural Experiment Station Journal Article No. 8003.*

ISBN 0-12-265250-9

opportunities to experience them in the past (Bureau of Outdoor Recreation, 1973). The demand for outdoor recreation and particularly the use of wildlands (national and state parks, forests and related areas) has increased throughout the world (Brockman and Merriam, 1973; Forster, 1973; Lavery, 1974; B.O.R., 1975), as has the attendance at national and state parks in the United States (Fig. 1). In state parks the rate of visitation in the past 20 years increased almost 425%. The national parks had an even greater visitor increase from 33.2 million in 1950 to 217.4 million in 1974, or a gain of more than 650%! This outdoor recreation "boom" and "return-to-nature" movement have brought man into closer contact with biting arthropods and enzootic diseases which are indigenous to wild areas not usually encountered by man.

However, we do not feel this has been the only reason for the increased importance of insect problems in recreational lands. In addition to the increased demand for outdoor recreation, there have been substantial changes in people's recreation habits and patterns. Recent government publications have tended to treat recreation in present day terms and to ignore the recreational activities of 50 years ago (Douglass, 1975). In 1910, the United States had a rural economy with approximately 55% of its population living in rural areas (Fig. 2). Today, nearly 75% of the population lives in urban centers while only 25% remains in rural areas (Fig. 2). Projections indicate that by the year 2000, about 85% of the expected 300 million people will be urban dwellers (Williamson, 1969). Rural living in the past, as today, involved close association with the natural environment and outdoor activities and contact with nature were part of everyday life. With the major shift from rural to urban and suburban living, man's attitudes toward outdoor recreation along with his patterns and habits have changed. Urbanites today are less conditioned than their rural counterparts of yesterday to the rigors and challenges of outdoor recreational activity (e.g., biting mosquitoes and blackflies). This is reflected in their selection of outdoor experiences such as camping in recreational vehicles (Catton *et al.*, 1969; Hendee, 1969). The use of trailers, campers and motor coaches is increasing at the rate of 25% each year (Anon., 1973a). People from cities and suburbs enjoy getting away from home, but they carry so much paraphernalia that life on the road is not really too different from that at home. When confronted with pestiferous insects in recreational lands, the modern-day camper demands immediate action and effective control measures for the complex and long-term pest problems that exist there (Ryan, 1976). Significant advances in recent years on disease epidemiology and surveillance, as well as diagnostic methods and techniques, have provided scientists with additional tools to study and identify previously unknown

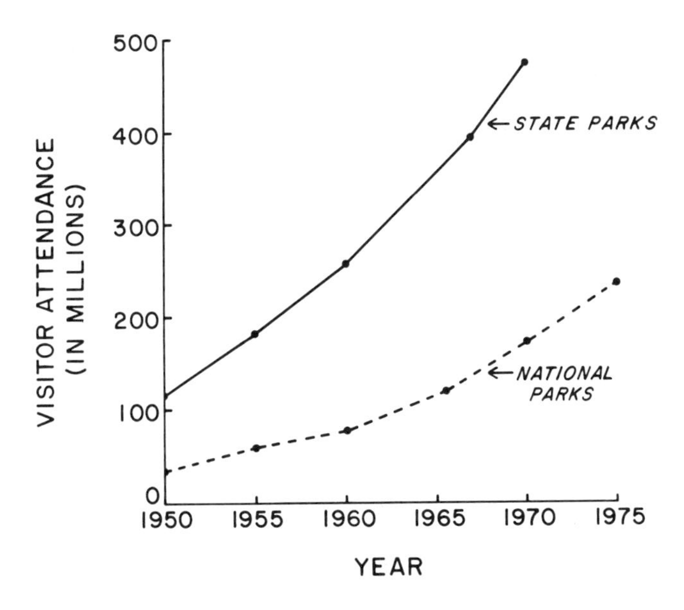

Fig. 1. Visitor attendance to national and state parks in the United States, 1950 to 1975.

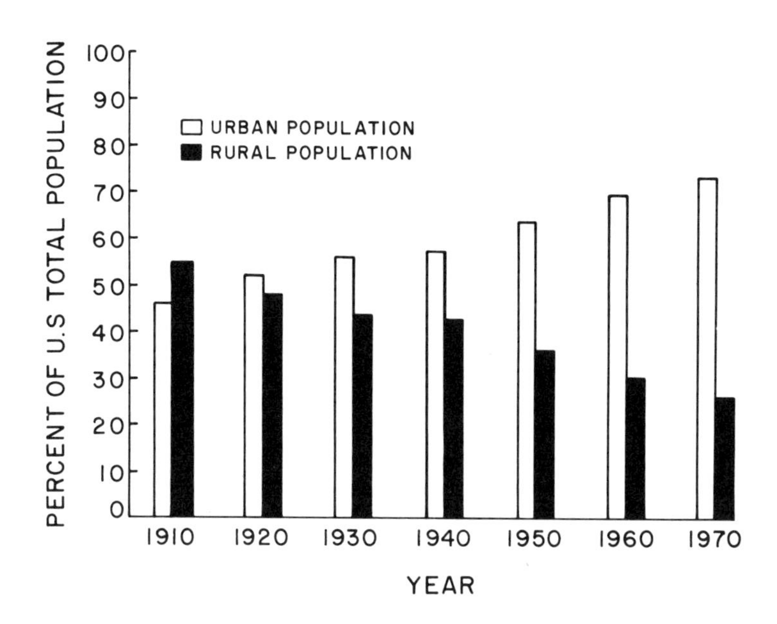

Fig. 2. Urban and rural population changes in the United States, 1910 to 1970.

etiologic agents of arthropod-borne diseases. Also, better communication and cooperation between physicians and state and county health departments have resulted in widespread dissemination of medical information to the public, and this has led to an increased awareness of the potential for disease transmission in recreation areas.

Arthropod problems associated with land and water-based recreation areas are now a major concern and some of the more important ones will be discussed in the following sections.

II. ARTHROPOD PROBLEMS ASSOCIATED WITH LAND-BASED RECREATIONAL ENVIRONMENTS

These problems are generally associated with state and national parks or forests and wilderness areas. Many of these areas not only contain pestiferous arthropods but also the natural zoonotic foci of various arthropod-borne diseases. A disease can be said to have a natural focus when the pathogen, its specific vector and its animal carrier have existed for generations, independently of man, under natural conditions in localized geographic areas (Audy, 1958; Pavlovsky, 1964). An intrusion by man into such a focus, in which the pathogen circulates between wild animals and arthropod vectors, can bring about transmission to man and cause infection and disease (zoonoses) (Fig. 3). Man is the side link or accidental host and not obligatory for the maintenance of the pathogen. Several diseases fit this category and are becoming increasingly important in recreational lands.

A. Rocky Mountain Spotted Fever

Rocky Mountain Spotted Fever (RMSF) accounts for more than 90% of the reported cases of rickettsial disease in the United States and represents the most important disease of this kind with respect to fatalities (Hattwick, 1971; Hattwick *et al.*, 1973; Burgdorfer, 1975). The disease is endemic in the United States and some parts of Canada, Mexico and South America (Hoogstraal, 1967; James and Harwood, 1969). It is principally a wild animal infection (Fig. 3) maintained and distributed by a variety of ticks belonging to the family Ixodidae. The infective agent of RMSF is *Rickettsia rickettsi* (Wolbach). When man enters the natural enzootic transmission cycle he may contact the infection, either by the bite of a tick or by infective crushed tick tissues or fluids, coming in contact with abraded skin or the conjunctivae of the eyes (Burgdorfer, 1975).

In the United States approximately 500 cases were reported

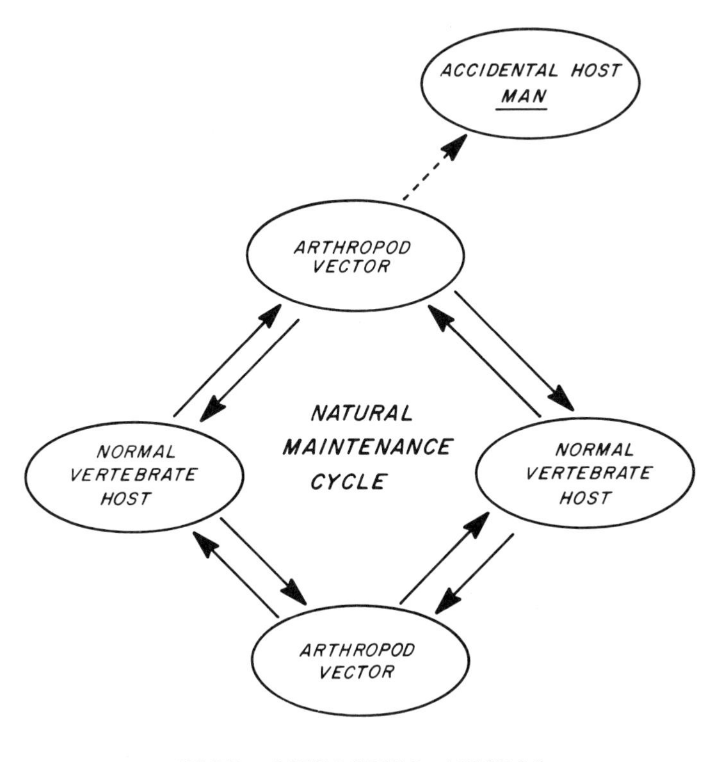

Fig. 3. Generalized zoonotic disease cycle in nature including man.

annually until 1950 when the availability of broad-spectrum antibiotics reduced reported cases sharply (ibid). However, since 1960 there has been a steady increase in reported cases, especially from the Atlantic seaboard and south central states (Hattwick, 1971; Peters, 1971; Hattwick *et al.*, 1973). In 1976 over 900 cases were reported (CDC, 1977).

Epidemiology of the disease is two-dimensional and complex. In the western United States the principal vector of RMSF is the Rocky Mountain wood tick, *Dermacentor andersoni* (Stiles), which is common in cutover mountainous areas, and in sagebrush thickets along streams (Bishopp, 1935; Hoogstraal, 1967). In this region RMSF is considered a rural occupational and recreational disease occurring most frequently among people engaged in outdoor work (e.g., farmers, foresters, road construction workers), or those enjoying outdoor recreational activities such as hunting, fishing, hiking and camping (Hoogstraal, 1967; Burgdorfer, 1976). Most cases are reported during April, May and June when adult *D. andersoni* are active (Burgdorfer, 1975). In the eastern United States, on the other hand, spotted fever is prevalent in both rural and urban areas during summer months when the primary vector, *Dermacentor variabilis* (Say), the American dog tick, exhibits its greatest activity (Rothenberg and Sonenshine, 1970; Sonenshine *et al.*, 1972). This species, in contrast to *D. andersoni*, is most common in small woodlots, abandoned farmland and old field-forest ecotones (Sonenshine and Stout, 1970; Sonenshine and Levy, 1972). In addition to being an important disease in certain recreational areas (Cooney and Burgdorfer, 1974), RMSF in the eastern and south central United States also occurs in rural and urban communities which are in the initial stages of urban or suburban development (Smadel, 1959; Linnemann *et al.*, 1973; Burgdorfer, 1975; Sonenshine, 1976). Exposure to the pathogen occurs predominantly in the environment of the home, and cases frequently occur in women and children because of their close association with tick-infested household pets, especially dogs (Hattwick *et al.*, 1973; Rothenberg and Sonenshine, 1970; also see Nelson, Chapt. 5, this volume).

Control efforts have generally involved the use of acaricides and mechanical clearing of natural trails and campgrounds to reduce the number of ticks in recreational areas (Collins and Nardy, 1951; Cooney and Pickard, 1972; Burgdorfer, 1975). With reference to prophylaxis, a commercial vaccine is available, yet its efficacy is uncertain (Calia *et al.*, 1970), and the incidence of RMSF is too low to justify general use (Bell and Stoenner, 1961). However, immunization of high risk persons is recommended (Burgdorfer, 1975). Treatment of the disease involves early diagnosis followed by prompt therapy with broad spectrum antibiotics. Human exposure will certainly be a continuing problem, as long as man continues to intrude into the natural zoonotic transmission cycle.

B. Colorado Tick Fever

Colorado Tick Fever (CTF) is an acute viral disease which occurs primarily in the western mountainous areas of the United States and Canada and is transmitted to man be the bite of the wood tick, *Dermacentor andersoni*. Is is similar to RMSF in that it is a zoonosis (Fig. 3) and is a localized disease correlated with the presence or absence of its normal rodent hosts, particularly the golden-mantled ground squirrel, (*Citellus lateralis* (Say) (Burgdorfer and Eklund, 1959; 1960). Recently it has been shown that other animals also are probably important in maintaining and spreading CTF virus in certain areas (Clark *et al.*, 1970). Clinically, physicians have frequently mistaken CTF for RMSF, which may account for the low number of reported cases (Spruance and Bailey, 1973). In a review of 115 confirmed cases of CTF in Utah from 1960 through 1969, it was shown that 78% of the patients were males, probably reflecting their greater outdoor activity and exposure to tick bites (ibid). All age groups were involved and the seasonal occurrence was from April to August. CTF is becoming more common in the West and will be of increasing importance in forest recreational areas.

C. Tick-borne Relapsing Fever

Tick-borne relapsing fever has a wide geographic distribution and is endemic throughout areas of the Old World and the Americas. In the western United States occasional sporadic outbreaks occur in campers and hikers in mountainous recreational areas. The infectious agent is a spirochete, *Borrelia* sp., and the chief vectors in the transmission of relapsing fever are soft-bodied argasid ticks, especially those of the genus *Ornithodoros*. Rodents and insectivores are the usual sylvan hosts, but when man enters the natural transmission cycle he may become infected. It is not unusual for patients to be unaware of having been bitten because ticks belonging to this genus cause painless bites and leave their hosts shortly after the blood meal (Felsenfield, 1971).

Although it is usually considered a sporadic endemic disease (Wynns, 1942; Gelman, 1961), large outbreaks can occur. In the United States the most recent outbreaks occurred in 1968 in northeastern Washington and in 1973 at the North Rim of the Grand Canyon, Arizona. The first instance involved a group of boy scounts using old wooden log cabins for camping near Spokane, Washington, (Thompson *et al.*, 1969). The disease affected 10 out of 20 persons who were sleeping in the cabins infested with rodents, particularly chipmunks. Infected ticks were also found inside the cabins and in nesting material of the rodents. The outbreak occurred during cold weather in March. Undoubtedly, the fact that fires were built inside the cabins was an important factor in activating the ticks since only one scout out of 22 sleeping outside contracted the disease (Thompson *et al.*, 1969). This area of Washington state was known to be a natural focus for tick-borne relapsing fever.

The second outbreak, which occurred in Grand Canyon National Park, involved 45 confirmed cases and was the largest outbreak of relapsing fever from a single focus in the United States (Maupin, 1974). The North Rim area attracts approximately 600,000 tourists from June to October each year and provides rustic cabin units for habitation in a ponderosa pine forest. Field and laboratory investigations revealed a rare instance of interaction between two zoonotic diseases, plague and relapsing fever. During the previous year (1972) plague was responsible for depleting the rodent population to such a degree that the infected tick vector, *O. hermsi*, had to rely on tourists and employees the following year (1973) for blood meals, thus transmitting the pathogen to man. The ability of this group of ticks to survive long periods without feeding and still retain their infectivity also was a contributing factor. Thus, large numbers of visitors to this park and unusually large rodent populations greatly enhanced the potential for human involvement in what otherwise may have remained an enzootic disease cycle.

D. Tularemia

Tularemia occurs throughout North America and in many parts of continental Europe, USSR and Japan. It is a plague-like disease caused by the rod-shaped bacterium, *Francisella tularensis* (McCoy and Chapin). The maintenance cycle in nature is from infected to susceptible rabbits via ticks; however, the pathogen is transmitted to man primarily by the bites of infected ticks or deer flies (Tabanidae), and by direct inoculation of the skin or conjunctival sac through skinning, dressing or otherwise handling infected mammals, ticks or flies. Approximately 70% of the human cases of tularemia in North America result from contact with hares and rabbits (McDowell *et al.*, 1964). An average of 180 human cases per year were reported in the United States over the last 10 years (CDC, 1977).

Seasonal peaks of tularemia are associated with hunting and outdoor recreation periods, and with large populations of infected deer flies and ticks. In the western United States human infections are more common in summer when ticks and deer flies are abundant and jack rabbits are hunted. In the eastern states infections are more common in winter when cottontails are hunted (McDowell *et al.*, 1964).

Even though the number of reported cases of tularemia have not dramatically increased during the past decade, it continues to constitute a potential hazard to those persons involved in outdoor recreational activities, particularly hunting. Preventive measures involve minimizing human contact with infected rabbits, ticks and deer flies during peak tularemia periods through education and thoughtful scheduling of hunting seasons and recreational activities (ibid).

E. Sylvatic Plague

Plague is widespread throughout the world and endemic in

many countries, including the Americas. Although only small numbers of sporadic cases of plague are reported annually in the United States (Caten and Kartman, 1968; CDC, 1977), these are not valid indicators of the importance of this disease. Widespread sylvatic plague is of increasing significance in the western third of the United States as more and more people turn to outdoor recreation and man continues to encroach on enzootic areas (Link, 1955; CDC, 1970, 1971). The causative agent of plague, *Yersinia (Pasteurella) pestis* (Yerskin and Kitasato), circulates in a natural reservoir of small rodents such as white-footed mice and meadow mice, causing little mortality. Occasionally, plague infected populations of more susceptible small mammals such as ground squirrels, chipmunks, wood rats, gophers and prairie dogs experience massive and observable die-offs. Epizootics of this type are particularly common in the southern Sierras and northeastern areas of California and involve forest and foothill rodent complexes (Stark and Kinney, 1969; Murray, 1971; Nelson and Smith, 1976). During such events people in resort areas may be exposed to bites of infective fleas, which have separated from their normal host animals, or to direct infection by handling sick or dead animals (CDC, 1970; Nelson and Smith, 1976).

In the past two decades over 80% of the human cases of plague occurred in the Rocky Mountain states, with very few in the Pacific states (CDC, 1970). The majority of these were reported in New Mexico and Arizona; several were associated with epizootics among prairie dogs. An extensive plague die-off among prairie dogs in 1971 at a major fishing and camping resort area in south central Colorado necessitated closing the facility for most of the season (Beadle, 1972). More recently (1976), a large epizootic affecting chipmunks (*Eutamias* spp.) and golden-mantled ground squirrels occurred in a newly developed U.S. Forest Service campground at Lake Davis, Plumas Co., California (Nelson, 1976). In this instance, the resulting population build-up of susceptible rodents was attributed to tourists who constantly fed and protected them in a resort area which was an enzootic plague focus. Major current and future problems focus around continued coexistence of highly susceptible plague hosts, such as prairie dogs or chipmunks, and highly mobile tourist populations (Beadle, 1972; Newson, 1977). The greatest potential danger to the public exists when domestic rats are exposed to infection from wild mammals in areas adjacent to human communities (Link, 1955). An outbreak of plague among these rats would expose many persons in urban centers to infective fleas (see Nelson, Chapt. 5, this volume).

Control programs for plague have been designed primarily to protect the visiting public from contact with plague-infected fleas or rodents. In recreation areas, good control of fleas on wild rodents has resulted from the dusting of rodent burrows with insecticides (Ryckman *et al.*, 1954; Barnes *et al.*, 1972), and the use of insecticide bait stations whereby rodents are attracted to stations where they treat themselves with insecticidal dust while taking bait (Kartman, 1958; Barnes and Kartman, 1960; Barnes *et al.*, 1974). Recently, rodent baits impregnated with either systemic insecticides (Miller *et al.*, 1975), or fumigants (Cole *et al.*1976) have shown promising results for flea control.

F. Problems Associated with Biting and Stinging Arthropods

In addition to playing an important role as vectors of disease agents to man, ticks are responsible for causing other human disorders such as dermatosis, envenomization and tick paralysis. Tick species that may feed on humans vary considerably between recreational areas, and their distribution is generally spotty and correlated with the population of wild or domestic animals that serve as their normal hosts. A review of those species which feed on man and cause annoyance and complications is given by James and Harwood (1969). Persons engaged in outdoor recreational activities can minimize their contact with ticks by: 1) impregnating clothing with repellents, 2) wearing protective clothing which will prevent ticks from coming in contact with the skin, and 3) avoiding areas which are known to have heavy tick infestations. Tick control efforts in recreational areas have generally involved the use of chemicals or habitat management techniques (McDuffie and Smith, 1955; Mount *et al.*, 1971, 1976; Cooney and Pickard, 1972). Biological control of ticks has been attempted using hymenopteran parasites (Cole, 1965) and fungi (Boicev and Rizvanor, 1960), but they have not been very successful.

Imported fire ants, *Solenopsis* spp., were introduced into the southeastern United States and have spread rapidly to become an important pest in several areas of this region (Lofgren *et al.*, 1975). Recent studies indicate that fire ants are assuming greater importance as a public health hazard because of allergic symptoms exhibited by some individuals when stung (ibid). In the past, fire ants were mainly regarded as pests of grazing and crop land but now are regarded also as important pests in recreational areas.

Yellowjackets, or members of the Vespidae (Hymenoptera), cause general annoyance to visitors in recreation areas because of their attraction to human foodstuffs (Lewallen, 1968). They also produce painful stings which may result in allergic reactions. Estimates of losses attributable to the presence of yellowjackets at some private California resorts range up to $5,000 annually (Poinar and Ennik, 1972, see also Davis, Chapt. 7, this volume). Five percent of all hospital injuries reported to the U.S. Forest Service each year are directly attributable to stinging Hymenoptera (Akre, 1976).

III. ARTHROPOD PROBLEMS ASSOCIATED WITH WATER-BASED RECREATIONAL ENVIRONMENTS

About one quarter of all U.S. outdoor recreation is dependent on the availability of water (Hofe, 1973), and the most common problems associated with pestiferous arthropods are encountered in water-based recreational environments. Major arthropod pests of man in these areas include mosquitoes, black flies, punkies, stable flies and nonbiting midges and gnats. They breed in a variety of habitats, from sewage oxidation ponds and man-made lakes to tree holes and snow pools. In an attempt to associate pestiferous insects with their characteristic breeding habitats, we have used the following classification system.

A. Lentic Habitats

1. Impoundments (artificial ponds and lakes)

Insect problems emanating from man-made lakes, reservoirs and ponds usually involve nonbiting midges and mosquitoes. As income and leisure time have increased, there has been a growing demand for water-related recreational facilities in both rural and residential areas. In addition, reservoirs and impoundments, which were originally created for hydroelectric power or flood control, are now experiencing multiple recreational use (Stroud, 1966). Recreation is now recognized as a legitimate purpose in output of water resource development projects (Badger, 1973).

A relatively recent phenomenon is the construction of residential-recreational lakes, which cater largely to middle income citizens from metropolitan areas who enjoy swimming, boating, waterskiing and fishing (Mulla *et al.*, 1971; Mulla, 1974). These shallow, warm-water lakes are increasing in number in areas of the western United States, particularly in southern California where a warm semiarid climate is conducive to year-round outdoor activity (Anderson *et al.*, 1965; Mulla, 1974). Homes, sporting facilities and other recreational establishments are generally situated directly on the shoreline, where drainage from fertilized lawns, gardens and inadequate septic systems introduce large amounts of organic nutrients into the lake. This accelerated process of enrichment (cultural eutrophication) causes undesirable changes in plant and animal life, resulting in increases in phytoplankton production or "algal blooms" which create problems of odor, taste and sight (Mackenthun *et al.*, 1964; Hasler, 1969). When algae and other aquatic plants die or are chemically treated, organic matter from this crop sinks and rapidly causes an oxygen deficit due to oxidation of such materials by microorganisms (Grodhaus, 1963; Warren, 1971). This situation leads to an oxygen-poor and nutrient-rich bottom habitat that is ideal for Chironomidae and/or Chaoboridae.

During periods of adult emergence these insects may occur in such numbers as to interfere or restrict various sorts of recreational and other outdoor activities. Annoyance consists mainly of personal discomfort; however, large numbers of adult midges have been responsible for creating traffic hazards, ruining paint on buildings, fouling swimming pools, causing a nuisance inside homes, increasing the density of spiders and webs on houses and leading to unpleasant odors when piles of dead midges begin to rot (Grodhaus, 1963; Anderson *et al.*, 1965; Magy, 1968; Beck and Beck, 1969; Mulla, 1974; also see review by Grodhaus, 1975). They also occasionally affect the health of man by causing allergy problems (Henson, 1966; Shulman, 1967).

Chironomid midge problems in recreational lakes are not a recent development nor are they restricted to small lakes and reservoirs. Enormous numbers of adult chironomids emerged from two large lakes (Fountain and Willow) which were dredged from a salt marsh for the 1938 World's Fair in New York (Felton, 1940). The salt marsh sod originally present on the bottom began decomposing as soon as it was flooded. In addition, raw sewage flowed into the lake and fertilizers applied on grass and ornamental plants near the shores were washed into the lakes, creating a favorable habitat for larval midges. Chironomids also created nuisance problems in Winter Haven, Florida, where two adjacent lakes received raw sewage and treated effluent from the city (Provost, 1958), and similar problems have occurred in other large man-made impoundments throughout the United States (Grodhaus, 1963; Cook, 1967).

The periodic summer appearance of large numbers of gnats, *Chaoborus astictopus* Dyar and Shannon, has been a long-term problem to the residents of Clear Lake, Lake Co., California, and has had an adverse effect on resort business (Hunt and Bischoff, 1960). Clear Lake has an area of over 16,842 ha, and it was determined that the total seasonal production in one 114 km^2 section of the lake approximated 712 billion gnats or 391 Mt of organisms (Mackenthun, 1969)! Efforts to chemically control this gnat have been ongoing for many years and have produced one of the best examples of a persistent pesticide buildup in an aquatic ecosystem (Hunt and Bischoff, 1960; Cook, 1965). A similar biological magnification of insecticides in carnivorous birds was reported on the Long Island tidal marshes in New York, where mosquitoes spraying with DDT had been practiced for some 20 years (Woodwell *et al.*, 1967).

Recent attempts to control the Clear Lake gnat with methyl parathion (Apperson *et al.*, 1976) have produced resistant populations (Apperson, 1976). Application rates now needed to achieve 90% control (populations still in excess of nuisance thresholds) pose serious biological hazards for non-target organisms in the lake. Abate(R), a substitute organic insecticide which offers an acceptable safety margin, is too costly when applied at an effective rate, having a projected cost of over one million dollars per treatment (ibid). Several organic larvicides, including some insect growth regulators, have shown favorable results against chironomid midges in smaller recreational lakes (Mulla *et al.*, 1971, 1975, 1976). Management of chironomid populations with predators and pathogens has been reviewed by Bay (1974) and Grodhaus (1975). More research is needed in the latter area as well as in the basic design and operation of recreational lakes (Magy, 1968).

Mosquitoes are generally considered the most widespread and serious pests in recreational areas and have delayed and complicated the development and operation of recreational facilities wherever they have been present in large numbers. Mosquito problems associated with impounded water have resulted from larvae breeding in shallow areas, usually in the upper reaches and tributary embayments of the impoundment. Typically, these contain emergent floating vegetation and are subject to prolonged or intermittent flooding during the mosquito breeding season (Surtees, 1971; Harmston and Ogden, 1975). Seepage habitats below dams and water impounded behind dikes and levees are also typical areas which produce mosquitoes. Potential mosquito problems associated with impounded waters in the southeastern United States were obviated or minimized in the construction and operation of a chain of reservoirs constructed by the Tennessee Valley Authority (Bishop and Gartrell, 1944; Gartrell *et al.*, 1972). The most important public health problem in these reservoirs was the control of malarial mosquitoes through reduction of mosquito-producing habitats (Hess and Hall, 1943; Elliot, 1973). Even though transmission of malaria by indigenous mosquitoes is a rarity in the United States today, focal outbreaks are a definite possibility if *Anopheles* mosquitoes have the opportunity to feed upon individuals with *Plasmodium* infections. Such an occurrence might involve military returnees or tourists from malarious areas visiting recreational sites and campgrounds (Brunetti *et al.*, 1953; Fontaine *et al.*, 1953). Mosquito problems associated with man-made recreational impoundments in other areas of the United States have been documented (Edman, 1964; Magy, 1968; Harmston and Ogden, 1975), and there now is concern about the relationship between sporadic outbreaks of encephalitis among humans and the production of vector mosquitoes in man-made reservoirs and lakes (Surtees, 1971; Harmston and Ogden, 1975).

Careful planning, design and construction of impounded water projects as well as the implementation of source reduction measures following impoundage appear to be the best management techniques for good mosquito control (Harmston and Ogden, 1975). Where measures of this type are not taken, significant mosquito problems can be expected to develop and effective control programs must be employed if these recreational areas are to be enjoyed and fully used. If a site is located within an established mosquito control district, several hundred of which are in operation throughout most states in this country, then adequate technical assistance is readily available. In many cases, however, parks, campgrounds and other recreational facilities established near water impoundments are far removed from the densely populated areas where most mosquito control districts are located. In such situations owners or operators of these facilities either must provide a satisfactory level of mosquito control, often at considerable expense, or be assured of markedly reduced patronage at their resorts.

Construction of large, man-made dams in Africa, such as the Aswan and Kariba, and subsequent reservoir developments have had serious public health implications in recent times (Lowe-McConnell, 1966; Farvar and Milton, 1972; Ackerman *et al.*, 1973). Although primarily constructed for hydroelectric power, these water projects are also economically important in terms of sports fisheries and other types of water recreation activities (Jackson, 1966; Kalisti, 1973). In addition to major unforseen complications that these engineering "miracles" have had on the spread of infectious diseases such as schistosomiasis, malaria, trypanosomiasis and onchocerciasis (Waddy, 1966; Heyneman, 1971; Hughes and Hunter, 1972; Dasmann *et al.*, 1973), mass outbreaks of nonbiting midges and mosquitoes have also caused severe annoyance and allergy problems (Rzoska, 1964; Lewis, 1966). These "ecological backlashes" brought about by environmental modification will continue to arise so long as planners think only in engineering terms and not in the context of the total ecosystem.

2. *Natural Lakes*

Although the majority of insect pest problems in permanent bodies of water are associated with man-made lakes, similar problems also occur in large natural lakes as human encroachment upon their shores increases. Serious outbreaks of nonbiting midges in Lake Winnebago and several other natural Wisconsin lakes have plagued residents in these areas for years

(Burrill, 1913; Johnson and Munger, 1930; Hilsenhoff, 1966). Also, large numbers of mayflies have become an expensive nuisance in towns and cities near large lakes; newly emerged adults have been reported to settle in great drifts over roads, bridges and streets, causing severe traffic and odor problems (Needham, 1920; Burks, 1953; Leonard and Leonard, 1962).

The stable fly, *Stomoxys calcitrans* (L.) (also called the dog fly or beach fly), a well known livestock pest, is an important man-biting pest in certain recreation areas. For years stable flies have bothered campers and fishermen in some reservoirs constructed by the Tennessee Valley Authority. Studies have shown that *Stomoxys* breed there in moist windrows of fine flotage mixed with mayfly bodies left undisturbed when water levels were lowered during scheduled draw-downs (Pickard, 1968). A different situation exists along the shores of Lake Superior where periodic appearances of *Stomoxys* force visitors to vacate beach areas. While these occasions occur sporadically, rather than continuously during the summer, large numbers of these flies make life intolerable for swimmers, fishermen and lakeside campers (Gill, 1976). Gill and his associates in northern Michigan found that in most instances shoreline debris was unable to support full development of stable fly larvae. Supportive data indicated that the primary source was conventional breeding sites (manure and urine mixed with straw) on the farms and stables of nearby areas (Love and Gill, 1965). They hypothesized that flies from these breeding areas migrate toward large lakes during favorable weather conditions (Love and Gill, 1965; Voegtline *et al.*, 1965). Flight range studies of *Stomoxys* support this hypothesis (Eddy *et al.*, 1962; Gill, 1976) but further research is needed to establish, with certainty, the breeding source of these flies.

Sterile male releases have been suggested as part of a pest management scheme for control of stable flies (LaBrecque *et al.*, 1975; Patterson *et al.*, 1975); however, more research on the ecology of *S. calcitrans* and a complete cost-benefit analysis should be conducted before an expensive program is instituted for a sporadic pest such as this.

3. *Wastewater Oxidation Ponds and Lakes*

In recent years an alternative to conventional methods of sewage treatment has been wastewater management systems that will clean municipal wastewater and recycle scarce nutrients by using solar energy to drive the system, thereby reducing need for fossil or nuclear fuels. Construction and operation of these types of sewage reclamation systems are increasing in many areas of the United States and their effluents are often used to fill ornamental and recreation lakes and irrigate crops, golf courses, parks and freeway landscapes (Mackenthun,

1969; Kardos, 1970; Anon., 1973b; Sullivan *et al.*, 1973). One such system is the Michigan State University Water Quality Management Project (Fig. 4) which consists of a combination of oxidation lakes, marshes and irrigated terrestrial areas for the removal and recycling of dilute wastes in domestic wastewater secondary effluent (Tanner, 1972).

Series of shallow oxidation lakes, common to most wastewater management facilities, provide ideal habitats for large numbers of nonbiting midges and gnats, as well as mosquitoes. Lakes such as these are often designed for some recreational use (fishing, boating, etc.) and are usually adjacent to urban centers. Intolerable numbers of chaoborid gnats originating from new sewage lagoons in Redding, California, led to threat-

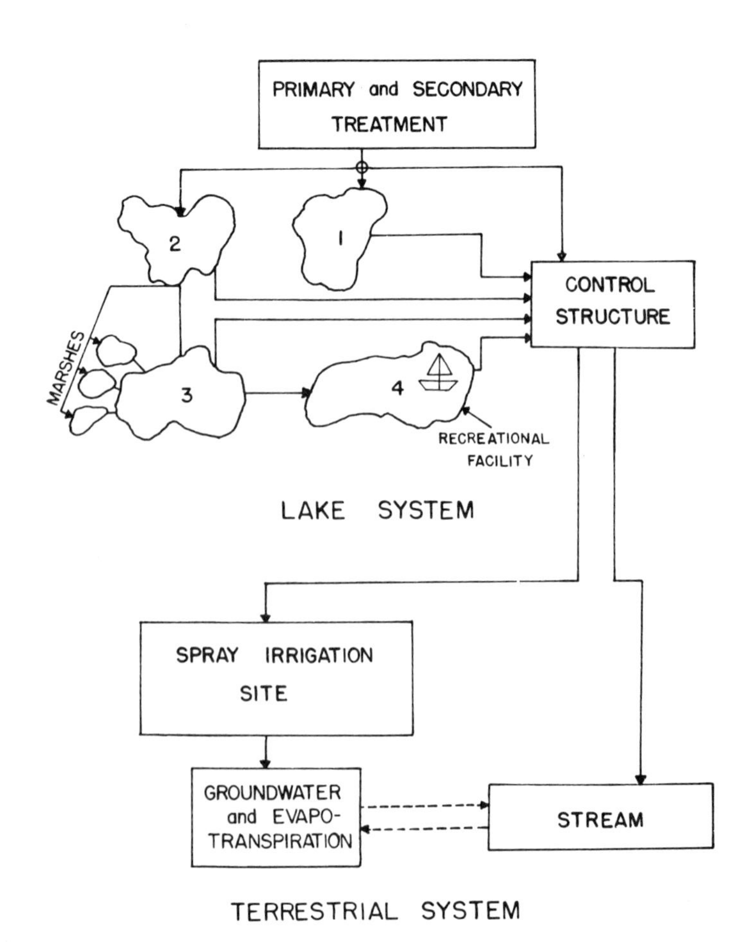

Fig. 4. Schematic diagram of the Michigan State University Water Quality Management Project.

ened lawsuits and the city spending nearly $200,000 to purchase surrounding property (Hazeleur, 1970; Lusk and Willis, 1970). There have been numerous reports of midge problems associated with wastewater lakes or lagoons in the United States and abroad (Bay, 1964; Bay *et al.*, 1965; Spiller, 1965; Brumbaugh *et al.*, 1969. Grodhaus (1975), in his review of chironomid midge nuisance and control, noted that wastewaters were the most important source of midge problems today. Mosquito problems have also been found in sewage oxidation ponds (Beadle and Rowe, 1960; Kimball, 1964; Mulla *et al.*, 1970); however, these have been associated with the presence of vegetation in and around the edges (Surtees, 1971).

New, promising measures for control of midges and mosquitoes in wastewater oxidation lakes, as well as in man-made recreational lakes, involve the use of insect growth regulators (Mulla *et al.*, 1974; Pelsue *et al.*, 1974; Mulla and Darwazeh, 1975). A number of these materials have shown high activity against target midges with little effect on non-target species. Beyond direct control of pest insects, proper design and operation of wastewater facilities are also of primary importance.

4. *Coastal Areas, Marshes, and other Wetlands*

Several families of biting flies have been associated with pest problems in coastal recreation areas. These problems have intensified with the increase in tourism and land development projects. The stable fly, or dog fly, has been an important pest during the summer and fall in coastal resort areas along the Atlantic and Gulf coasts (King and Lenert, 1936; Simmons and Dove, 1941; Hansens, 1951). Larvae breed in fermenting tidal deposits of bay grasses and marine algae which storm tides have laid down high above the normal tide lines on bay shores. Grasses deposited on beaches by normal tides are subject to daily saltwater inundations, thus preventing fermentation, and are unsuitable for larval development (Simmons and Dove, 1941; Simmons, 1944). Early control techniques involved spraying larval breeding areas in marine grasses (Blakeslee, 1945); however, in past years this method has not provided adequate control. More recent attempts include thermal fogging with insecticides (Mount *et al.*, 1966) or application of insect growth regulators to marine plants (Wright, 1974).

Members of the family Ceratopogonidae, often called sandflies, punkies, or no-seeums, have gained notoriety as pests on beaches and in coastal marshes and swamps throughout the world (Wirth and Blanton, 1974). It has been stated that these biting insects were largely responsible for the lack of early development of the southern Atlantic seaboard (Dove *et al.*, 1932). Visitors to mountainous recreational areas have also been plagued by these biting flies (Jamnback, 1969). Bites of these insects produce long-lasting and recurrent irritations, resulting in welts and lesions that persist for many hours and are the main discomfort caused by these flies (James and Harwood, 1969; Linley and Davies, 1971). On some resort islands in the Caribbean it is not uncommon for more than 100 biting *Culicoides* to be collected from a single arm or leg over a 15 minute period (Kettle and Linley, 1969). Many species of adult ceratopogonids are crepuscular in synchrony with the pattern of outdoor activity of many tourists in tropical and subtropical resorts. Larvae breed in a wide range of habitats including intertidal zones, salt marshes, mangrove swamps, wet clay or mud and other organic material (Kwan and Morrison, 1974). Linley and Davis (1971) thoroughly discussed the relationship of sandflies to tourism in Florida, the Bahamas and Caribbean area. They contended that, in developing tourism in these areas, sandfly problems should be avoided whenever possible, rather than trying to control them. In addition to life history information, they discussed several important entomological principles and their economic impact on resort development in sandfly areas. A recent bibliography on this economically important group of insects was provided by Atchley *et al.* (1975).

Due to the diversity of larval breeding habitats, management techniques for ceratopogonids are sometimes difficult and complex. Larvicidal treatment of breeding grounds has been widely employed (LaBreque and Goulding, 1954; Clements and Rogers, 1968); however, the breeding area may be extensive and problems involving insecticide resistance (Smith *et al.*, 1959) or the effects of these compounds on non-target organisms should be considered. Linley and Davies (1971) believe that of all control methods currently available, impoundment (permanent flooding) or the alteration and destruction of the breeding sites is the best and most successful. Protection from adult sandflies for limited periods can be gained through the use of insect repellents (Davies, 1967), and conventional window screens painted or sprayed with insecticide (Jamnback, 1961).

In addition to sandflies and dog flies, two other important groups of blood-sucking Diptera breed in coastal and estuarine marshes throughout the world. These are mosquitoes, primarily *Aedes* spp. and the Tabanidae, deerflies (*Chrysops*) and horseflies (*Tabanus*) (also called salt marsh greenheads). These groups are most abundant in the summer, and during the peak of their activity make some seashore areas virtually uninhabitable. In the United States, the Atlantic seaboard and Gulf Coast resort and recreational areas have had major problems because of these biting flies (Provost, 1949, 1959; Hansens, 1952; Mulrennan and Sowder, 1954; Jamnback and Wall, 1959; Wall and Doane, 1960; Jones and Anthony, 1964; Jamnback, 1969). These insect groups are not restricted to coastal areas and have been reported as pests at inland recreational areas with suitable breeding areas such as freshwater swamps and bogs (Defoliart *et al.*, 1967; Evans and McCuiston, 1971; Gojmerac and Devenport, 1971). More concern over these pests in recreational areas has come about with the awareness of their role as actual or potential vectors of Eastern Equine Encephalitis in states along the Eastern Seaboard and Gulf Coast (Hayes *et al.*, 1962; Reeves, 1965; Chamberlain, 1968).

Methods of management and control of salt marsh mosquitoes and tabanids are numerous. Effective control of salt marsh mosquitoes has been obtained by organophosphate larvicides and adulticides (Sutherland, 1970). Other promising chemicals are the aliphatic amines and similar compounds (Wilton and Fay, 1969). Because of the larval habitats, chemical control of horseflies and deerflies has been difficult. Applications to restricted areas have been of little value and wide area control programs have been only partially effective, while increasing the danger to non-target organisms (Pechuman, 1972; Wall and Marganian, 1973). Trapping devices such as the canopy and Manitoba traps baited with dry ice (Hansens *et al.*, 1971; Axtell *et al.*, 1975) have been used successfully to reduce tabanid populations on beaches and golf courses in coastal areas.

Biological control studies of mosquitoes have increased sharply in recent years as resistance to insecticides has increased. The most effective method to date has involved various species of predator fish (Geverich and Laird, 1968; Legner *et al.*, 1974). Invertebrate predators of mosquitoes are beginning to receive more attention as their role in natural control becomes clearer. Legner *et al.* (1974) recently reviewed the literature on biological control of medically important arthropods.

Traditionally, mosquito management in tidal wetlands has been based on water management or source reduction. In past years tidal marshes have been regarded as mosquito-producing wastelands suitable only for sanitary landfills and housing developments (Ferrigno, 1969; Cookingham, 1971). Today, people are more aware of the ecological value of marshes and estuaries (Odum, 1971). Mosquito control now is regarded as an integral part of wetlands management and studies have indicated that so long as there is cooperation between the different agencies involved, good mosquito control may, in some instances, increase marsh productivity (Catts *et al.*, 1963; Ferrigno, 1969; Ferrigno and Jobbins, 1966; Cookingham, 1971; Provost, 1973).

5. *Temporary Standing Water Habitats*

Some of the most annoying insect pests in or near forested recreational areas are the snow-pool breeding *Aedes* mosquitoes. These are major pests during the spring and summer of each year in the western mountains, northern midwest and northeastern areas of the United States (Nielson, 1959; Carpenter, 1966; DeFoliart *et al.*, 1967; Jamnback, 1969; Wagner and Newson, 1975). In Michigan, Wagner and Newson (1975) studied the effects of snow-pool mosquito biting activity on recreation and found that when mean biting counts were equal to or greater than two per minute, park visitors avoided recreational sites located in forested regions (i.e., nature trails) and stayed in those localities that possessed good prevailing winds and lacked a forest canopy (beach areas). When mosquito counts were less than or equal to one per minute the forested areas were utilized by park visitors.

Most snow-pool mosquito species are univoltine and females are principally daytime biters, remaining in rather close proximity to their larval habitats (Carpenter, 1961). The most comprehensive studies on the ecology and distribution of mosquitoes in forest recreational areas were conducted by Carpenter and published in *California Vector Views* from 1961 through 1974. These woodland mosquitoes have assumed even greater importance with the discovery that several species are vectors of California encephalitis virus (CEV), a zoonotic pathogen of increasing medical importance (Sudia *et al.*, 1971), and that some subtypes of CEV undergo transovarial transmission in certain species (Watts *et al.*, 1973; LeDuc *et al.*, 1975). Further research is needed on the ecology and behavior of species present in areas where development of new recreational facilities are planned.

Insect control practices in forest recreational areas have created some unusual problems in the past. One interesting example occurred in the Lake Tahoe region on the California-Nevada border. This area is one of the most renowned mountain recreational regions in the United States and consists of several towns and many hotels, motels, second homes, campgrounds, golf courses and marinas. It has one of the highest population densities, particularly during the summer months, of any recreational region (Carpenter and Gieke, 1974). Snow-pool *Aedes* are problems in this region. From 1963 to 1968 weekly fogging with malathion in residential areas was the method of control (Roberts, 1971). In the summer of 1968 a heavy infestation of the pine needle scale, *Chionaspis (Phenacaspis) pinifoliae* (Fitch), was found in lodgepole and Jeffrey pines within the residential area of South Lake Tahoe (Dahlsten *et al.*, 1969). Follow-up studies showed that fogging not only had eliminated mosquitoes, but also the predator-parasite complex of the scale insects, thereby allowing scale populations to increase unchecked (Luck and Dahlsten, 1975). In 1969 fogging was cut back dramatically, and the scale population declined due to an increase in parasitism and predation (Roberts *et al.*, 1973; Luck and Dahlsten, 1975). Failure to consider the effects of control strategies on the total ecosystem is probably not unique to Lake Tahoe and should be a concern of other developing and expanding forest recreational areas faced with decisions regarding management and control of nuisance insects.

B. Lotic Habitats

1. *Streams and Rivers*

Next to mosquitoes, the most notorious insect pests of recreational areas are the simuliids or black flies. Immature stages of most simuliids develop in running water, ranging from semipermanent trickles to large rivers. Adults emerge in large numbers during the spring and summer and are active biters during the day. They are discriminatory in their choice of hosts since many species feed only on birds (ornithophilic) yet they are frequently annoying because they swarm around one's head and crawl into hair, nostrils, ears, under clothing and are sometimes even inhaled (Jamnback, 1969, 1973). Unlike many mosquitoes, their flight range is fairly extensive (approximately 11 km) (Baldwin *et al.*, 1975), and they may frequent camping and recreational areas some distance from their breeding sites.

The bite of these flies usually produces an itching wheal, and the area may become inflamed and edematous. In sensitized individuals there also may be a general systemic reaction with headache, fever, nausea, and adenitis. An outbreak of the black fly *Boophthora erythrocephala* De Geer in Yugoslavia resulted in the medical treatment of over 2,000 persons for cutaneous lesions caused by the bites (Zivkovic and Burany, 1972). Recently, a "Black Fly Control Committee" was established in the state of Maine to determine control measures for a species of black fly not previously recorded there, which is now making life extremely miserable for residents and tourists in recreational areas (Sleeper, 1975). It has been suggested that effective antipollution measures to clean up Maine's rivers and streams may be partially responsible for these recent outbreaks. Serious black fly problems are now recognized in Alaska, much of Canada, the United States, Europe, Africa and many parts of Mexico and Central America (Nicholson and Mickel, 1950; Sailer, 1953; Davies and Peterson, 1956; Anderson and DeFoliart, 1961; Rzoska, 1964; Fredeen, 1969; Jamnback, 1969; WHO, 1971; Burany *et al.*, 1972; Mulla and Lacey, 1976).

Because of their specific and restricted breeding habitats, black flies have been particularly susceptible to control by persistent larvicides (Jamnback, 1973). However, unlike lakes which are generally viewed as closed systems, rivers and streams are open systems and compounds added to one part of the stream to control black fly larvae may also affect other organisms as far as 160 km downstream (Fredeen, 1975). In addition, the use of persistent insecticides has resulted in their accumulation in the food chain and produced resistance problems, therefore more research now is being conducted on nonpersistent chemicals with low toxicity to non-target organisms, and non-chemical methods of control (Cumming and McKague, 1973; Molloy and Jamnback, 1975). Recent developments in the control of black flies were reviewed by Jamnback (1973). Black fly problems resulting from the modification of existing streams and rivers are often easier to manage than those in more pristine habitats since larvae are usually localized near water impoundments or in dam spillways and constructed water courses, and are more easily found and treated. However, in many remote recreational areas where adult black flies are a problem, topical repellents and protective clothing have been and will continue to be the best means of temporary relief.

Other insects which have become nuisance problems primarily along major rivers are the mayflies (Ephemeroptera) and caddisflies (Trichoptera). Although these insects do not bite or sting, during their summer emergence they swarm in large numbers around lights, store windows, restaurants, resort establishments, and interfere with traffic and planned outdoor activities (Fremling, 1960; Jamnback, 1969; Mackenthun, 1969). Large numbers of scales, hairs and other fragments may evoke allergic reactions such as rhinitis, eczema and asthma in hypersensitive individuals (Shulman, 1967). In the United States, these insects have been major problems in cities along the Mississippi River and the St. Lawrence drainage system (Peterson, 1952; Burks, 1953; Fremling, 1960).

IV. CONCLUSIONS

One of the major goals of this report was to examine the impact of human attitudes and practices on urban insect problems in recreational areas. During its preparation, there were two concepts that became quite obvious. These are: 1) that proper planning and management of recreational facilities should include biological considerations as well as the traditional aspects such as business, engineering and construction factors; and 2) human tolerance to pestiferous insects are extremely variable.

In the words of Linley and Davies (1971), "If an extensive development is projected and if it involves the disposition of various facilities, entomological planning should be incorporated into the overall design at the outset." There are numerous examples, some cited earlier in this paper, in which entomological problems in recreational facilities or the outbreak of arthropod-borne diseases in or near recreational sites have had a drastic economic and social impact on the areas involved. Despite this, new recreational facilities are now being planned or constructed in areas where significant entomological problems either now exist or are certain to appear as a result of the development activities. Establishing land-based recreational facilities in areas where enzootic disease transmission (e.g., Rocky Mountain spotted fever or plague) is known to exist, or placing water-based recreational sites near or adjacent to swamps or salt marshes that produce large populations of pestiferous insects is a virtual guarantee of future management problems unless adequate control provisions are incorporated into the routine operations of these resorts. Even where no entomological problems now exist, faulty design, construction or management of man-made water impoundments can produce problems that are as serious and intractable as any that may occur naturally. The developers of new recreational facilities would be well advised to utilize appropriate biological specialists from the outset in the planning, site selection, construction and operation of new resort areas. In addition, municipal zoning commissions should require that rezoning applications involving the construction of new recreational areas or the expansion of existing ones include biological information concerning the actual or potential insect problems which could result from development in these areas. In this way the biological impact of planned activities could be anticipated and future insect problems avoided or minimized. Evidence supports the contention that this approach is not only less expensive, but usually is more effective than attempting to correct such problems after-the-fact.

The threshold of human tolerance for insect pests in recreational areas is a difficult concept to deal with because of problems in analyzing the wide range of factors involved. Many of these are socioeconomic in nature and not readily assessable. For instance, residents of Clear Lake, California, who are accustomed to 10,000 gnats swarming around their patio may consider a reduction of 90% to be a noticeable relief. However, city residents moving to a suburban development with "minor" gnat problems may consider 10 gnats on the patio annoying and 100 intolerable (Cook, 1967). In recent studies, Gerhardt *et al.* (1973) and Washino (1973) used public opinion surveys in attempting to assess threshold levels of tolerance to medically important insects. It is conceivable that this approach might be useful to determine attitudes and tolerances of potential tourists and residents for which a new recreational area is being planned, and that the information thus obtained could be used in planning design features and control operations needed to keep pest insect populations below objectionable levels for user groups. Polling methods and evaluation techniques must be markedly improved, however, before they can be expected to produce reliable information that would be useful in the planning and operation of recreation areas.

It is anticipated that demands for development of more recreation areas will continue to increase and, with this increase, insect control practices will cause additional stress on the environment. Land available for recreational development is limited and much of it already has significant insect problems or has the potential for their development. It is possible in some of these areas, despite the most careful planning, that consideration for the environment will make it impossible to achieve satisfactory levels of insect control. Indeed, this is the situation now in some existing recreation facilities. In such cases, if insect control practices are to be compatible with the maintenance of desired environmental conditions, man must accept a certain degree of discomfort and show a greater willingness to tolerate, rather than dominate, the rigors of nature.

ACKNOWLEDGMENTS

We would like to thank the following individuals for their helpful comments and suggestions during the preparation of this manuscript: R. D. Akre, J. R. Anderson, W. Burgdorfer, M. Chubb, G. D. Gill, R. S. Lane, B. C. Nelson, D. Sonenshine, I. B. Tarshis and R. K. Washino.

REFERENCES

Ackerman, W. C., G. F. White, and E. B. Worthington. (Eds.) 1973. Man-made lakes: Their problems and environmental effects. Geophys. Monogr. Ser. 17. Am. Geophys. Union, Washington, D.C.

Anderson, J. R., and G. R. DeFoliart. 1961. Feeding behavior and host preference of some black flies (Diptera: Simuliidae) in Wisconsin. *Ann. Entomol. Soc. Am. 54*: 716-729.

Anderson, L. D., E. C. Bay, and M. S. Mulla. 1965. Aquatic midge investigations in southern California. *Proc. Calif. Mosq. Contr. Assoc. 33*: 31-33.

Anonymous. 1973a. Roughing it the easy way. *Time Magazine 102*: 60-61.

Anonymous. 1973b. The wastewater tide ebbs slowly. *Environ. 15*: 34-42.

Apperson, C. S. 1976. Personal communication. Lake County Mosquito Abatement District, Lakeport, California.

Apperson, C. S., R. Elston, and W. Castle. 1976. Biological effects and persistence of methyl parathion in Clear Lake, California. *Environ. Entomol. 5*: 1116-1120.

Akre, R. D. 1976. Personal communication. Dept. of Entomology, Washington State University, Pullman, Washington.

Atchley, W. R., W. W. Wirth, and C. T. Gaskins. 1975. A bibliography and a keyword-in-context index of the Ceratopogonidae (Diptera) from 1758 to 1973. Texas Tech Press, Lubbock, Texas.

Audy, J. R. 1958. The localization of disease with special reference to the zoonoses. *Trans. Roy. Soc. Trop. Med. Hyg. 52*: 308-334.

Axtell, R. C., T. D. Edwards, and J. C. Dukes. 1975. Rigid canopy trap for Tabanidae (Diptera). *J. Georgia Entomol. Soc. 10*: 67-73.

Badger, D. D. 1973. Economic impact of water-based recreation. *In* Man-made lakes: Their problems and environmental effects (W.C. Ackermann, G.F. White, and E.B. Worthington, eds.). Geophys. Monogr. Ser. 17. Am. Geophys. Union, Washington, D.C., pp. 775-782.

Baldwin, W. F., A. S. West, and J. Gomery. 1975. Dispersal pattern of black flies (Diptera: Simuliidae) tagged with 32p. *Canad. Entomol. 107*: 113-118.

Barnes, A. M., and L. Kartman. 1960. Control of plague vectors on diurnal rodents in the Sierra Nevada of California by use of insecticide bait boxes. *J. Hyg. 58*: 159-167.

Barnes, A. M., L. J. Ogden, and E. G. Campos. 1972. Control of the plague vector *Opisocrostis ludovicianus*, by treatment of prairie dog (*Cynomys ludovicianus*) burrow with 2% carbaryl dust. *J. Med. Entomol. 4*: 330-333.

Barnes, A. M., L. J. Ogden, W. S. Archibald, and E. Campos. 1974. Control of plague vectors on *Peromiscus maniculatus* by use of 2% carbaryl dust in bait stations. *J. Med. Entomol. 11*: 83-87.

Bay, E. C. 1964. California chironomids. *Proc. Calif. Mosq. Contr. Assoc. 32*: 82-84.

Bay, E. C. 1974. Predator-prey relationships among aquatic insects. *Ann. Rev. Entomol. 19*: 441-453.

Bay, E. C., L. D. Anderson, and J. Sugerman. 1965. The abatement of a chironomid nuisance on highways at Lancaster, California. *Calif. Vector Views 12*: 29-32.

Beadle, L. D. (Ed.) 1972. Plague activity of major concern - 1971. Vector Contr. Briefs, Feb. 1972: 11.

Beadle, L. D., and J. A. Rowe. 1960. Sewage lagoons and mosquito problems. *In* Proc. Symp. on Waste Stabilization Lagoons, Kansas City, Mo. U.S. Publ. Health Serv., Dept. Health Educ. Welfare, pp. 101-104.

Beck, E. C., and W. M. Beck, Jr. 1969. The Chironomidae of Florida. II. The nuisance species. *Florida Entomol. 52*: 1-11.

Bell, E. J., and J. G. Stoenner. 1961. Spotted fever vaccine: Potency assay by direct challenge of vaccinated mice and

toxin of *R. rickettsiae*. *J. Immunol. 87*: 737-746.
Bishop, E. L., and F. E. Gartrell. 1944. Permanent works for the control of anophelines on impounded waters. *J. Nat. Malaria Soc. 3*: 211-219.
Bishopp, F. C. 1935. Ticks and the role they play in the transmission of diseases. Smithsonian Rep. for 1933: 389-406.
Blakeslee, E. B. 1945. DDT surface sprays for control of stable flies breeding in shore deposits of marine grass. *J. Econ. Entomol. 38*: 548-552.
Boicev, D., and K. Rizvanov. 1960. Relation of *Botrytis cinerea* Pers. to ixodid ticks (in Russian). *Zool. Zhur. 39*: 462.
Brockman, C. F., and L. C. Merriam, Jr. 1973. Recreational use of wild lands. McGraw-Hill Co., New York.
Brumbaugh, L. R., J. L. Mallars, and A. V. Viera. 1969. Chironomid midge control with quick breaking emulsions in wastewater stabilization lagoons at Stockton, CA. *Calif. Vector Views 16*: 1-9.
Brunetti, R., R. F. Fritz, and A. C. Hollister, Jr. 1953. An outbreak of malaria in California, 1952-1953. *Am. J. Trop. Med. Hyg. 3*: 779-788.
Burany, B., D. Miskov, V. Zivkovic, M. Morovic, and V. Stojanovic. 1972. *Boophthora erythrocephala* (Diptera, Simuliidae) as a new health problem in the places along the River Tisa in Yugoslavia. *Acta Parasitol. Jugoslavica 3*: 117-128.
Bureau of Outdoor Recreation. 1973. Outdoor recreation - A legacy for America. U.S. Dept. Interior, Washington, D.C.
Bureau of Outdoor Recreation. 1975. Assessing demand for outdoor recreation. U.S. Dept. Interior, Washington, D.C.
Burgdorfer, W. 1975. A review of Rocky Mountain spotted fever (tick-borne typhus), its agent, and its tick vectors in the United States. *J. Med. Entomol. 12*: 269-278.
Burgdorfer, W. 1976. Personal communication. U.S. Dept. of Health, Education and Welfare, Public Health Serice, National Institute of Health, Rocky Mountain Laboratory, Hamilton, Montana.
Burgdorfer, W., and C. M. Eklund. 1959. Studies on the ecology of Colorado tick fever in western Montana. *Am. J. Hyg. 69*: 127-137.
Burgdorfer, W., and C. M. Eklund. 1960. Colorado tick fever. I. Further ecological studies in western Montana. *J. Infect. Dis. 107*: 379-383.
Burks, B. 1953. The mayflies or Ephemeroptera of Illinois. Bull. Illinois Nat. Hist. Survey, Vol. 26.
Burrill, A. C. 1913. Economic and biologic notes on the giant midge, *Chironomus (Tendipes) plumosus* Meigen. *Bull. Wisconsin Nat. Hist. Soc. 10*: 124-163.

Calia, F. M., P. J. Bartelloni, and R. W. McKinnery. 1970. Rocky Mountain spotted fever. Laboratory infection in a vaccinated individual. *J. Am. Med. Assoc. 211*: 2012-2014.

Carpenter, S. J. 1961. Observations on the distribution and ecology of mountain *Aedes* mosquitoes in California. I. Species and their habitats. *Calif. Vector Views 8*: 49-53.

Carpenter, S. J. 1966. Observations on the distribution and ecology of mountain *Aedes* mosquitoes in California. X. Mosquito problems at Sierra Nevada recreational areas. *Calif. Vector Views 13*: 7-13.

Carpenter, S. J., and P. A. Gieke. 1974. Observations on the distribution and ecology of mountain *Aedes* mosquitoes in California. XXII. Mosquito problems in Lake Tahoe recreational region in the Sierra Nevada. *Calif. Vector Views 21*: 1-8.

Caten, J. L., and L. Kartman. 1968. Human plague in the United States 1900-1966. *JAMA 205*: 333-336.

Catton, W. R., Jr., J. C. Hendee, and T. W. Steinburn. 1969. Urbanism and the natural environment: An attitude study. Institute for Social Res., Univ. of Washington, Seattle, Mimeograph ISR 69-14.

Catts, E. P., F. H. Lesser, R. F. Darsie, O. Florschutz, and E. E. Tindall. 1963. Wildlife usage and mosquito production on impounded tidal marshes in Delaware, 1956-62. Trans. 28th N. Am. Wildl. and Nat. Res. Conf. March, 1963.

Center for Disease Control. 1970. Plague surveillance, report no. 1. July, 1970.

Center for Disease Control. 1971. Plague surveillance, report no. 2. July, 1971.

Center for Disease Control. 1977. Veterinary public health notes. April, 1977.

Chamberlain, R. W. 1968. Arboviruses: The arthropod-borne animal viruses. *Current Topics Microbiol. Immunol. 47*: 37-58.

Clark, G. M., C. M. Clifford, L. Fadness, and E. K. Jones. 1970. Contributions to the ecology of Colorado tick fever virus. *J. Med. Entomol. 7*: 189-197.

Clements, B. W., Jr., and A. J. Rogers. 1968. Tests of larvicides for control of salt marsh sand flies (*Culicoides*), 1967. *Mosq. News 28*: 529-534.

Cole, M. M. 1965. Biological control of ticks by the use of hymenopterous parasites, a review. WHO/EBL/43.65.

Cole, M. M., W. C. Bennett, G. N. Graves, J. R. Wheeler, B. E. Miller, and P. H. Clark. 1976. Dichlorvos bait for control of fleas on wild rodents. *J. Med. Entomol. 12*: 625-630.

Collins, D. L., and R. V. Nardy. 1951. Effect of spray residues on larvae of the tick *Dermacentor variabilis* Say. *J. Econ. Entomol. 43*: 861-863.

Cook, S. F., Jr. 1965. The Clear Lake gnat: Its control, past, present, and future. *Calif. Vector Views 12*: 43-48.
Cook, S. F., Jr. 1967. The increasing chaoborid midge problem in California. *Calif. Vector Views 12*: 43-48.
Cookingham, R. A. 1971. Mosquito control - an important part of wetlands management. *Proc. N. J. Mosq. Exterm. Assoc. 58*: 34-39.
Cooney, J. C., and E. Pickard. 1972. Comparative tick control field tests - Land-Between-the-Lakes. *Down to Earth 28*: 9-11.
Cooney, J. C., and W. Burgdorfer. 1974. Zoonotic potential (Rocky Mountain spotted fever and tularemia) in the Tennessee Valley region. I. Ecologic studies of ticks infesting mammals in Land-Between-the-Lakes. *Am. J. Trop. Med. Hyg. 23*: 99-108.
Cumming, J. E., and B. McKague. 1973. Preliminary studies of effects of juvenile hormone analogues on adult emergence of black flies (Diptera: Simuliidae). *Canad. Entomol. 105*: 509-511.
Dahlsten, D. L., R. Garcia, J. E. Prine, and R. Hunt. 1969. Insect problems in forest recreation areas. *Calif. Agric. 23*: 4-6.
Dasmann, R. F., J. P. Milton, and P. H. Freeman. 1973. Ecological principles for economic development. John Wiley & Sons, London.
Davies, J. B. 1967. A review of research into the biology and control of the biting sandflies of Jamaica, with recommendations for future control measures. Report to the Ministry of Health, Jamaica, January 1967.
Davies, D., and B. Peterson. 1956. Observations on the mating, feeding, ovarian development and oviposition of adult black flies (Simuliidae: Diptera). *Canad. J. Zool. 34*: 615-655.
DeFoliart, G. R., M. R. Rao, and C. D. Morris. 1967. Seasonal succession of bloodsucking Diptera in Wisconsin during 1965. *J. Med. Entomol. 4*: 363-373.
Douglass, R. W. 1975. Forest recreation. Pergamon Press, New York.
Dove, W. E., D. G. Hall, and J. B. Hull. 1932. Salt marsh sandfly problem (Culicoides). *Ann. Entomol. Soc. Am. 25*: 505-527.
Eddy, G. W., A. R. Roth, and F. W. Plapp, Jr. 1962. Studies on the flight habits of some marked insects. *J. Econ. Entomol. 55*: 603-607.
Edman, J. D. 1964. Control of *Culex tarsalis* Coquillett and *Aedes vexans* (Meigen) on Lewis and Clark Lake (Gavins Point Reservoir) by water level management. *Mosq. News 24*: 173-185.

Elliott, R. A. 1973. The TVA experience: 1933-1971. *In* Man-made lakes: Their problems and environmental effects (W.C. Ackermann, G.F. White, and E.B. Worthington, eds.). Geophys. Monogr. Ser. 17. Am. Geophys. Union, Washington, D.C., pp. 251-255.

Evans, E. S., Jr., and L. G. McCuiston. 1971. Preliminary mosquito survey of the Wharton State Forest - Summer 1970. *Proc. N. J. Mosq. Exterm. Assoc. 58*: 118-125.

Farvar, M. T., and J. P. Milton. (Eds.) 1972. The careless technology: Ecology, and international development. Nat. History Press, New York.

Felsenfeld, O. 1971. Borrelia, strains, vectors, human and animal borreliosis. W.H. Green, Inc., St. Louis, Missouri.

Felton, H. L. 1940. Control of aquatic midges with notes on the biology of certain species. *J. Econ. Entomol. 33*: 252-264.

Ferrigno, F. 1969. Ecological approach for improved management of coastal meadowlands. *Proc. N. J. Mosq. Exterm. Assoc. 56*: 188-203.

Ferrigno. F., and D. M. Jobbins. 1966. A summary of nine years of applied mosquito-wildlife research on Cumberland County, N. J., salt marshes. *Proc. N. J. Mosq. Exterm. Assoc. 53*: 97-112.

Fontaine, R. E., H. F. Gray, and T. Aarons. 1953. Malaria control at Lake Vera, California, in 1952-53. *Am. J. Trop. Med. Hyg. 3*: 789-792.

Forster, R. R. 1973. Planning for man and nature in national parks. International Union for Conservation of Nature and Natural Resources. IUCN, Publ. New Series No. 26. Morges, Switzerland.

Fredeen, F. J. H. 1969. Outbreaks of the black fly *Simulium arcticum* Malloch in Alberta. *Quaest. Entomol. 5*: 341-372.

Fredeen, F. J. H. 1975. Effects of a single injection of methoxychlor black-fly larvicide on insect larvae in a 161-km (100-mile) section of the North Saskatchewan River. *Canad. Entomol. 107*: 807-817.

Fremling, C. R. 1960. Biology and possible control of nuisance caddisflies of the upper Mississippi River. *Iowa State Univ. Agric. Home Econ. Exp. Sta. Bull. 483*: 856-879.

Gartrell, F. E., W. W. Barnes, and G. S. Christopher. 1972. Environmental impact of mosquito water resource management projects. *Mosq. News 32*: 337-342.

Geberich, J. B., and M. Laird. 1968. Bibliography of papers relating to the control of mosquitoes by the use of fish, an annotated bibliography for the years 1901-1966. F.A.O. Fisheries Tech. Paper No. 75 (1).

Gelman, A. C. 1961. The ecology of relapsing fevers. *In* Studies in disease ecology (J.M. May, ed.). Hafner Publ.

New York, pp. 113-141.

Gerhardt, R. R., J. C. Dukes, J. M. Falter, and R. C. Axtell. 1973. Public opinion on insect pest management in coastal North Carolina. N.C. Agric. Ext. Serv. Misc. Publ. No. 97.

Gill, G. D. 1976. Personal communication. Dept. of Biology, Northern Michigan University, Marquette, Michigan.

Gojmerac, W. L., and E. C. Devenport. 1971. Tabanidae (Diptera) of Kegonsa State Park, Madison, Wisconsin: Distribution and seasonal occurrence as determined by trapping and netting. *Mosq. News 31*: 572-575.

Grodhaus, G. 1963. Chironomid midges as a nuisance. II. The nature of the nuisance and remarks on its control. *Calif. Vector Views 10*: 27-37.

Grodhaus, G. 1975. Bibliography of chironomid midge nuisance and control. *Calif. Vector Views 22*: 71-81.

Hansens, E. J. 1951. The stable fly and its effect on seashore recreational areas in New Jersey. *J. Econ. Entomol. 44*: 482-487.

Hansens, E. J. 1952. Some observations on the abundance of salt marsh greenheads. *Proc. N. J. Mosq. Exterm. Assoc. 39*: 93-98.

Hansens, E. J., E. M. Bosler, and J. M. Robinson. 1971. Use of traps for study and control of salt-marsh greenhead flies. *J. Econ. Entomol. 64*: 1481-1486.

Harmston, F. C., and L. J. Ogden. 1975. Mosquito problems associated with man-made impoundments in western and midwestern United States. *Proc. Calif. Mosq. Contr. Assoc. 43*: 97-99.

Hasler, A. D. 1969. Cultural eutrophication is reversible. *BioScience 19*: 425-431.

Hattwick, M. A. W. 1971. Rocky Mountain spotted fever in the United States, 1920-1970. *J. Infect. Dis. 124*: 112-114.

Hattwick, M. A. W., A. H. Peters, M. R. Gregg, and B. Hanson. 1973. Surveillance of Rocky Mountain spotted fever. *J. Am. Med. Assoc. 225*: 1338-1343.

Hayes, R. O., L. D. Beadle, A. D. Hess, O. Sussman, and M. J. Bonese. 1962. Entomological aspects of the 1959 outbreak of eastern encephalitis in New Jersey. *Am. J. Trop. Med. Hyg. 11*: 115-121.

Hazeleur, W. C. 1970. Gnat control operations in the Shasta mosquito abatement district. *Proc. Calif. Mosq. Contr. Assoc. 37*: 114-115.

Hendee, J. C. 1969. Rural-urban differences reflected in outdoor-recreation participation. *J. Leisure Res. 1*: 333-341.

Henson, E. B. 1966. Aquatic insects as inhalant allergens. A review of American literature. *Ohio J. Sci. 66*: 592-632.

Hess, A. D., and T. F. Hall. 1943. The intersection line as a factor in anopheline ecology. *J. Nat. Malaria Soc. 3*: 93-98.

Heyneman, D. 1971. Mis-aid to the third world: Disease repercussions caused by ecological ignorance. *Canad. J. Pub. Health. 62*: 303-313.

Hilsenhoff, W. L. 1966. The biology of *Chironomus plumosus* (Diptera: Chironomidae) in Lake Winnebago, Wisconsin. *Ann. Entomol. Soc. Am. 59*: 465-473.

Hofe, G. D., Jr. 1973. Summary: Outdoor recreational use of man-made lakes. *In* Man-made lakes: Their problems and environmental effects (W.C. Ackermann, G.F. White, and E. B. Worthington, eds.). Geophys. Monogr. Ser. 17., Am. Geophys. Union, Washington, D.C., pp. 769-774.

Hoogstraal, H. 1967. Ticks in relation to human diseases caused by *Rickettsia* species. *Ann. Rev. Entomol. 12*: 377-420.

Hughes, C. C., and J. M. Hunter. 1972. The role of development in promoting disease in Africa. *In* The careless technology: Ecology and international development (M.T. Farvar and J.P. Milton, eds.). Nat. History Press, New York, pp. 69-101.

Hunt, E. G., and A. I. Bischoff. 1960. Inimical effects on wildlife of periodic DDD applications to Clear Lake. *Calif. Fish and Game 46*: 91-106.

Jackson, P. B. N. 1966. The establishment of fisheries in man-made lakes in the tropics. *In* Man-made lakes. (R.H. Lowe-McConnell, ed.). Inst. of Biol. Symp. No. 15. Academic Press, London, pp. 53-74.

James, M. T., and R. F. Harwood. 1969. Herms' medical entomology (6th ed.). Macmillan Co., New York.

Jamnback, H. 1961. The effectiveness of chemically treated screens in killing annoying punkies, *Culicoides obsoletus*. *J. Econ. Entomol. 54*: 578-580.

Jamnback, H. 1969. Bloodsucking flies and other outdoor nuisance arthropods of New York State. Mem. 19, New York State Mus. and Sci. Serv., Albany.

Jamnback, H. 1973. Recent developments in the control of black flies. *Ann. Rev. Entomol. 18*: 281-304.

Jamnback, H., and W. Wall. 1959. The common salt marsh Tabanidae of Long Island, New York. New York State Mus. Sci. Serv. Bull. 375.

Johnson, M. S., and F. Munger. 1930. Observations on the excessive abundance of the midge *Chironomus plumosus* at Lake Pepin. *Ecol. 11*: 110-126.

Jones, C. M., and D. W. Anthony. 1964. The Tabanidae (Diptera) of Florida. USDA Tech. Bull. 1295.

Kalitsi, E. A. K. 1973. Volta Lake in relation to the human population and some issues in economics and management.

In Man-made lakes: Their problems and environmental effects (W.C. Ackermann, G.F. White, and E.B. Worthington, eds.). Geophys. Monogr. Ser. 17. Am. Geophys Union, Washington, D.C., pp. 77-85.

Kardos, L. T. 1970. A new prospect. *Environ. 12*: 10-21.

Kartman, L. 1958. An insecticide-bait-box method for the control of sylvatic plague vectors. *J. Hyg. Camb. 56*: 455.

Kettle, D. S., and J. R. Linley. 1969. The biting habits of some Jamaican *Culicoides*. II. *C. furens* (Poey). *Bull. Entomol. Res. 558*: 729-753.

Kimball, J. H. 1965. Integration of sewage disposal and mosquito control in Orange County, California. *Calif. Vector Views 12*: 5-7.

King, W. V., and L. G. Lenert. 1936. Outbreaks of *Stomoxys calcitrans* (L.) ("dog flies") along Florida's northwest coast. *Florida Entomol. 19*: 33-39.

Kwan, W. E., and F. O. Morrison. 1974. A summary of published information for field and laboratory studies of biting midges, *Culicoides* species (Diptera: Ceratopogonidae). *Ann. Entomol. Soc. Quebec 19*: 127-137.

LaBrecque, G. C., and R. L. Goulding. 1954. Tests with granulated BHC and dieldrin for controlling sand fly larvae. *Mosq. News 14*: 20-22.

LaBrecque, G. C., D. W. Meiffert, and D. E. Weidhaas. 1975. Potential of the sterile-male technique for the control or eradication of stable flies, *Stomoxys calcitrans* Linnaeus. *In* Sterility principle for insect control. Internat. Atomic Energy Agency, Vienna, IAE-SM-186/55, pp. 449-459.

Lavery, P. 1974. The demand for recreation. *In* Recreational geography (P. Lavery, ed.). John Wiley and Sons, New York, pp. 21-51.

LeDuc, J. W., W. Suyemoto, B. F. Eldridge, P. K. Russell, and A. R. Barr. 1975. Ecology of California encephalitis viruses on the Del Mar Va peninsula. II. Demonstration of transovarial transmission. *Am. J. Trop. Med. Hyg. 24*: 124-126.

Legner, E. F., R. D. Sjogren, and I. M. Hall. 1974. The biological control of medically important arthropods. *In* CRC critical reviews in environmental control, Vol. 4 (1). CRC Press, Cleveland, pp. 85-113.

Leonard, J. W., and F. A. Leonard. 1962. Mayflies of Michigan trout streams. Cranbrook Inst. Sci., Bloomfield Hills, Michigan.

Lewallen, L. L. 1968. Preliminary toxicity studies on the yellowjacket *Vespula pensylvanica* (Saussure) in California. *Calif. Vector Views 15*: 1-2.

Lewis, D. J. 1966. Nile control and its effects on insects of medical importance. *In* Man-made lakes (R.H. Lowe-McCon-

nell, ed.). Inst. Biol. Symp. No. 15. Academic Press, London, pp. 43-45.
Link, V. B. 1955. A history of plague in the United States. Publ. Health Monogr. No. 26. U.S. Dept. Health Educ. Welfare.
Linley, D. R., and D. B. Davies. 1971. Sandflies and tourism in Florida and the Bahamas and Caribbean areas. *J. Econ. Entomol. 64*: 264-278.
Linnemann, C. C., Jr., P. Jansen, and G. M. Schiff. 1973. Rocky Mountain spotted fever in Clermont County, Ohio. Description of an endemic focus. *Am. J. Epidemiol. 97*: 125-130.
Lofgren, C. S., W. A. Banks, and B. M. Glancey. 1975. Biology and control of imported fire ants. *Ann. Rev. Entomol. 20*: 1-30.
Lowe-McConnell, R. H. (Ed.) 1966. Man-made lakes. Inst. Biol. Symp. No. 15. Academic Press, London.
Love, J. A., and G. D. Gill. 1965. Incidence of coliforms and enterococci in field populations of *Stomoxys calcitrans* (Linnaeus). *J. Invert. Pathol. 7*: 430-436.
Luck, R. F., and D. L. Dahlsten. 1975. Natural decline of a pine needle scale (*Chionaspis pinifoliae* [Fitch]), outbreak at South Lake Tahoe, California, following cessation of adult mosquito control with malathion. *Ecol. 56*: 893-904.
Lusk, E. E., and J. W. Willis. 1970. The development of midge problems in the Shasta mosquito abatement district. *Proc. Calif. Mosq. Contr. Assoc. 37*: 114.
Mackenthun, K. M. 1969. The practice of water pollution biology. U.S. Dept. Int., F.W.P.C.A., Washington, D.C.
Mackenthun, K. M., W. M. Ingram, and R. Porges. 1964. Limnological aspects of recreational lakes. P.H.S.P. No. 1167, U.S. Dept. Health Educ. Welfare, Washington, D.C.
Magy, H. I. 1968. Vector and nuisance problems emanating from man-made recreational lakes. *Proc. Calif. Mosq. Contr. Assoc. 36*: 36-37.
Maupin, G. O. 1974. An outbreak of tick-borne relapsing fever at Grand Canyon National Park. *In* Proc. 10th Biennial Publ. Health Vector Contr. Conf., Ft. Collins, Colorado, February 19-21, 1974. CDC and Inst. of Rural Environ. Health, Colorado State Univ., pp. 31-32.
McDowell, J. W., H. G. Scott, C. J. Stojanovich, and H. B. Weinburgh. 1964. Tularemia. U.S. Dept. Health Educ. Welfare. CDC, Atlanta, Georgia.
McDuffy, W. C., and C. N. Smith. 1955. Recommended current treatments for tick control. *Public Health Rep. 70*: 327-330.
Miller, B. E., W. C. Bennett, G. N. Graves, and J. R. Wheeler. 1975. Field studies of systemic insecticides. I. Eval-

uation of phoxim for control of fleas on cotton rats. *J. Med. Entomol. 12*: 425-430.
Molloy, D., and H. Jamnback. 1975. Laboratory transmission of mermithids parasitic in blackflies. *Mosq. News 35*: 337-342.
Mount, G. A., C. S. Lofgren, and J. B. Gahan. 1966. Malathion, naled, fenthion and Bayer 39007 thermal fogs for control of the stable fly (dog fly) *Stomoxys calcitrans* (Diptera: Muscidae). *Florida Entomol. 49*: 170-173.
Mount, G. A., N. W. Pierce, and C. S. Lofgren. 1971. Effectiveness of twenty-two promising insecticides for control of the lone star tick. *J. Econ. Entomol. 64*: 262-263.
Mount, G. A., R. H. Grothaus, J. T. Reed, and K. F. Baldwin. 1976. *Amblyomma americanum*: Area control with granules or concentrated sprays of diazinon, propoxur, and chlorpyrifos. *J. Econ. Entomol. 69*: 257-259.
Mulla, M. S. 1974. Chironomids in residential-recreational lakes. An emerging nuisance problem - measures for control. *Entomol. Tidskr. 95(Suppl.)*: 172-176.
Mulla, M. S., H. A. Darwazeh, and D. R. Peters. 1970. Mosquito control in sewage oxidation ponds with drip and pour-in larvicides. *Mosq. News 30*: 456-460.
Mulla, M. S., R. L. Norland, D. M. Fanara, H. A. Dawrazeh, and D. W. McKean. 1971. Control of chironomid midges in recreational lakes. *J. Econ. Entomol. 64*: 300-307.
Mulla, M. S., R. L. Norland, T. Ikeshoji, and W. L. Kramer. 1974. Insect growth regulators for control of aquatic midges. *J. Econ. Entomol. 67*: 165-170.
Mulla, M. S., and H. A. Darwazeh. 1975. Evaluation of insect growth regulators against chironomids in experimental ponds. *Proc. Calif. Mosq. Contr. Assoc. 43*: 164-168.
Mulla, M. S., D. R. Barnard, and R. L. Norland. 1975. Chironomid midges and their control in Spring Valley Lake, California. *Mosq. News 35*: 389-395.
Mulla, M. S., and L. A. Lacey. 1976. Biting flies in the lower Colorado River Basin: Economic and public health implications of *Simulium* (Diptera: Simuliidae). *Proc. Calif. Mosq. Contr. Assoc. 44*: 130-133.
Mulla, M. S., W. L. Kramer, and D. R. Barnard. 1976. Insect growth regulators for control of chironomid midges in residential-recreational lakes. *J. Econ. Entomol. 69*: 285-291.
Mulrennan, J. A., and W. T. Sowder. 1954. Florida's mosquito control system. *Publ. Health Rep. 69*: 613-618.
Murray, K. F. 1971. Epizootic plague in California, 1965-1968. Unpubl. rep. Calif. Dept. Pub. Health, Bur. Vector Contr.
Needham, J. G. 1920. Burrowing mayflies of our larger lakes and streams. *Bull. Bur. Fish. 36*: 269-290.

Nelson, B. C. 1976. Personal communication. California State Dept. of Public Health, Vector Control Section, Berkeley, California.

Nelson, B. C., and C. R. Smith. 1976. Ecological effects of a plague epizootic on the activities of rodents inhabiting caves at Lava Beds National Monument, California. *J. Med. Entomol. 13*: 51-61.

Newson, H. D. 1977. Arthropod problems in recreation areas. *Ann. Rev. Entomol. 22*: 333-353.

Nichelson, H., and C. Mickel. 1950. The black flies of Minnesota (Simuliidae). Minnesota Agric. Exp. Sta. Tech. Bull. 199.

Nielsen, L. T. 1959. Seasonal distribution and longevity of Rocky Mountain snow mosquitoes of the genus *Aedes*. *Proc. Utah Acad. Sci. 36*: 83-87.

Odum, E. P. 1971. Fundamentals of ecology (3rd. ed). W.B. Saunders Co., Philadelphia.

Patterson, R. W., G. C. LaBrecque, and D. F. Williams. 1975. Use of the sterile-male technique to control stable flies on St. Croix. Study in progress July 74-77 at USDA ARS Federal Exp. Sta., St. Croix, U.S. Virgin Islands.

Pavlovsky, E. N. 1964. Natural nidality of transmissible disease with special reference to the landscape epidemiology of zooanthroponoses. Transl. by F.K. Plovs, Jr., edited by N.D. Levine. 1966. Univ. Illinois Press, Urbana.

Pechman, L. L. 1972. The horse flies and deer flies of New York (Diptera: Tabanidae). *Search (Cornell Univ. Agric. Exp. Sta.) 2*: 1-72.

Pelsue, F. W., G. C. McFarland, and C. Beesley. 1974. Field evaluations of two insect growth regulators against chironomid midges in water spreading basins. *Proc. Calif. Mosq. Contr. Assoc. 42*: 157-161.

Peters, A. H. 1971. Tick-borne typhus (Rocky Mountain spotted fever). Epidemiologic trends, with particular reference to Virginia. *J. Am. Med. Assoc. 216*: 1003-1007.

Peterson, D. 1952. Observations on the biology and control of pest Trichoptera at Fort Erie, Ontario. *Canad. Entomol. 84*: 103-107.

Pickard, E. 1968. *Stomoxys calcitrans* (L.) breeding along TVA reservoir shorelines. *Mosq. News 28*: 644-645.

Poinar, G. O., Jr., and F. Ennik. 1972. The use of *Neoaplectana carpocapsae* (Steinernematidae: Rhabditoidea) against adult yellowjackets (*Vespula* spp., Vespidae: Hymenoptera). *J. Invert. Pathol. 19*: 331-334.

Provost, M. W. 1949. Mosquito control and mosquito problems in Florida. *Proc. Calif. Mosq. Contr. Assoc. 17*: 32-35.

Provost, M. V. 1958. Chironomids and lake nutrients in Florida. *Sew. and Ind. Wastes 30*: 1417-1419.

Provost, M. W. 1959. Current research in mosquito biology and control at Florida's Entomological Research Center. *Proc. N. J. Mosq. Exterm. Assoc. 46*: 64-69.

Provost, M. W. 1973. Salt marsh management in Florida. *Proc. Tall Timbers Conf. Ecol. Animal Control by Habitat Manag. 5*: 5-17.

Reeves, W. C. 1965. Ecology of mosquitos in relation to arboviruses. *Ann. Rev. Entomol. 10*: 25-46.

Roberts, F. C. 1971. The evolution of mosquito control at South Lake Tahoe. *Proc. Calif. Mosq. Contr. Assoc. 39*: 44-46.

Roberts, F. C., R. F. Luck, and D. L. Dahlsten. 1973. Natural decline of a pine needle scale population at South Lake Tahoe. *Calif. Agric. 27*: 10-12.

Rothenberg, R., and D. E. Sonenshine. 1970. Rocky Mountain spotted fever in Virginia: Clinical and epidemiological features. *J. Med. Entomol. 7*: 663-669.

Ryan, T. 1976. Recreation Vehicle Industry Assoc., Elkhart, Indiana.

Ryckman, R. E., C. T. Ames, C. C. Lindt, and R. D. Lee. 1954. Control of plague vectors on the California ground squirrel by burrow dusting with insecticides and the seasonal incidence of fleas present. *J. Econ. Entomol. 47*: 604-607.

Rzoska, J. 1964. Mass outbreaks of insects in the Sudanese Nile basin. *Verh. Internat. Limnol. 15*: 194-200.

Sailer, R. I. 1953. The blackfly problem in Alaska. *Mosq. News 13*: 232-235.

Shulman, S. 1967. Allergic responses to insects. *Ann. Rev. Entomol. 12*: 323-346.

Simmons, S. W. 1944. Observations on the biology of the stable fly in Florida. *J. Econ. Entomol. 37*: 680-686.

Simmons, S. W., and W. E. Dove. 1941. Breeding places of the stable fly, or "dog fly", *Stomoxys calcitrans* (L.) in northwestern Florida. *J. Econ. Entomol. 34*: 457-462.

Sleeper, F. 1975. Visit from a small monster. *Sports Illus. 43*: 46-49.

Smadel, J. E. 1959. Status of rickettsioses in the United States. *Ann. Intern. Med. 51*: 421-435.

Smith, C. N., A. N. Davis, D. E. Weidhaas, and E. L. Seabrook. 1959. Insecticide resistance in the salt-marsh sand fly *Culicoides furens*. *J. Econ. Entomol. 52*: 352-353.

Sonenshine, D. E. 1976. Personal communication. Dept. of Biological Sciences, Old Dominion University, Norfolk, Virginia.

Sonenshine, D. E., and I. J. Stout. 1970. A contribution to the ecology of ticks infesting wild birds and rabbits in the Virginia North Carolina piedmont (Acarina: Ixodidae). *J. Med. Entomol. 7*: 645-54.

Sonenshine, D. E., and G. F. Levy. 1972. Ecology of the American dog tick, *Dermacentor variabilis*, in as study area in Virginia. 2. Distribution in relation to vegetative types. *Ann. Entomol. Soc. Am. 65*: 1175-1182.
Sonenshine, D. E., A. H. Peters, and G. F. Levy. 1972. Rocky Mountain spotted fever in relation to vegetation in the eastern United States, 1951-1971. *Am. J. Epidemiol. 96*: 59-69.
Spiller, D. 1965. Methods used for chironomid larvae surveys of sewage oxidation ponds and natural waters at Auckland, New Zealand. *Calif. Vector Views 12*: 9-15.
Spruance, S. L., and A. Bailey. 1973. Colorado tick fever. *Arch. Intern. Med. 131*: 288-293.
Stark, H. E., and A. R. Kinney. 1969. Abundance of rodents and fleas as related to plague in Lava Beds National Monument, California. *J. Med. Entomol. 6*: 287-294.
Stroud, R. H. 1966. American experience in recreational use of artificial waters. *In* Man-made lakes (R.H. Lowe-McConnell, ed.). Academic Press, London, pp. 189-200.
Sudia, W. D., W. F. Newhouse, C. H. Calisher, and R. W. Chamberlain. 1971. California group arboviruses: Isolations from mosquitoes in North America. *Mosq. News 31*: 576-600.
Sullivan, R. H., M. M. Cohn, and S. S. Baxter. 1973. Survey of facilities using land application of wastewater. U.S Govt. Print. Office, Washington, D.C.
Surtees, G. 1971. Urbanization and the epidemiology of mosquito-borne disease. *Abstr. Hyg. 46*: 121-134.
Sutherland, D. J. 1970. Newer chemicals for mosquito control. *Proc. N. J. Mosq. Exterm. Assoc. 57*: 177-185.
Tanner, H. A. 1972. Michigan State University's project for recycling of waste water and associated nutrients. *In* Environmental quality: Now or never (C.L. San Clemente, ed.). Continuing Educ. Serv., Michigan State Univ., East Lansing, Michigan, pp. 111-115.
Thompson, R. S., W. Burgdorfer, R. Russell, and B. J. Francis. 1969. Outbreak of tick-borne relapsing fever in Spokane County, Washington. *JAMA 210*: 1045-1050.
Voegtline, A. C., G. W. Ozburn, and G. D. Gill. 1965. The relation of weather to biting activity of *Stomoxys calcitrans* (Linnaeus) along Lake Superior. *Paper Michigan Acad. Sci. Arts and Letters 50*: 107-114.
Waddy, B. B. 1966. Medical problems arising from the making of lakes in the tropics. *In* Man-made lakes (R.H. Lowe-McConnell, ed.). Inst. of Biol. Symp. No. 15. Academic Press, London, pp. 87-94.
Wagner, V. C., and H. D. Newson. 1975. Mosquito biting activity in Michigan state parks. *Mosq. News 35*: 217-222.
Wall, W. J., Jr., and O. W. Doane, Jr. 1960. A preliminary study of the bloodsucking Diptera on Cape Cod, Massachu-

setts. *Mosq. News 20*: 39-44.

Wall, W. J., Jr., and V. M. Marganian. 1973. Control of salt marsh *Culicoides* and *Tabanus* larvae in small plots with granular organophosphorus pesticides, and the direct effect on other fauna. *J. Econ. Entomol. 33*: 88-93.

Warren, C. E. 1971. Biology and water pollution control. W. B. Saunders Co., Philadelphia.

Washino, R. K. 1973. Tolerance thresholds in mosquito pest management. *Proc. Calif. Mosq. Contr. Assoc. 41*: 105.

Watts, D. M., S. Pantuwatana, G. R. DeFoliart, J. M. Yuill, and W. H. Thompson. 1973. Transovarial transmission of La Crosse virus (California encephalitis group) in the mosquito, *Aedes triseriatus*. *Science 182*: 1140-1141.

WHO. 1971. Blackflies in the Americas. WHO/VBC/71.283.

Williamson, F. S. L. 1969. Population pollution. *BioScience 19*: 979-983.

Wilton, D. P., and R. Fay. 1969. Action of amine ovicides on *Aedes aegypti* mosquitoes. *Mosq. News 29* : 361-365.

Wirth, W. W., and F. S. Blanton. 1974. The West Indian sandflies of the genus *Culicoides*. USDA Tech. Bull. No. 1474.

Woodwell, G. M., C. F. Wurster, Jr., and P. A. Isaacson. 1967. DDT residues in an east coast estuary: A case of biological concentration of persistent insecticide. *Science 156*: 821-824.

Wright, J. E. 1974. Insect growth regulators, juvenile hormone analogs for control of the stable fly in marine plants in Florida. *Mosq. News 34*: 160-163.

Wynns, H. L. 1942. The epidemiology of relapsing fever. *In* A symposium on relapsing fever in the Americas (F.R. Moulton, ed.). Am. Assoc. Adv. Sci., Washington, D.C., pp. 100-105.

Zivkovic, V., and B. Burany. 1972. An outbreak of *Boophthora erythrocephala* (Diptera: Simuliidae) in Yugoslavia in 1972. *Acta Vet. 22*: 133-142.

YELLOWJACKET WASPS IN URBAN ENVIRONMENTS

Harry G. Davis

Yakima Agricultural Research Laboratory
Agricultural Research Service, USDA
Yakima, Washington

I. INTRODUCTION AND MAN'S ATTITUDE

A review of the literature relating specifically to yellowjacket wasps and their ecology in urban environments reveals a paucity of information on this topic. Although the subject is treated indirectly by workers such as Duncan (1939), Kemper and Dohring (1967), and Spradbery (1973a) in their extensive publications on vespine biology and ecology, only a few researchers have studied and published specifically on these social insects as they occur in urban ecosystems. The term yellowjacket has considerable variation in meaning among vespine researchers. Therefore, for the purpose of this paper, I am defining the term yellowjacket to include the truly social

ISBN 0-12-265250-9

wasps that belong to the genera *Vespula* (ground nesters) and *Dolichovespula* (aerial nesters) of the subfamily Vespinae. Yellowjacket is not appropriate for members of the genus *Vespa*, the hornets, nor the important paper wasps which belong to the genus *Polistes*. However, certain species of these genera occupy important ecological niches in urban environments so they will occasionally be included with my comments on yellowjackets.

In general, man's attitude toward yellowjackets - or indeed to all wasps - is a negative one. The response is to eliminate the wasps immediately with no thought given to population levels or actual problems encountered. Wagner and Reierson (1969) stated that the decision as to when a population of yellowjackets can be considered a problem is entirely subjective and is influenced by the degree of exposure and the mental attitude of the person involved. National Park Service personnel in the State of Washington have determined that seven to 10 yellowjackets per hour annoying people constitute a nuisance level, and at this level they receive many complaints from park users. Actually, the attitude of people toward yellowjackets probably contributes to the problem. They become frightened and attempt to kill the yellowjackets, which often results in unnecessary stings.

The prevalent negative attitude toward yellowjackets can be confirmed from people's reactions to researchers collecting vespine wasps in the Pullman, Washington area. Almost without exception, people did not want the wasps on their property, and their primary concern was being stung. Undoubtedly, this attitude exists among most people in all localities where yellowjackets occur. An attitude such as this reflects ignorance of the subject and only continuous education of the public will alter it. It is especially important to stress, however, that control practices become mandatory when wasp populations reach levels that are intolerable.

II. TAXONOMY AND DISTRIBUTION

The genus *Vespula* is composed of two species groups, the *Vespula rufa* (L.) group and the *Vespula vulgaris* (L.) group. These are separated mainly by differences in nest structure, colony duration, and foraging behavior although there are also morphological differences between members of the two groups in the occipital carina and the structure of the first abdominal segment.

As noted by Mac Donald *et al.* (1974), members of the *V. vulgaris* group are well-studied in Europe and in North America, probably because of their large colonies and pestiferous workers. The group includes the Holarctic *V. vulgaris*,

Palearctic *V. germanica* (Fab.), and the Nearctic *V. pensylvanica* (Saussure) and *V. maculifrons* (Buysson). *Vespula germanica* has also become established in New Zealand (Thomas, 1960), Tasmania (Spradbery, 1973b), Cape Peninsula in South Africa (Whitehead and Prins, 1975), Chile (Mac Donald, 1976), and the northeastern coastal states of the United States (Menke and Snelling, 1975). Recently, *V. germanica* has extended its range to Ohio and Indiana (Garnett, 1976).

The *V. rufa* group is comprised of the Holarctic *V. austriaca* (Panzer), Palearctic *V. rufa*, and Nearctic *V. atropilosa* (Sladen), *V. acadica* (Sladen), *V. intermedia* (Buysson), *V. vidua* (Saussure), *V. consobrina* (Saussure), *V. squamosa* (Drury), and *V. sulphurea* (Saussure). With the exception of *V. squamosa*, members of this group form small colonies which decline earlier than those of the *V. vulgaris* group, and the workers are not pests of man (Mac Donald *et al.*, 1974). Although *V. squamosa* morphologically belongs to the *V. rufa* group, it appears biologically similar to the *V. vulgaris* group because it often consists of large colonies which may remain active into the fall months (Mac Donald, 1976). For example, Tissot and Robinson (1954) reported finding large colonies of this species in Florida, some of which had overwintered and were into a second year of development.

III. ECOLOGY AND BEHAVIOR

Yellowjackets appear to be increasingly conspicuous members of the insect fauna in urban localities in various parts of the world. Unfortunately, the detrimental attributes of a few pestiferous vespine species are well documented and remembered. However, the beneficial contributions made by the vast majority of yellowjacket species have been little studied by scientists, or appreciated by the general public, or even by some entomologists. Most vespine species are highly beneficial because they are predators of various noxious insect species which occur in their foraging areas. They commonly prey upon such insects as grasshoppers, caterpillars, moths, lygus bugs, and adult flies. Furthermore, even workers of pestiferous species become a problem only during a brief period - usually in late summer and early fall when the colonies start to decline.

The problem of yellowjacket invasion into sites of human activities, especially in urban-suburban areas, is probably in part the result of ever-increasing human encroachment into previously undisturbed nesting and foraging areas utilized by these insects. Natural foraging areas have been greatly reduced by urban sprawl that includes new housing developments, industrial parks, playgrounds, parks, golf courses, etc. In addition, these "modified" environments provide the scavenging species with additional sources of food which become readily available to them in the form of garbage and litter commonly associated with urban and recreational conditions. Moreover, these pest species compete with man as their own natural food sources become depleted later in the summer. This period coincides with colony decline and associated behavioral changes in workers. Larvae are no longer adequately attended and many are pulled from the comb and killed which results in reduced foraging responsibilities for the workers. At this time the workers become more aggressive and have an increased tendency to sting.

In some urban communities man has changed the environment in such a way to enhance the establishment of yellowjacket colonies. For example, Mac Donald *et al.* (1974) found that *V. pensylvanica* and *V. atropilosa* establish many nests in pastures and golf courses in southeastern Washington, near Pullman, where soil moistures are high and water sources remain available during the dry summer and fall months (July, August, September). They also noted that most nests of *V. pensylvanica* (over 95%) and *V. atropilosa* (90%) were subterranean and typically associated with rodent burrows. Few nests were found in dry agricultural lands planted to wheat or peas in the surrounding area. Thus without the urban environment, nesting sites would probably not be available in this area. Similar choices of nesting sites for *V. pensylvanica* have been observed by the author at playgrounds and nature study areas located in the center of Portland, Oregon.

The bibliography of the yellowjacket literature of the world compiled by Akre *et al.* (1974) reveals numerous publications devoted to the biology, ecology, and behavior of these insects. In actuality, however, few of these references represent in-depth studies, and most relate only to a few species pestiferous to man. From this published information, it can be inferred that other species, especially those closely related, may have similar ecological and behavioral characteristics. Such inferences are proving to be inaccurate as additional information accumulates from recent research. For example, Greene *et al.* (1976) reported the biology and ecology of *Dolichovespula arenaria* (Fab.) to be vastly different from that of *Vespula* spp. Moreover, virtually nothing is known about *V. acadica*, a member of the *V. rufa* group, and only very limited information is available concerning other important vespine species such as *D. maculata* (L.).

Although large gaps remain in our knowledge about yellowjackets, information on nesting, foraging habits, food sources, effects of weather, population fluctuations, queen hibernation, etc., nevertheless provides a better understanding of why, how, and where wasp populations develop. A few of these topics merit attention in this paper.

A. Nesting: Establishment of Nests

Behavioral observations indicate there is competition among yellowjacket queens for nests. For example, Akre *et al.* (1976) reported that very few overwintered queens of *V. pensylvanica* and *V. atropilosa* establish nests; some continue to search for nesting sites throughout July and into August. They concluded that during the initial phases of nest construction there is probably intense intraspecific competition for nests. Although they did not observe queens fighting for nests, they frequently found the remains of one to three queens lying in the entrance tunnel or in the cavity under the nests. Mac Donald and Matthews (1975) found evidence that queens of *V. squamosa* regularly usurp established colonies of *V. maculifrons*. More recent data (Mac Donald and Matthews, 1976) showed about 80% of the *V. squamosa* colonies represented a previous *V. maculifrons* colony. Since *V. maculifrons* is a scavenging species and *V. squamosa* probably is not, this usurpation definitely influences the abundance of the more pestiferous *V. maculifrons*. Interestingly, Thomas (1960) noted that despite apparently high mortality among spring queens of *V. germanica* in New Zealand, only a small percentage of queens need to establish nests to maintain the same population densities from year to year.

B. Foraging: Distance from Nests

In recent years, several researchers have established distances that yellowjackets travel during their foraging flights. For example, Akre *et al.* (1975), using metal labels and magnets, determined that about 80% of the workers of *V. pensylvanica* foraged within 335 m of the nest. This distance compares favorably with data obtained by Rogers (1972) with dye-marked wasps (*V. pensylvanica* and *V. vulgaris*) recovered at distances ranging from 6 m to 402 m from the nest; none was recovered by traps at 805 m. Arnold (1966, cited in Spradbery, 1973a) found that workers of *V. rufa* and *D. sylvestris* (Scolpoli) foraged 900 m from the nest. However, numbers of workers collected during this study were extremely small. Unfortunately, no definitive studies of foraging distances have been conducted for other species.

C. Foraging: Food Preferences

Although yellowjackets apparently have food and prey preferences (Duncan, 1939; Howell, 1973; Mac Donald *et al.*, 1974), Spradbery (1973a) noted that the wide range of protein and carbohydrate sources utilized by foragers undoubtedly enable these wasps to overcome problems that develop with the depletion of a preferred food source. Duncan (1939) listed many foods of high sugar content that are especially utilized by workers including honey, honeydew, manufactured sweets (jams, jellies, molasses, etc.), and juices and pulp of well-ripened fruit such as apples, peaches, prunes, and grapes. He also listed various kinds of meat (fresh, cured, or cooked) that are collected by foraging wasps as sources of protein for the developing larvae.

D. Adaptability of Foraging Behavior

Spradbery (1973a) reported that flexibility in behavior is characteristic of vespine wasps. They are far less constant to a particular duty than are honey bees, *Apis mellifera* L., (Kalmus, 1954) and frequently switch from one activity to another (Brian and Brian, 1952). This may involve foraging for fluids (carbohydrates and water), protein (meat, insect prey, etc.), collecting wood pulp for nest construction, or performing sanitation duties in the nest. Brian and Brian (1952) also noted there may be behavioral patterns associated with certain periods of the day. For example, *Vespula* (= *Dolichovespula*) *sylvestris* workers foraged and collected more flesh in the evening, wood pulp collections became predominant in the middle part of the day, and fluid was gathered throughout the day.

Other examples of flexibility are cited by Akre *et al.* (1976) who found that some workers specialized in prey capture while others rarely attacked prey. All foraged for honey and water. Interestingly, workers that visited a honey dish for less than 12 seconds usually continued on to collect fiber or protein (presumably for their own energy requirements), whereas workers that remained at the dish for more than 12 seconds usually returned immediately to the nest to feed nestmates.

E. Foraging: Finding the Food

Brian and Brian (1952) reported that wasps which search for carbohydrates respond primarily to odors as they move from flower to flower or examine the leaves where honeydew producers are present. Gaul (1952) observed that wasps fly upwind to honey baits but are unable to detect glucose (odorless) from a distance. Akre *et al.* (1975) reported that 10 to 20% of the *V. pensylvanica* workers in their study were not constant to a particular location when scavenging for protein but searched in widely separated areas. They concluded also that learning and sight played a part, as workers investigated any platform placed in yellowjacket foraging areas, whether or not it was supplied with fish. Moreover, they suggest that learning may help explain why workers arrive so quickly at a new protein source, such as a dead animal, and why they may shift their foraging to human food at picnic and camping grounds where they become a nuisance.

It is important to point out that when finding and locating food sources wasps have the ability to learn (Spradbery, 1973a). Evans and Eberhard (1970) noted that workers returned time after time to the same food, water, or pulp sources after initially performing orientation flights. Later, they observed that when the source is exhausted, workers "forgot" that site and moved to another place to start performing a different task. To quote Evans and Eberhard (1970): "In general then, a social wasp worker seems to be a non-ovipositing Jack-of-all-trades capable of changing her speciality from one minute to the next regardless of her size or age."

F. Foraging: Influence of Weather

Herold (1952) observed that light rain or moderate winds will not prevent social wasps from foraging. Blackith (1958) found that cool weather, high winds, and rain may reduce the foraging population temporarily but had little effect on the overall activity of the wasps throughout the day. The author has found this to be the case on several occasions while observing the foraging activities of *V. pensylvanica* workers during light rainstorms in the Willamette Valley of Oregon.

G. Communication

Kalmus (1954) showed that workers of yellowjackets such as *V. germanica*, unlike honey bees, lack communication to alert their colony mates when food sources are located. However, Ishay and Schwartz (1973) reported that hungry larvae of *Vespa orientalis* Fab. scrape the sides of their cells to produce sounds (or vibrations) that attract adult hornets. Akre *et al.* (1976) reported that small groups of *V. pensylvanica* larvae twitched back and forth almost in unison every four to six seconds. This twitching was especially noticeable in the queen nest when the queen was away, became less pronounced when workers appeared, and never involved a major proportion of the larval population at any one time. It appeared to attract both queen and workers. Akre and co-workers believe that perhaps foraging may be correlated to the stimulation of workers by "twitching" larvae.

H. Colony Drift

This phenomenon, commonly associated with honey bees, involves the leaving of one hive by some individuals to join another. Data gathered by Akre *et al.* (1976) indicates that workers of *V. pensylvanica* and *V. atropilosa* may also exhibit colony drift or at least join other nearby colonies under certain circumstances. This suggests that if workers were away from their nests at the time of its destruction, they could become members of a second nest, a factor to consider when developing a control program.

I. Wasp Populations: Densities

Mac Donald (1976) noted that some species of yellowjackets become particularly pestiferous when population densities are high, especially in urban and recreational areas. For example, he found 28 nests of *V. maculifrons* and 10 nests of *V. squamosa* in a stretch of land about 275 m along a stream in Athens, Georgia. Akre (1976) collected 60 to 70 nests of *Vespula* spp. (*V. pensylvanica* and *V. atropilosa*) in Pullman, Washington, during 1974 and 1975. About half of these nests were on or near a nine-hole golf course. Workers from these nests caused many problems in the surrounding areas, and stings resulting from nest disturbance were frequent. In another study area in Idaho he observed, during the severe outbreak of *V. pensylvanica* and *V. vulgaris* in 1973, from four to seven workers flying per square meter of forest floor. Preiss (1967) found 151 *V. maculifrons* nests within a 14 ha woodlot over a two year period (1965-1966) at the University of Delaware at Newark. Moreover, an article that appeared recently in a newspaper at Christchurch, New Zealand, reported that 17 nests of *V. germanica* were found within 274 m of a school located on the west coast of the country. The wasp problem was so severe that officials were contemplating closing of the school.

J. Wasp Populations: Influence of Weather

For years researchers have speculated on the effect of weather in influencing wasp populations. Dohring (1960) concluded, after a 15-year study of wasp populations in the city of Berlin, that cold, wet winters were unimportant, but that sudden cold periods in April and May severely reduced colony survival. After six years of study, Akre (1976) concluded that weather in the form of a very cold, wet spring in April, May, and early June (when yellowjacket colonies are in a "critical stage" with the queen and the first five to seven workers) is the single most important factor determining yellowjacket abundance in the Pullman area of southeastern Washington, although other factors undoubtedly are involved.

K. Summary: Ecology and Behavior

For the general public to understand and appreciate the importance of yellowjacket wasps, and in some cases the problems that arise with these insects, will require considerable education on its part. Moreover, this will come about only after extensive research enables us to fill the gaps in information that now exist for many of the ecological and behavioral aspects just covered. Furthermore, this information will greatly facilitate the development of sound, efficient, and intelligent control programs whenever they become necessary.

IV. IMPORTANT PEST SPECIES IN URBAN AREAS

As previously mentioned, several species of yellowjackets often become a nuisance in urban areas in various parts of the world. Some of these pests affect man directly by building their nests in or near his dwellings, lawns, shrubs, trees, and undeveloped lands along the margins of his property. As a result of this, foraging by these insects in these "man-made" environments necessarily brings them into close contact with man. It is appropriate also to point out that the term urban is difficult to define because man often congregates, lives, and/or works in numbers, large or small, in villages, parks, apiaries, vineyards, orchards, berry fields, mink ranches, and other areas that provide the wasps with environments not unlike those found in strictly urban areas. For this reason, I have elected to use the term urban in its broadest sense.

Some brief comments that pertain to a few of the more important *Vespula* and *Dolichovespula* species known to be pestiferous in urban localities follow:

A. *Vespula germanica*

This is perhaps the most widely studied *Vespula* species of the world, probably because it is a nuisance to man in various localities throughout its large geographic range.

1. V. germanica *in New Zealand*

In his comprehensive study of *V. germanica* in New Zealand, Thomas (1960) described the establishment, distribution, biology, and behavior of this species since its introduction into the country in 1945. He found that under favorable conditions aerial colonies of *V. germanica* could survive throughout the winter and spring thus enabling some of them to reach huge proportions. For example, one nest attached to a large totara tree nearly six meters above ground measured 4.6 x 1.5 x 0.6 m. Similarly, some of the terrestrial nests were equally enormous when compared with those found in Europe. Thomas (1960) reported that one of the largest was 119 x 102 x 97 cm with 27 comb levels and a total weight of approximately 45.4 kg. Moreover, he noted that the most important economic factors associated with this wasp were losses suffered by beekeepers; their hives were often robbed with considerable loss of honey and/or loss of the colonies. He also described *V. germanica* workers as a great nuisance in the autumn when they searched for sweets in dwellings, breweries, and honey houses, or in places where they were stored or handled. He also noted they sometimes bothered persons picking and handling fruit. Moreover, a survey conducted by Walton and Reid (1976) in New Zealand in 1974 and 1975 revealed that *V. germanica* was considered by 88.6% of the beekeepers to be a nuisance and by 73.6% as a cause of financial loss. Perrott (1975), also in New Zealand, noted that forestry workers and construction gangs on roads and power lines were bothered by *V. germanica* workers, resulting in changed work schedules. In some rural areas children were forbidden to play on grass turfs because of *V. germanica*, and the wasps also spoiled the enjoyment of people attending holiday resorts.

2. V. germanica *in Tasmania*

Spradbery (1973b) reported that this species, introduced in 1959, was most abundant in suburban and coastal areas and along the major river valleys. He reported it as established throughout 70% of the 67,340 km^2 state and in all the major vegetational regions except the mountainous woodlands. He believes the major factors responsible for the successful invasion and establishment of *V. germanica* in Tasmania are climate, abundance of nesting sites and forage, absence of natural enemies, and the capacity to requeen colonies, overwinter, and increase reproduction. Although he noted that nests with more than a million cells have been found, which undoubtedly result in high wasp populations in these localities, he did not indicate that this species is a pest of man in Tasmania.

3. V. germanica *in South Africa*

Whitehead and Prins (1975) recently reported that a worker of *V. germanica* was collected in Cape Town, Cape Peninsula, South Africa, and there is evidence that this species bas been present in that country for two years. They noted, however, that the only economic threat this species apparently poses is the robbing of weak beehives and the nuisance they may cause in recreational areas in autumn.

4. V. germanica *in Chile*

Mac Donald (1976) in communication with researchers in Chile reported that *V. germanica* had become established.

5. V. germanica *in Europe and U.S.A.*

In Europe, Biegel (1953) found that *V. germanica* built mostly subterranean nests (nearly 74%) near Erlangen, Germany, and Spradbery (1971) noted that over 80% of the nests in England were subterranean. Spradbery indicated, however, that many colonies that had been discovered but not included in his analysis were aerial nests found in roofs, attics and cavity walls. Menke and Snelling (1975) reported that pest control operators in the Washington, D.C., area in the United States commonly found *V. germanica* nests in attics and walls of houses.

Kemper and Dohring (1967) found that workers of *V. germanica* are a problem in bakeries, markets, and butcher shops in several German cities. Kemper (1962) showed that food gathered by workers in Berlin consisted mainly of sweets, protein from flies and other living insects, and meat and fish . Morse (1976) found from surveys conducted the past two years in the Ithaca, New York, area that *V. germanica* constituted 87% of the population of yellowjackets found around garbage cans near several restaurants and at one local park. Morse noted that it "thrives" on foods such as hamburgers, hot dogs, soda pop, beer, etc. Menke and Snelling (1975) reported that *V. germanica* was common in some parts of New York, Pennsylvania, New Jersey, Delaware, Maryland, and Washington, D.C. Caron (1976) found that in parts of Maryland, *V. germanica* workers create problems in public areas such as parks and zoos. He receives calls each year from persons hosting outside parties and cookouts inquiring about what can be done about the "bees," which nearly always turn out to be *V. germanica*. Calls for assistance are also received from people at golf courses, etc., where outside soft drink dispensing machines are located. The wasps are attracted and become a nuisance and threat to the patrons.

Menke and Snelling (1975) reported that although this wasp was collected in the United States on several occasions (1891, a few from 1920-1950), this insect has become abundant only since 1968. This would lead to speculation why this species was unable to become established prior to this time. Perhaps the populations that are now established represent a new or different biotype that thrives in environments greatly modified by man's presence. Moreover, will their behavior be such that they will become increasingly dependent upon man, and, for example, scavenge entirely for their food or utilize his structures even more frequently for nest sites?

B. *Vespula vulgaris*

As described by Miller (1961) *Vespula vulgaris* is a Holarctic species that is transcontinentally distributed in the Nearctic region. He noted that in areas where this species and *V. maculifrons* appear together the generic barrier between populations is not complete and hybridization may occur.

Studies by Biegel (1953) in Erlangen showed that 70.9% of the nests of *V. vulgaris* were subterranean and 29.1% were established above ground. Spradbery (1971) found about the same ratio for this species among nests he analyzed in England.

In several German cities Kemper and Dohring (1967) found *V. vulgaris* workers, as well as those of *V. germanica*, a nuisance in bakeries, markets, and butcher shops. Løken (1964) described this insect as being the most annoying to human beings of any species in Norway. He described it as quick-tempered, aggressive, and apparently irritated by strong scents, such as perfume. In the United States it commonly becomes a pest of man probably as a result of its scavenging habits. In the western United States it frequently builds nests in rotted logs and stumps in coniferous forests. It is also found however, in urban-suburban areas in many parts of the country where it nests (usually underground) in various types of vegetation. For example, Ennik (1973) found it nests in oak and madrone woodlands in the San Francisco Bay area where it becomes a nuisance to residents. Grant *et al.* (1968) also reported that most *V. vulgaris* nested in wooded areas in San Mateo County (a part of the San Francisco Bay area), and that the wasps become such a severe nuisance that organized control programs are required almost every year.

In 1974, nests of *V. vulgaris* were reported in Hawaii by Miyahira (1974). During the same year the heaviest populations of this species on record were reported (and observed by the author) in the Anchorage-Palmer regions of southern Alaska. They became extremely annoying and inflicted numerous stings on people residing or working in urban areas and on those in agricultural and recreational areas.

C. *Vespula maculifrons*

This species is restricted to the Austral region of eastern North America (Miller, 1961). Lord (1976) stated that *V. maculifrons* is abundant in wooded sections of suburban Delaware were it becomes a major pest in playgrounds, parks and picnic areas. The nests are usually subterranean and constructed in mouse holes usually located under tree bases, fallen limbs, etc., or they may sometimes nest on open ground unassociated with any protective object. The workers forage for food from many sources including trash cans, picnic tables, dead carcasses, etc., as well as preying upon other insects. However, Lord noted that staff members at the University of Delaware have observed over a five year period that *V. maculifrons* workers seldom, if ever, become aggressive unless the nest is disturbed.

According to Mac Donald (1976), *V. maculifrons* is the most pestiferous species in Athens, Georgia, because of its scavenging habits. He noted that this species, as well as *V. squamosa*, nests abundantly in parks, and that they exhibit a pronounced propensity for nesting in yards of homes. Nearly 300 colonies were located over two seasons (1974/1975) in northern Georgia and western North Carolina. In addition, he pointed out that the impact of yellowjackets on man in the south-eastern part of the United States depends on their high population densities in urban and recreational areas where they are particularly pestiferous. Mac Donald and Matthews (1976) are currently preparing manuscripts on the biology of both *V. squamosa* and *V. maculifrons* which will contribute to our knowledge of these two important species in urban and recreational areas. Moreover, Howell *et al.* (1974) reported that jellowjackets are considered to be one of the most serious pests in Georgia's outdoor recreational areas and that *V. maculifrons* and *V. squamosa* constitute the bulk of the populations.

D. *Vespula pensylvanica*

This yellowjacket, commonly called the western yellowjacket, is restricted in its distribution to the Canadian and

Transition zones in western North America (Miller, 1961). As described by Duncan (1939), their nests are mostly terrestrial; however, there are exceptions. Akre and co-workers (unpublished data) found, for example, a nest in the roof under cedar shakes at a home in Pullman, Washington, and a second nest between the walls of a house. Also, Buckell and Spencer (1950) reported finding two nests of this species between the walls of houses in British Columbia.

The western yellowjacket, because of its scavenging habits, is in close contact with man in urban environments in many parts of its range. Wagner (1961) reported workers of *V. pensylvanica* often became so numerous at a large park in the Los Angeles area that it was difficult to eat in comfort and safety. He also pointed out that large numbers of adults and children were stung every year by these wasps and that even animals in the nearby zoo were molested.

In 1973, a year of unusually high *V. pensylvanica* populations throughout the Pacific Northwest, the wasps were a hazard in places such as school playgrounds, amusement parks, zoos, golf courses, building construction sites, canneries, sanitary landfills, backyards and numerous other places in urban-suburban environments. In agricultural areas hordes of workers invaded fruit orchards, berry fields, vegetable gardens, mink ranches, and pasture lands. Moreover, in many of the forested areas they were a serious problem for sawmill workers, personnel operating fire-lookout towers, fire fighters (some were stung 20 to 30 times as they walked by smoke-filled nests), foresters, and visitors at the recreation parks and compgrounds.

According to Akre (1976) aggressiveness of *V. pensylvanica* can be linked to the season; in the fall during colony decline the workers are most unpredictable and aggressive. During June and July he and his co-workers were sometimes able to work among these yellowjackets without protective clothing and head veils; later, once the colony started to decline, the jellowjackets became aggressive and had a much greater tendency to sting. Akre also found that wasp colonies continuously fed honey (and house flies for protein) throughout the fall do not become aggressive. Therefore, much of this aggressive behavior in field colonies in late summer (during colony decline) can probably be attributed to the inability of workers to secure nourishment from the larvae.

E. Other Species

Other species of yellowjackets, most of them highly beneficial, become "incidental" pest in urban areas and occasionally require control because of the location of their nests.

1. Vespula atropilosa

This wasp is commonly found in similar ecological habitats as the pestiferous scavenger *V. pensylvanica* and, unfortunately, is not usually recognized by the general public as a highly beneficial insect.

2. Dolichovespula maculata

This is another highly beneficial species frequently found in urban areas that builds large aerial nests in shrubs and trees. Unfortunately, it often selects sites which are near sidewalks, porches, patios, etc., necessitating colony destruction. Wagner and Reierson (1971) noted that colonies of *D. maculata* that are not a direct hazard should be left alone because this species is an effective predator of flies and other insects.

3. Dolichovespula arenaria

This beneficial yellowjacket usually reaches its population peak before the pestiferous species such as *V. germanica* and *V. vulgaris* become troublesome. However, they do on occasion create problems. At Pullman, Washington, for example, Akre (1976) and co-workers annually remove, at the request of homeowners, 20 to 40 nests of this species from eaves of houses, garages, fences, etc. Also importantly, they point out that once their nests have been disturbed they remain aggressive and easily aroused, as do most yellowjackets. In southern Alaska in 1974, *D. arenaria* became especially troublesome in wooded areas that were being prepared for new housing. Thousands of nests were destroyed when the trees were cut down leaving untold numbers of wasps without shelter. As a result, their presence and irritability became vividly evident to many of the construction workers and residents of the area.

4. Vespa *spp.*

Hornets belonging to this genus are also known to create problems for man. Chan (1972) reported that three species of *Vespa* were troublesome in Singapore, but that *V. tropica leefmansi* van der Vecht appeared to be associated more closely with man than the other species, *V. affinis indosinensis* Perez and *V. analis analis* Fab. Its distribution was more or less related to that of houses. The number of stinging incidents during the past several years has prompted the implementation of control programs, mostly through nest destruction.

5. Vespa mandarinia *Smith*

The giant hornet was reported by Matsuura and Sakagami (1973) to annually destroy thousands of beehives in Japan. *Vespa mongolica* Andre is another species that causes serious damage to beehives. Choi (1968) pointed out that *Vespa mandarinia* is also the most important pest of honey bees in Korea and that beekeepers annually suffer large losses because of these insects.

6. Polistes *spp.*

Members of this important group are well known for their predation on other insects and for nonaggressive habits around man. They may become troublesome under certain circumstances, however, which requires their control. For example, a large outbreak of *Polistes fuscatus aurifer* Saussure in the Kwajalein Islands in 1973 created serious problems for the residents and interrupted work schedules for many of the personnel working at facilities on the islands. Numerous people were stung during the outbreak. Also, *Polistes* sp. were reported by Perez (1973) as responsible for damage to fully ripened fruit (grapes, mango, sapodilla) in Puerto Rico.

V. ECONOMIC IMPACT OF YELLOWJACKETS

Despite the highly important beneficial roles that yellowjackets play in ecosystems, they are responsible for considerable economic losses. Hawthorne (1969) reported that losses in 1968 in agricultural operations in California, due to the attacks of *Vespula* spp. on fruit pickers, feed lot workers, etc., were estimated at $200,000, due mostly to lost wages and medical expenses. Additionally, thousands of dollars are lost each year by beekeepers in many parts of the world because yellowjackets rob hives and destroy the colonies (Matsuura and Sakagami, 1973; Walton and Reid, 1976).

In recreational areas, losses are suffered by owners of resorts, parks, campgrounds, etc., in the form of decreased revenues; their patrons refuse to tolerate yellowjackets. Poinar and Ennik (1972) estimated that losses at some private resorts in California annually reach $5,000 per resort. Others in recreational areas who are affected by the wasps include loggers, road construction workers, fire fighters, etc., mostly from time lost from work; some years these losses are substantial. In 1973, for example, workers of *Vespula pensylvanica* and *V. vulgaris* were responsible for halting logging operations, closing sawmills, and delaying fire fighting operations in the Pacific Northwest.

In urban-suburban areas during "bad" wasp years, yellowjackets are responsible for severe losses in revenues at amusement parks, playgrounds, zoos, and athletic facilities. Also, laborers in cities and towns such as garbage collectors, landfill equipment operators, food cannery personnel, etc., can be severely affected. Furthermore, residents in some large cities are annoyed and disturbed by these insects and are unable to use their yards for cookouts and barbeques. Indeed, yellowjackets can be responsible for automobile accidents, some of which have resulted in fatalities. Other losses attributed to yellowjackets include money spent for medical bills, insurance (health and life), unemployment compensation, and property damage.

VI. MEDICAL IMPORTANCE OF YELLOWJACKETS

As mentioned previously, man's primary concern with yellowjackets is the threat of being stung. Fluno (1961) pointed out the public's concern in the United States about wasps and noted that the United States Department of Agriculture received over 10,000 requests each year for information about these insects and how to control them. About one-half of the requests were accompanied by statements that one or more members of the family had suffered severe reactions from stings. Present day wasp researchers continue to receive similar letters year after year.

Frankland (1968, cited in Spradbery, 1973a) has shown that all people react to insect stings and bites although the symptoms vary from person to person, and that some people may become allergic to bee, wasp, and hornet stings. As described by Fluno (1961), some victims experience only an intense immediate pain at the site of the wound with localized reddening and swelling, whereas others may experience swelling that extends considerably beyond the site of the sting, for example, an entire leg or arm. Unfortunately, still others (although few) experience lethal or near-lethal reactions from anaphylactic shock. This usually occurs within 10 to 20 minutes after receiving the sting, although it can occur 10 to 20 hours later.

Reisman (1975) suggested several simple measures which may help avoid stings. He noted that insects are attracted to perfumes, hair sprays, suntan lotion, cosmetics, and bright colors. He recommends that susceptible persons avoid outdoor cooking and eating when yellowjacket workers become troublesome. Allergic reactions are combated by: 1) adrenaline (epinephrine) by injection, 2) antihistamine tablets, and 3) an inhaler containing adrenaline to alleviate possible throat swelling.

VII. CONTROL OF YELLOWJACKETS

When wasp populations reach intolerable levels they must be controlled. Unfortunately, we do not know what levels must be reached before action is required; some people can tolerate annoying numbers of wasps while others become alarmed by the presence of a single worker.

The destruction of wasp nests remains one of the principal means of controlling these insects. This is especially true for the homeowner who finds nests on or in close proximity to his property. Subterranean nests are easily eliminated by pouring toxicants such as gasoline or kerosene into the entrance hole and plugging it. Control attempts should be done only at night when the insects have returned to the nest. The materials used should not be ignited. Aerial nests should also be destroyed at night and, fortunately, several aerosol products have been recently developed that propel a quick-knock-down insecticide from distances of approximately three to five meters.

Several control methods developed in recent years include the use of protein baits, specific for yellowjackets, mixed with low concentrations of insecticides. For example, Grant *et al.* (1968) published the results of a five year testing program that utilized cooked horsemeat and chlordane,[1] 1,2,4,5,6,7,8,8-octachloro-3*a*,4,7,7*a*-tetrahydro-4,7-methanoindane, and found that it effectively controlled *V. pensylvanica* and *V. vulgaris* populations in park and suburban areas in San Mateo County, California. Wagner and Reierson (1969) reported excellent control of *V. pensylvanica* in southern California by using fish-flavored cat food in conjunction with mirex, dodecachloro-octahydro-1,3,4-metheno-1*H*-cyclobuta[*cd*]pentalene, a slow-acting insecticide, plus a chemical wasp attractant.

Ennik (1973) reported the abatement of populations of *V. pensylvanica* and *V. vulgaris* using encapsulated formulations of diazinon, *O,O*-diethyl O-(2-isopropyl-4-methyl-6-pyrimidyl) phosphorothioate, and stirofos, 2-chloro-1-(2,4,5-trichlorophenyl)vinyl dimethyl phosphate, mixed with tuna fish catfood, at eight test sites in northern California. Foraging yellowjackets were reduced an estimated 75 to 95% within two days of exposure to the bait. More recently, Perrott (1975) found that a bait of canned fish poisoned with 0.5% of 1.0% mirex substan-

---

[1] *This paper reports the results of research only. Mention of a pesticide in this paper does not constitute a recommendation for use by the U.S. Department of Agriculture nor does it imply registration under FIFRA as amended.*

tially reduced populations of *Vespula germanica* in a resort area in New Zealand.

Davis *et al.* (1967) discovered that 2,4-hexadienyl butyrate was a highly specific, potent lure for *V. pensylvanica*. Later (1973) he and co-workers demonstrated that small carton traps, containing the synthetic lure heptyl butyrate, placed around the periphery of a 8.9 ha peach orchard in Oregon, effectively depressed worker populations to a level allowing harvesting to be resumed.

Biological control does not show much promise as a control method. Spradbery (1973a) described parasites, predators, and pathogens as having little effect on wasp populations. He states, "... except in weakened colonies, biotic agents have little effect on wasp density, levels of parasitism being generally low and the number of affected colonies small. To regulate wasp populations on an annual basis, colonies must be severely weakened or destroyed before queen production gets underway, or predation and parasitism must be directed at the queen brood or adults."

Finally, it should be emphasized that perhaps the best method of controlling yellowjackets in urban environments involves the proper management of man's garbage and litter. Wagner (1961) showed that the treatment of trash containers once a week with an aqueous spray containing 0.75% dichlorvos, 2,2-dichlorovinyl dimethyl phosphate, in picnic areas in southern California reduced *V. pensylvanica* worker populations to about 1% of those found in test areas without treatment. Similarly, Akre and Retan (1973) noted the effectiveness of dichlorvos (DDVP) strips attached to the inside of trash can lids.

VIII. CONCLUSIONS

Although several wasp abatement methods have been developed in recent years that have proven effective in various degrees (Grant *et al.*, 1968; Wagner and Reierson, 1969; Ennik, 1973; Davis *et al.*, 1973), we still lack effective control programs for most species (Mac Donald *et al.*,1976). One of the most promising control programs currently under investigation involves using a meat bait/encapsulated insecticide combination which is attractive to yellowjackets. However, these types of materials have been tested only in California on *V. pensylvanica*.

Until the necessary biological and behavioral information on other species in other areas is forthcoming, intelligent control programs cannot be developed.

ACKNOWLEDGMENT

The author gratefully acknowledges the assistance of Dr. Roger D. Akre, Department of Entomology, Washington State University, and Dr. John F. Mac Donald, Department of Entomology, Purdue University, for their review and helpful criticism of the manuscript.

REFERENCES

Akre, R. D. 1976. Personal communication. Dept. of Entomol., Washington State University, Pullman, Washington.

Akre, R. D., and A. H. Retan. 1973. Yellowjackets and paper wasps. Wash. State Univ. Coop. Ext. Serv. Bull. 643.

Akre, R. D., J. F. Mac Donald, and W. B. Hill. 1974. Yellowjacket literature (Hymenoptera: Vespidae). *Melanderia 18*: 67-93.

Akre, R. D., W. B. Hill, J. F. Mac Donald, and W. B. Garnett. 1975. Foraging distance of *Vespula pensylvanica* workers (Hymenoptera: Vespidae). *J. Kansas Entomol. Soc. 48*: 12-16.

Akre, R. D., W. B. Garnett, J. F. Mac Donald, A. Greene, and P. Landolt. 1976. Behavior and colony development of *Vespula pensylvanica* and *V. atropilosa* (Hymenoptera: Vespidae). *J. Kansas Entomol. Soc. 49*: 63-84.

Arnold, T. S. 1966. Biology of social wasps: Comparative ecology of the British species of social wasps belonging to the family Vespidae. M. Sc. Thesis, Univ. London. *Cited in* J.P. Spradbery. 1973a. Wasps. Univ. Washington Press, Seattle, Washington.

Biegel, W. 1953. Die Tagesaktivität der Arbeiter eines Nestes von *Vespa germanica* F. *Mitt. Schweiz. Entomol. Ges. 26*: 293-294.

Blackith, R. E. 1958. Visual sensitivity and foraging in social wasps. *Insectes Soc. 5*: 159-169.

Brian, M. V., and A. D. Brian. 1952. The wasp, *Vespula sylvestris* Scopoli: Feeding, foraging and colony development. *Trans. Roy. Entomol. Soc. London 103*: 1-26.

Buckell, E. R., and G. J. Spencer. 1950. The social wasps (Vespidae) of British Columbia. *Proc. Entomol. Soc. Br. Columbia 46*: 33-40.

Caron, D. M. 1976. Personal communication. Dept. of Entomol., University of Maryland, College Park, Maryland.

Chan, K. L. 1972. The hornets of Singapore: their identification, biology and control. *Singapore Med. J. 13*: 178-187.

Choi, S. Y. 1968. Personal communication. Dept. of Agric. Biol., Seoul National University, Suwon, Korea.

Davis, H. G., G. W. Eddy, T. P. McGovern, and M. Beroza. 1967. 2-4-Hexadienyl butyrate and related compounds highly attractive to yellowjackets (*Vespula* spp.). *J. Med. Entomol.* *4*: 275-280.

Davis, H. G., R. W. Zwick, W. M. Rogoff, T. P. McGovern, and M. Beroza. 1973. Perimeter traps baited with synthetic lures for suppression of yellowjackets in fruit orchards. *Environ. Entomol.* *2*: 569-571.

Dohring, E. 1960. Zur Häufigkeit, hygienischen Bedeutung und zum Fang sozialer Faltenwespen in einer Groβstadt. *Z. Angew. Entomol.* *47*: 67-79.

Duncan, C. D. 1939. A contribution to the biology of North American vespine wasps. Stanford Univ. Publ., Univ. Ser., Biol. Sci. 8.

Ennik, F. 1973. Abatement of yellowjackets using encapsulated formulations of diazinon and rabon. *J. Econ. Entomol.* *66*: 1097-1098.

Evans, H. E., and M. W. Eberhard. 1970. The wasps. Univ. Michigan Press, Ann Arbor, Michigan.

Fluno, J. A. 1961. Wasps as enemies of man. *Bull. Entomol. Soc. Am.* *7*: 117-119.

Frankland, A. W. 1968. Allergy and the bee-keeper. Report Cent. Assoc. British Beekeep. Assoc. *Cited in* J.P. Spradbery. 1973a. Wasps. Univ. Washington Press, Seattle, Washington.

Garnett, W. B. 1976. Personal communication. Raymond Walters General and Technical College, Univ. of Cincinnati, Cincinnati, Ohio.

Gaul, A. T. 1952. Additions to vespine biology. X. Foraging and chemotaxis. *Bull. Brooklyn Entomol. Soc.* *47*: 138-140.

Grant, C. D., C. J. Rogers, and T. H. Lauret. 1968. Control of ground-nesting yellowjackets with toxic baits - A five-year program. *J. Econ. Entomol.* *61*: 1653-1656.

Greene, A., R. D. Akre, and P. Landolt. 1976. The aerial yellowjacket, *Dolichovespula arenaria* (Fab.): Nesting biology, reproductive production, and behavior (Hymenoptera: Vespidae). *Melanderia* *26*: 1-34.

Hawthorne, R. M. 1969. Estimated damage and crop loss caused be insect/mite pests - 1968. Calif. Dept. Agric., Bureau Entomol., Unpubl. Report.

Herold, W. 1952. Beobachtungen über die Arbeitsleistung einiger Arbeiter von *Vespa germanica* F.-*Dolichovespula germanica* (F.). *Biol. Zentralbl.* *71*: 461-469.

Howell, J. O. 1973. Notes on yellowjackets as a food source for the bald faced hornet, *Vespula maculata* (L.). *Entomol. News* *84*: 141-142.

Howell, J. O., T. P. McGovern, and M. Beroza. 1974. Attractiveness of synthetic compounds to some eastern *Vespula* species. *J. Econ. Entomol* *67*: 629-630.

Ishay, J., and A. Schwartz. 1973. Acoustical communication between the members of the oriental hornet (*Vespa orientalis*) colony. *J. Acoust. Soc. Am. 53*: 640-649.

Kalmus, H. 1954. Finding and exploitation of dishes of syrup by bees and wasps. *Br. J. Anim. Behavior 2*: 136-139.

Kemper, H. 1962. Nahrung und Nahrungserwerb der heimischen sozialen Vespiden. *Z. Angew. Entomol. 50*: 52-55.

Kemper, H., and E. Dohring. 1967. Die sozialen Faltenwespen Mitteleuropas. Paul Parey, Berlin, Hamburg.

Løken, A. 1964. Social wasps in Norway (Hymenoptera: Vespidae). *Nor. Entomol. Tidsskr. 12*: 195-218.

Lord, W. D. 1976. Personal communication. Dept. of Entomol. and Applied Ecol., University of Delaware, Newark, Delaware.

Mac Donald, J. F. 1976. Personal communication. Dept. of Entomol., Purdue University, West Lafayette, Indiana.

Mac Donald, J. F., R. D. Akre, and W. B. Hill. 1974. Comparative biology and behavior of *Vespula atropilosa* and *V. pensylvanica* (Hymenoptera: Vespidae). *Melanderia 18*: 1-66.

Mac Donald, J. F., and R. W. Matthews. 1975. *Vespula squamosa:* A yellowjacket wasp evolving toward parasitism. *Science 190*: 1003-1004.

Mac Donald, J. F., and R. W. Matthews. 1976. Personal communication. Dept. of Entomol., Purdue University, West Lafayette, Indiana, and Univ. of Georgia, Athens, Georgia.

Mac Donald, J. F., R. D. Akre, and R. W. Matthews. 1976. Evaluation of yellowjacket abatement in the United States. *Bull. Entomol. Soc. Am. 22*: 397-401.

Matsuura, M., and S. F. Sakagami. 1973. A bionomic sketch of the giant hornet, *Vespa mandarinia,* a serious pest for Japanese apiculture. *J. Fac. Sci., Hokkaido Univ., Ser. 6, Zool. 19*: 125-162.

Menke, A. S., and R. Snelling. 1975. *Vespula germanica* (Fabricius), an adventive yellow jacket in the Northeastern United States (Hymenoptera: Vespidae). *USDA, APHIS, Coop. Econ. Insect Rep. 25*: 193-200.

Miller, C. D. F. 1961. Taxonomy and distribution of Nearctic *Vespula.* Canad. Entomol. Suppl. 22.

Miyahira, N. 1974. *USDA, APHIS, Coop. Econ. Insect Rep. 24*: 743.

Morse, R. A. 1976. Personal communication. Dept. of Entomol., Cornell University, Ithaca, New York.

Perez, R. P. 1973. Personal communication. Estacion Experimental Agricola, Recinto Universitario se Mayaquez, Universidad de Puerto Rico, Ponce, Puerto Rico.

Perrott, D. C. F. 1975. Factors affecting use of mirex-poisoned protein baits for control of European wasp (*Paravespula germanica*) in New Zealand. *New Zealand J. Zool. 2*: 491-508.

Poinar, G. O., and F. Ennik. 1972. The use of *Neoaplectana carpocapsae* (Steinernematidae: Rhabditoidea) against adult yellowjackets (*Vespula* spp., Vespidae: Hymenoptera). *J. Invert. Pathol. 19*: 331-334.

Preiss, F. J. 1967. Nest site selection, microenvironment and predation of yellowjacket wasps, *Vespula maculifrons* (Buysson), (Hymenoptera: Vespidae) in a deciduous Delaware woodlot. M.S. Thesis, Dept. of Entomol., University of Delaware, Newark, Delaware.

Reisman, R. E. 1975. Insect stings--Danger season opens. *U.S. News and World Rep. 78(16)*: 65.

Rogers, C. J. 1972. A method for marking ground nesting yellowjackets. *J. Econ. Entomol. 65*: 1487-1488.

Spradbery, J. P. 1971. Seasonal changes in the population structure of wasp colonies (Hymenoptera: Vespidae). *J. Anim. Ecol. 40*: 501-523.

Spradbery, J. P. 1973a. Wasps. Univ. Washington Press, Seattle, Washington.

Spradbery, J. P. 1973b. The European social wasp, *Paravespula germanica* (F.) (Hymenoptera: Vespidae) in Tasmania, Australia. International Union Study Social Insects, VII Internat. Congr. Proc. 7: 375-380.

Thomas, C. R. 1960. The European wasp (*Vespula germanica* Fab.) in New Zealand. New Zealand Sci. Ind. Res., Inform. Ser. No. 27.

Tissot, A. N., and F. A. Robinson. 1954. Some unusual insect nests. *Florida Entomol. 37*: 73-92.

Wagner, R. E. 1961. Control of the yellowjacket *Vespula pensylvanica* in public parks. *J. Econ. Entomol. 54*: 628-630.

Wagner, R. E., and D. A. Reierson. 1969. Yellowjacket control by baiting. 1. Influence of toxicants and attractants on bait acceptance. *J. Econ. Entomol. 62*: 1192-1197.

Wagner, R. E., and D. A. Reierson. 1971. Recognizing and controlling yellowjackets around buildings. *PCO News 6-7*: 30-32.

Walton, G. M., and G. M. Reid. 1976. The 1975 New Zealand European wasp survey. *New Zealand Beekeeper 38*: 26-30.

Whitehead, V. B., and A. J. Prins. 1975. The European wasp, *Vespula germanica* (F.), in the Cape Peninsula. *J. Entomol. Soc. S. Afr. 38*: 39-42.

URBAN APICULTURE

Michael Burgett
Department of Entomology
Oregon State University
Corvallis, Oregon

Dewey M. Caron
Department of Entomology
University of Maryland
College Park, Maryland

John T. Ambrose
Department of Entomology
North Carolina State University
Raleigh, North Carolina

ISBN 0-12-265250-9

I. INTRODUCTION

The association of humans and honey bees (*Apis mellifera* L.) is one whose history predates the written word. During the thousands of years of man's primitive husbandry of honey bees the usual procedure was to "rob" wild honey bee colonies, with little concern over the survival of the plundered colony. Man was, in essence, another vertebrate predator of the honey bee. Eventually, potential nest sites were provided by man for the bees' use. This was an important first step in improving the efficiency of beekeeping.

Apiculture has been defined by Michener (1974) as: "... the management of honey bees usually for honey or wax production or for pollination." The key word is management. Beekeepers practice applied insect husbandry, using the bee's biological requirements as guidelines for their management systems.

An important point in discussing the relationship between man and the honey bee is that the bee is not "domesticated" in the sense of other animals. Crane (1975) has noted that, "... however cleverly man exploits the bees, they are no more 'domesticated' today than before man existed." Man adjusts to the bee rather than adapting the bee to suit man's needs.

In the urban environment honey bees present a contradiction. Many consider honey bees pests that constitute a human health hazard because they sting. Their pest status is reinforced by the difficulty and expense of their removal when they nest in human habitations. Others view the honey bee as a valued member of the urban insect community; a creature that can be manipulated to provide honey, beeswax and pollination services with a minimum of interaction with humans. Within the urban setting the honey bee provides man an important avocational pastime and pollinates garden, landscape and wild plants.

Whether a friend or insect foe, there is a general dearth of literature on honey bees in the urban environment. Two articles on urban beekeeping have appeared in the popular beekeeping journals in the last five years (see Greve, 1974 and DeJong, 1975). Of the numerous extension publications on beekeeping, only two leaflets prominently address the urban beekeeper (see Caron, 1973 and Thurber and Johansen, 1974). Of numerous state and regional beekeeping manuals, at most only a paragraph or two is directed to urban apiculture. Yet there are important differences between urban and rural environments that affect the beekeeper. It is hoped that this chapter will coalesce the recommended management practices that should be followed by urban beekeepers.

II. WHO ARE THE BEEKEEPERS?

It has been estimated by Morse (1972) that the United States beekeeper population is nearly 300,000 persons. That a larger segment of the population is not involved in beekeeping is accredited by Morse (1975) to an acquired fear of stings on the part of most laypersons. This is probably true since stings are never pleasant and are always possible in the maintenance of bee colonies. Successful beekeepers accept stings as part of honey bee management.

Beekeepers can be roughly classified into three categories; namely commercial, sideliner, and hobbyist. A commercial beekeeper in the U.S. has, by United States Department of Agriculture (USDA) definition, in excess of 300 colonies. A workable number of colonies for a commercial beekeeper, who employs no permanent outside help, is from 500 to 1,000 colonies. Several commercial operations in excess of 10,000 colonies are found in the western U.S. The majority of commercial beekeepers derive all or the major proportion of their income from honey production and pollination rental fees. There are currently about 1,600 persons in the U.S. who fit the commercial category (U.S.I.T.C., 1976). In other parts of the world, the figure of 300 colonies or over as being commercial would also be accurate although Crane (1975) says that only 100 to 300 colonies could be handled by one person in much of Europe. Generally, commercial beekeepers would be found only in developed or developing countries. The beekeepers in the less developed countries would likely be farmers handling one or more animals or crops.

In the U.S. a concentration of commercial beekeepers is found in the midwest (North and South Dakota, Montana, Nebraska and Minnesota) as well as in California, Texas and Florida. Commercial beekeepers account for nearly half of the honey bee colonies in the U.S. and better than two-thirds of the total honey production, according to USDA statistics. Commercial beekeeping requires large acreages of entomophilous flowers, either wild or cultivated, for honey production or for crop pollination rental fees. Primarily for this reason, the commercial beekeeper rarely frequents urban environments.

The sideliners are beekeepers who have more than 25 but less than 300 colonies. This group, numbering about 10,000 persons in the U.S., supplement their regular income by maintaining honey bees for honey production or pollination rental. Sideliners usually have more than one apiary location. The sideliner level of beekeeping, like that of the commercial, has greater space and floral requirements than the urban environment can usually provide. The sideliner locates in rural areas, or in outlying suburban areas.

Sideliners exhibit a high degree of beekeeping skill.

They have time, fiscal resources, and energy to maximize productivity on an individual colony basis, whereas commercial operations manage colonies as apiary units (Crane, 1975). Sideliners experiment with and invent both equipment and techniques. They have contributed greatly to our knowledge about and management of honey bees.

In less developed and developing countries even a few bee colonies can be an important means of income supplement. Expenses may be kept to a minimum by utilization of natural nesting sites, and management is often primitive and not extensive. The full productivity is not frequently attained under such conditions but the income is still valuable to the individual.

The majority of beekeepers are hobbyists, owning less than 25 colonies. This last group will be found in every type of environment, and is nearly always the only type of beekeeper found in the urban area. Within this largest group of beekeepers is found the full complement of beekeeping skill, from those highly proficient to those termed "beehavers," possessing very little skill or biological knowledge of the honey bee. Hobbyists keep bees for many different reasons, and they are often more concerned with the aesthetics and atavisms of beekeeping rather than economic returns.

The number of honey bee colonies in the U.S. has been declining since the end of World War II (Anderson, 1969). The distribution pattern of colonies has also changed. Where every farm used to possess a few hives of bees for pollination and/or honey production, we now see concentrations of commercial colonies operated by relatively few full-time commercial beekeepers. The demographic change from a rural to an urban/suburban population has also changed the pattern of beekeeping in the U.S. (McGregor, 1976). Other parts of the world have witnessed similar changes but not to the degree of development of commercial beekeeping as in the U.S. (Crane, 1975). In the U.S.S.R. and other socialist countries, the individual beekeeper is often superseded by the state-run collective farm. A farm averages 80 colonies with some farms specializing in apiculture having 6,000 to 25,000 or more colonies (Crane, 1975). Beekeeping in much of Asia and Africa has not yet begun to develop.

III. EQUIPMENT AND STARTING WITH BEES

Beekeeping is an adaptation by man to the honey bee's biology. It is not based on man-made changes of the bee. Beekeeping equipment therefore conforms to a few basic biological requirements. Bees require a dark, protected cavity which allows for the construction of multiple and parallel combs.

They also need an adequate food and water supply to permit wax elaboration and storage of food surpluses for periods of food unavailability. In addition to providing for the basic needs of the honey bee, man has developed equipment that permits the manipulation of the bees so that he may share in the stored food surpluses.

Beekeeping may be divided into two distinct periods: Modern beekeeping after 1851 and beekeeping prior to 1851. The year 1851 is significant because in that year L. L. Langstroth (Langstroth, 1853) discovered the principle of "bee space" which allowed him to develop the modern, top-opening, movable-frame hive. Prior to this discovery the beekeeper had to physically cut or otherwise remove sections of comb from the hive to obtain any honey. This process frequently resulted in great disruption and much damage to the colony. It was most often messy, time consuming and necessitated the killing of all or some of the bees.

Langstroth's discovery that honey bees naturally construct their combs with a space of approximately 9.5 mm between them enabled him to develop movable frames or combs which could be placed into a bee hive and later removed *in toto* for honey collection by the beekeeper. After removing the honey, the combs could be returned to the hive. By 1861 the movable frame hive was in general use throughout the U.S. and the standard Langstroth 10-frame hive is the most popular unit in the U.S. today. Other countries have one or more hives of different dimensions, but nearly all beekeepers in the developed and many beekeepers in the developing countries keep bees in hives that utilize the principle of bee space.

The basic man-made domicile for a honey bee colony is called a hive. A hive basically consists of a series of boxes, usually rectangular rather than square. Inside each box, parallel beeswax combs are suspended by, and contained within, a frame of wood that can be handled without damaging the fragile comb. One or more top boards cover the boxes and they usually rest on a bottom board that frequently provides for the exterior opening.

Bee hives may be placed on the ground or in some building (with opening to the exterior for the bees) or on a building. The site where bees may be kept is termed an apiary and one to several hundred colonies may be placed at one apiary site.

Beekeeping equipment may be purchased or constructed by hand. Most hives are made of wood but plastics are coming into greater use. Developed countries have bee equipment manufacturers that supply everything a beekeeper might need. In the U.S. there are five major bee equipment supply dealers that distribute free catalogues and another dozen or so who make and sell a more limited line of equipment. Two major supply manufacturers (A. I. Root Co., and Dadant and Sons) have local dealers throughout the U.S. and local dealers are available for beekeepers in Europe, England, Canada, Australia, New Zealand and a few other countries. Obtaining equipment in developing and less developed countries is extremely difficult and the situation is a deterrent to the development of the bee industry there.

The best method for a beginning beekeeper to determine what equipment he or she needs to begin beekeeping is to obtain a catalogue from any one of the major bee supply manufacturers. Most of these catalogues list a "Beginner's Outfit" which includes the basic equipment for the novice beekeeper. Generally, these "outfits" contain a standard 10-frame hive body (called a brood chamber or super) to house the bees, 10 wooden frames and wax foundation for the bees to use in food storage and brood production, and a top cover and bottom board for the hive body. In addition, the kits generally contain a bee veil to protect the beekeeper's face, gloves, a metal smoker which is used to quiet the bees when colonies are being examined, and a hive tool which is used for prying and manipulating hive parts.

Beginners kits usually contain only the equipment to begin beekeeping, and they do not include the most important ingredient, the bees. Bees may be obtained in a number of ways. One may attempt to collect a swarm or remove a feral or unmanaged colony from the wall of a house or other structure. Neither of these methods is suggested for the novice. The best alternative is to purchase the bees from a commercial queen breeder and package bee producer. The usual recommendation is to buy 1.4 kilos of bees (about 10,500 bees) and a queen to start a colony.

After purchasing the "beginner's outfit" and the necessary bees, the beekeeper will need additional equipment for expansion of the brood area and for honey storage by the bees. For this purpose additional supers (hive bodies), frames and wax foundation should be obtained. Additional supers may be the same depth as the original brood super; however, most beekeepers prefer "shallow" or "three-quarters" supers which are easier to handle because of the reduced depth and weight. The most important consideration is that the beekeeper purchase or construct "standard" 10-frame equipment, which allows for easy interchange of equipment between hives.

An alternative to buying new equipment and a quantity of bees is to purchase an already established hive. This is generally less expensive than buying new equipment, but there are potential problems. It is very difficult for the novice beekeeper to recognize the various bee diseases. It is beneficial for any beekeeper to request a disease inspection certificate, prior to purchase, which states that the bees and/or equipment are free from infectious disease. Used bee equipment may not be in top condition and it should be examined carefully.

One problem the beginning beekeeper must face is which kind of honey to produce. Extracted honey (liquid honey removed from the comb) requires a substantial investment in equipment to separate and remove the honey from the wax comb. The basic equipment includes a heated knife, a centrifugal force extractor and a settling tank. This material can cost in excess of $200. Because of this many hobby beekeepers attempt to produce section comb honey or a variety of section honey, cut comb honey. Unfortunately for the novice beekeeper, both of these types of honey production are complicated and require intense bee management and an excellent honey flow. For these reasons the beginning beekeeper is often unsuccessful in producing a high quality product. The best alternative is extracted honey with the hobbyist renting or borrowing extracting equipment from an established beekeeper or taking the honey combs to the established beekeeper for extraction.

Hobbyists, as a general rule, tend to use more types and pieces of equipment than do commercial beekeepers, and they also tend to keep their equipment in better condition. The commercial beekeeper is a businessman and endeavors to keep expenses at a minimum, whereas the hobbyist is often willing to spend more time, energy and money than may be economically sound.

The hobby beekeeper, in either the rural or urban environment, who is interested in bees from an educational or aesthetic viewpoint and not honey production, should consider the use of an observation beehive (Gary, 1968; Connor, 1974; Gary and Lorenzen, 1976) instead of the standard full-size hive. Observation hives are exceptionally versatile in that they may be located almost anywhere in the home or office and, depending upon the style, the hive may consist of one or multiple frames. Such hives have the additional value of being easily transportable and may be used in beekeeping lectures at schools, garden clubs and civic associations. The only serious limitation is that bees in observation hives are not able to store enough honey to survive the entire winter in most temperate climates.

IV. BEE COLONY LOCATIONS

In the urban environment the one consideration that should be paramount in bee colony location is that of fellow humans. Urban beekeepers need to consider the human safety factor. It is estimated that from four to eight percent of the population can be expected to have an allergy to bee stings (Settipane *et al.*, 1972; Frazier, 1976). Many more people believe they have an allergy to bees. An allergic response to envenomation may be an increasingly severe reaction to each bee or wasp sting or an anaphylactoid shock which may result from a single sting. Urban beekeepers also need to consider psychological fears associated with insects. Entomophobia is an important factor in the location of urban apiaries.

Honey bees can be, and are, kept just about everywhere. A bee colony location, termed an apiary, is far from uniform in either the urban or rural environment. In rural areas an apiary may be located on a farm, in an open field, in the wood lands, *ad infinitum*, whereas the potential location of urban apiaries is much more limited, both physically and often legally. However, criteria for selecting an apiary location are basically the same in both urban and rural areas, and the eventual choice is usually based on a compromise between the needs of the bees and those of the beekeeper.

Morse (1972) gives the following description of a proper location for an apiary. "A good apiary is secluded, exposed to full sunlight, has good air circulation and water drainage, a source of fresh water, and is in close proximity to a multitude of flowering plants." This description should probably be modified to include easy access for the beekeeper's vehicle (Peer, 1976), and the need for midday shade in hotter climates.

An apiary should be secluded because this will reduce the probability of vandalism and the possibility that the bees may become a nuisance to nearby residents (Greve, 1974). The latter is of particular importance in urban areas. Kauffeld and Knutson (1976) suggest that it is a good policy to place the apiary at least 15 m from where the bees could interface with human or mechanical traffic. Seclusion may also be obtained by placing the hives so that nearby hedges, trees, fences or buildings will force the bees to fly over the heads of pedestrians (Caron, 1976a).

Another technique for secluding an apiary, particularly relevent in urban areas, is to place the hives on a roof or in the attic of a building (a modification of a bee house). Elevating colonies is effective in reducing vandalism and reducing the possibility of low flying bees, but it does have several disadvantages; the most obvious is that the beekeeper must carry equipment up to the bees and eventually carry the honey back down. A more serious problem is that the summer temperatures may overheat hives enclosed in attics. Construction of artificial shading on a roof would help to alleviate any overheating problem (Owens and McGregor, 1964). In spite of the inherent difficulties, colonies have been and are still being placed on rooftops (Phillips, 1928; Eckert and Shaw, 1960; DeJong, 1975; Kelley, 1975).

Another good method for apiary seclusion is the use of a bee house or house apiary (Root, 1974). The use of bee houses is not very common in the U.S. but it is widespread in Europe. In such structures the hives are usually placed on shelves next to the buildings inner walls and the bees are allowed egress through openings in the walls. There are several advantages to such a system. The house may hold a large number of hives, and the beekeeper can manipulate bees inside the structure which reduces the potential for disturbed bees annoying neighbors. The obvious disadvantage is the cost of such a building, but this may be offset by the advantages, particularly in metropolitan areas.

One criterion of a good apiary site, exposure to full sunlight, may not always be possible in urban environments. Such exposure is generally deemed important because it will enable the bees to fly earlier in the morning and to continue foraging later in the afternoon, thus increasing the temporal foraging period during nectar flows. In certain geographical areas the midday temperatures can become excessive enough to cause thermoregulatory problems in the hives. Under such conditions, artificial shading should be considered (Owens and McGregor, 1964). Shaded locations and bee hives painted camouflage colors may help concealment (Greve, 1974).

Good air circulation and water drainage are necessary apiary criteria for it is important that the hives remain dry. A wet hive or wet bottom board causes temperature control problems and can also lead to greater disease susceptibility problems. This is particularly important during the winter months when moisture (and not cold winter temperatures) is the greatest threat to the survival of a honey bee colony.

Honey bees need a fresh water supply because they use water to both dilute the stored honey before feeding it to larvae and to cool the interior of the hive during periods of temperature stress. Morse (1975) stated that a bee colony may require up to three liters of water per day during the hottest portions of the year. If the beekeeper does not insure that a nearby water source is available, bees may then become a nuisance by obtaining water from such man-made receptacles as swimming pools or bird baths. Finally, apiary sites need to be in the proximity of a multitude of flowering plants. Even though it is generally difficult for an urban apiary location to compete with a rural apiary in terms of nectar and pollen producing plant abundance, there are usually enough ornamental shrubs, trees, and various melliferous flora for the beekeeper to keep at least a few colonies at any given location. The urban apiary is usually limited to 10 or fewer hives because of the limited nectar and pollen resource. However, it should also be noted that a small urban apiary is less likely to win the neighbor's disapproval than a large apiary.

A description of urban bee colony locations would be incomplete without some reference to the presence of unmanaged or feral colonies. Sound data are lacking but such colonies are presumed present in all urban centers. Burgett (unpub. data) estimates a minimum population of 25 feral colonies in

Corvallis, Oregon, a municipality of 40,000 persons. Feral colonies may be found in numerous structures, but these locations may be grouped into four general types: 1) Outside or unprotected (not in a cavity); 2) inside wall voids of structures; 3) in human artifacts such as automobiles, refrigerators, unoccupied bird houses, boxes, *ad infinitum*, and 4) tree cavities. In all of these situations, the inhabitants of the feral colonies are generally considered to be pests and indeed they are not manageable to any degree, even by experienced beekeepers.

V. BEEKEEPING AND THE LAW

Of specific concern to the urban beekeeper are the statutory restrictions and regulations promulgated by municipal authorities. Such regulations usually result from complaints of relatively few citizens regarding actions of one or a few beekeepers.

Nearly every adult can relate at least one experience that involved a stinging insect, though often the person is unable to positively identify the species. Culpability is usually assigned to a "bee." Envenomation by stinging Hymenoptera is not a particularly pleasant sensation. The initial pain and lingering discomfort, which varies greatly with individuals, often causes the development of an avoidance syndrome which associates any flying, buzzing insect with displeasure (entomophobia) (Crane, 1976). This entomophobic reaction is too often transferred to honey bees and occasionally, beekeepers. Associated with this is the fear of a fatal reaction resulting from envenomation. According to Parrish (1963) the average number of persons dying in the U.S. each year from bee, wasp and hornet stings, was 22.5 for the 10 year period 1950 to 1959. This statistic is often contrasted with the average number of people dying from snake envenomation during the same period (13.8). Additionally, news media presentations of the "Killer Bees" invading the U.S. from South America have done little to relieve the layman's fear of stinging insects (Jaycox, 1976).

Historically the banning of honey bees from urban environments has met with little success (Pellett, 1938; Root, 1974). The usual scenario is that the inconsiderations of an individual beekeeper cause his or her immediate neighbors to petition the local statutory body for restrictive legislative relief. Most often the local beekeeping association will appeal and biologically more suitable regulations result (Luver, 1973).

The problem as seen by the non-beekeeper is one of entomological trespass. The beekeeper is unable to prevent the foraging patterns of his or her bees from becoming established on adjacent properties. This is, more often than not, viewed as an infringement upon the rights of the non-beekeeping neighbors.

Several comprehensive reviews of beekeeping from the legalistic standpoint have been written during the past 50 years (Amer. Honey Producers' League, 1924; Blanchard, 1956; Burr, 1959). These works provide considerable detail on the pertinent legal cases that have influenced the present view in which the courts hold bees and beekeeping.

The legalities of urban beekeeping often hinge on the interpretation of the term "nuisance." Bailey (1975) points out that: "An occupant or (land) owner is not permitted to use his property in a manner that seriously interferes with a neighbor's legal right to use his own property as he also sees fit." When such an interference takes place the law terms it a nuisance. The nuisance is not the honey bee as a biological entity; it is most often the manner in which the honey bee is managed. The honey bee *per se* is legally considered *ferae naturae*, i.e., a wild animal. For bees, or more specifically beekeeping, to be declared a nuisance, the alleged inconvenience caused must be proven real and substantial (Bailey, 1975) and recurring (Blanchard, 1956). A classic case occurred in 1889 when the Arkansas State Supreme Court struck down a city ordinance which automatically declared the bee a nuisance (Root, 1974). In reversing a lower court's decision the Supreme Court declared that neither bees nor beekeeping were in themselves nuisances, but the management on the part of the beekeeper could be so declared under varying circumstances.

That peace can exist between the beekeeper and the layman in urban areas is seen in the existence of city ordinances which are satisfactory to all parties. Seattle, Washington, the largest urban center in the Pacific Northwest, possesses such an entomologically amenable law (Luver, 1973). It has been viewed as a model city ordinance and can be found in Appendix I, given with the permission of the Seattle Mayoralty.

It is appropriate to discuss other examples of urban ordinances and their apicultural ramifications. There are beekeepers and bee colonies in the middle of even the largest cities. New York, Washington, D.C., and Long Island will serve as examples. In New York, one beekeeper opposite Central Park has several hives. MacMillan (1938), a roof top beekeeper, stated that there were "several hundred" beekeepers in the Borough of The Bronx, and he felt there was room for many more. At the time there was an active Bronx County Beekeepers association. MacMillan (1938) knew of 115 colonies within a 3.2 km radius in the city.

Today, there is an active beekeeping organization on Staten Island, while in Brooklyn an association with the pretentious name "Brooklyn Botanic Beekeeping Benevolent Brotherhood" meets regularly. The brotherhood maintains several colonies at the Brooklyn Botanic Gardens. One member has maintained two hives in Brooklyn, about 1.6 km from Wall Street, for 15 years and averages 113 kg of honey per year (Davidson, 1973). Morse (1972) stated that New York City is a poor beekeeping area. A harvest of 23 to 34 kg of honey per year might be possible, but only for a limited number of colonies.

A city ordinance prohibits beekeeping in Washington, D.C. Despite this, there are several hundred colonies actively maintained in the city, and they do rather well. A bee colony has been an exhibit in the Smithsonian Institution since 1926. Mrs. Calvin Coolidge, although a frequent Smithsonian bee exhibit visitor, insisted that a bee colony established from a swarm on the White House grounds be removed (Anon., 1972a). In 1973 bees in a tree hollow were ordered removed from the White House grounds for security reasons.

A third example of successful urban beekeeping is Long Island. Two county organizations, now a single Long Island Beekeepers Club, have been strong, active groups with large memberships. They publish a newsletter and have prepared an informative and useful leaflet on urban beekeeping (Muller, 1974). The by-laws of the Long Island Beekeepers Club list six points for being a good beekeeping neighbor. The six tenets are: 1) No more than four hives for each 0.1 ha or less, 2) no hive within three meters of a boundary line, 3) a 1.8 m hedge or partition fence between hives and neighbors if a hive is within three meters of and facing a neighbor, 4) no hive without an adequate water supply within six meters of the hive March through October (the active foraging season), 5) no hive inspected less than four times between March and November with a written record of each inspection, and 6) no hive maintained in a residential area in such a manner so that it shall constitute a substantial nuisance.

The Long Island Beekeepers Club has been active in counteracting restrictive regulations against beekeeping in Long Island communities. In a recent article, Peabody (1976), the current club president, provided a chronology of past successes and failures. The club failed in 1966 when the city of Glen Cove instituted an ordinance that made it nearly impossible to keep bees within the city boundaries. In 1973 the group was successful in helping to obtain an ordinance favorable to beekeepers in Islip, Long Island. This ordinance has served as a model for other municipalities on Long Island including Oyster Bay, Hempstead, and most recently Patchoque (Peabody, 1976). Peabody provides valuable insight into the reasons for the club's success and suggests how others may gain favorable and equitable ordinances.

VI. URBAN BEE COLONY MANAGEMENT

Honey bees do not require frequent attention to insure their survival. Up to a point, management input will correspondingly increase returns. Hobbyists as a group tend to "play" with colonies, disturbing them more frequently than necessary and then often not performing the proper management at the best time.

The traditional objectives of securing the largest possible honey crop and/or providing efficient crop pollination may be of little concern to the urban apiculturist. The urban beekeeper's objectives may be to simply enjoy and experience the bee's biology while obtaining only enough honey to satisfy a single household and for distribution to friends and associates. Large honey surpluses may present problems and detract from the total enjoyment of the hobby. At some point the urban beekeeper may wish to switch from hobbyist to sideliner, which would mean different management procedures to meet different objectives.

The primary goal in bee colony management is to have the greatest population of bees of foraging age at the time of the major honey flow. Although seemingly simple, this ideal is difficult to achieve. Progressive beekeeping involves anticipation rather than reaction and hence the greatest management skills are required when anticipatory time is the shortest. Whether urban or rural, the principles of bee colony management are the same, but the application may differ.

It is necessary to master several management skills to control the population increase inherent in a healthy honey bee colony. Colonies can be stimulated to build more rapidly in population in a number of ways. Early spring feeding of sugar water (carbohydrate) and pollen supplement (protein) will stimulate a higher oviposition rate by the queen. Reversal of hive bodies or the addition of empty supers will permit the bees to expand upward (their preferred direction of expansion) leading to larger populations. Perhaps the single most important factor in colony population dynamics is the queen's ability to produce eggs quantitatively. The skillful beekeeper learns to gauge a queen's egg laying capacity with regular inspections.

A difficult skill to acquire is that of equalizing colony strength to simplify management and to keep large colonies intact and productive. As colonies normally expand in the spring, conditions become favorable for asexual colony division (swarming). Although the causative factors in swarming are not clear, an important feature is overcrowding in the brood area of the hive (Caron, 1972). No reliable detection methods exist to determine which colonies are preparing to swarm until swarming preparations are well underway. Techniques for reversing this behavior are time consuming and employ concepts difficult to grasp. Many beekeepers lack the necessary understanding or manipulative skills to adequately deal with swarming. In the urban environment, management is ironically complicated by the fact that effective swarm control may result in large colonies, which are more difficult to handle, or a greater number of colonies, resulting in an over-population of the urban apiary.

A swarm of honey bees, although normally not aggressive, is generally viewed with alarm by most laypersons (Haff, 1974). Swarms are usually not very difficult to capture (Avitabile and Kasinskas, 1974). Some urban beekeepers take great pride in performing civic duties in swarm retrieval. They leave their names and telephone numbers with police, firemen, extension agents and others who receive calls from citizens with uninvited swarms on their property. Beekeepers usually collect swarms at no charge. However, as Ebeling (1975) points out, beekeepers sometimes remove their names from swarm retrieval lists after capturing one or a few swarms due to space limitations and other factors. Late season swarms may go uncollected for want of a beekeeper willing to assist.

An increased public awareness of pesticides has also led to some problems for beekeepers desiring to perform public service in swarm removal. In Maryland a legal problem surfaced in 1976 regarding the state pesticide applicators law. Maryland law is more broadly written than federal law and regulates pest control advice and devices as well as pesticides. Beekeepers who took swarms, gave advice or charged fees, were included under this law, and they needed to be registered, pay a fee, take a test and have insurance coverage. Recently, a committee of state beekeeping association members received a favorable ruling to exclude beekeepers who capture swarms or remove stinging insects from houses, trees, etc., without pesticides (Caron, 1976b). Several beekeepers can meet this requirement by using a vacuum cleaner for pest removal or by the use of traditional methods involving exposure of feral nests and transfer of comb.

In addition to swarming there are other aspects of urban bee management that differ from the more traditional rural management.

A. Water Sources for Bees

Bees cannot be kept confined in their hives because of their need to visit flowers for the collection of food materials - pollen and nectar. The nectar they collect is "ripened" into honey. Before using honey, which is less than 18.6% water, for feeding larval bees, the adult bees dilute it with water which they collect outside the hive. Unlike honey and pollen, water is not stored in the hive. It must be collected at the time it is needed. Water is a continuous necessity during the active foraging season, and it is a biological factor that must be considered by the urban beekeeper.

It is extremely difficult to modify water collecting behavior of bees once it is established. Thus is not always possible to change water collection from nearby bird baths or swimming pools to more desired sites. If trays or pans of water are placed near bee colonies in early spring, bees will usually forage at these close sites rather than fly to farther water locations (Greve, 1974; DeJong, 1975). Floats in open pools should be provided to assist in minimizing the number of bee drownings (Owens and McGregor, 1964).

B. Gentleness of the Bee

It is imperative to maintain gentle stock in the urban environment (Root, 1974). Although our knowledge of bee genetics is not extensive, especially as it pertains to aggressiveness (Kerr, 1974), we do know that bee colonies vary greatly in their individual temperament. Temperament is both genetically and environmentally influenced. In the U.S., two hybrids have been developed by a commercial company which are advertised as gentle. Both of these proprietary hybrids, "Starline" and "Midnight," as well as the standard Italian, Caucasian and Carniolan varieties, are generally considered gentle and nonaggressive under most circumstances.

When a bee colony becomes aggressive in the urban environment it should be removed or requeened with a queen from known gentle stock. Requeening is a difficult management skill to master even when dealing with gentle colonies. Numerous procedures and techniques are available for requeening (Johanssen and Johannsen, 1971; Caron, 1975), and the urban beekeeper should insure that only colonies of gentle temperament are kept in urban apiaries.

C. Colony Inspections

The urban beekeeper has a responsibility to practice the best available management techniques. There is no one way to manage bees - there are several good methods with hundreds of variations. The Long Island Beekeepers Club requires that its membership inspect a hive not less than four times between March 1 and October 31 each season (Muller, 1974). In addition to a thorough inspection for food stores, brood pattern and the absence of brood diseases each spring and fall, beekeepers need to control swarming and add or remove honey supers. Additional inspections can be performed to insure that the queen is healthy and brood rearing is progressing normally.

During colony inspections the urban beekeeper should insure that the bees are not agitated. Colony inspections should only be done on sunny days when the bees are actively foraging. When opening and examining colonies proper use of the smoker is important. The smoker is a vital tool, as smoke disrupts normal defensive behavior of the guard bees in a colony. Smoking causes a large number of bees to engorge on honey. This displacement activity normally results in gentle bees (Newton, 1968, 1969). Too much smoke as well as too little is undesirable.

Gloves are used by many urban beekeepers, but their use can lead to clumsiness and carelessness which will result in more stings. With each sting, alarm pheromones are released which aggrevate the stinging behavior on the part of other bees. Koover (1971) and others suggest the use of plastic gloves (such as the gloves used for washing dishes) that will give better control while still offering some sting protection. Bee stings serve the beekeeper with a biological warning that the colony is being mishandled. It is recommended that gloves not be used by urban beekeepers.

Urban beekeepers should always endeavor to control robbing. Robbing involves the foraging for honey from exposed combs or from other colonies. Bees that rob and bees in the colony being robbed become very aggressive, and searching behavior by robbing bees can quickly lead to a problem with neighbors. Bee combs with honey left exposed during periods of pollen and nectar dearth can quickly bring about robbing. Removal of honey, which involves exposing honey combs, can be an especially critical management practice. Any technique that avoids exposure of honey to robbing bees should be used when examining colonies and particularly when removing honey from colonies.

Another management technique that the urban beekeeper will perform is that of feeding supplemental carbohydrate (sugar) to bee colonies that are low on food stores. Feeding of sugar syrup or a dilute honey solution stimulates a colony to rear brood. Feeding may be necessary to avoid starvation or as a stimulant to colony growth. It is also important to feed colonies newly established from packages of bees. The feeder usually recommended to beginning beekeepers is the Boardman; however, it is not the best type to use in an urban situation. The Boardman feeder inserts into the front entrance of the hive. Its use leads to searching behavior by bees when nectar sources are not usually plentiful, and this may cause problems for the urban beekeeper (Greve, 1974). In urban apiaries, it is best to use other systems for feeding sugar, such as a feeder pail that is placed above the frames with an empty hive body placed around it. This system does not lead to searching behavior outside the colony.

D. Vandalism

Vandalism is an increasing problem for beekeepers everywhere. Theft, knocking over the bee colonies, stealing of honey and similar colony distrubances are all of concern to urban beekeepers. Such colony disturbances create the potential for robbing and aggravating bees. Vandalized colonies are difficult to realign, and bees from distrubed hives can alarm and sting neighbors before the problem is corrected.

Colony or honey theft may be a problem for the urban beekeeper who attempts to maintain suburban or rural apiaries. Such locations are often not as well protected or looked after as the home residence apiary. Anyone with a little knowledge of bees can easily take honey from a colony or steal entire hives in a very short period of time. Marking equipment may not prevent such pilferage, but it may help prevent casual or compulsive thievery. On the other hand, honey bees may help avoid vandalism problems in the urban environment. In a survey of Maryland beekeepers, Caron (1970) found that a small percentage of beekeepers listed protection as a reason for starting with bees. In suburban Washington, D.C., geologists used active and inactive bee hives to protect field equipment used in experiments in a wooded park area. Unprotected equipment was continually vandalized, while active bee hives protected data recording instruments and helped make record collection possible (Phipps, 1972).

E. Public Relations

Good beekeeping by the urban apiculturist involves some public relations work. Martin (1975) and others believe that beekeepers who permit their bees to become a nuisance force communities to institute restrictive ordinances. Haydak (1968) recommended consulting neighbors before starting with bees. Neighbors should receive honey each year (Greve, 1974). The story of the value of the honey bee should be related to civic, community and school groups (Connor, 1976; Peabody, 1976). Observation bee hives at fairs, schools or at other areas of public activity draw a good deal of attention and can be valuable educational and public relations tools.

Public relations start at home with the immediate neighbors and should extend outward as time and energy permit. Many persons with a rural background know and can appreciate the value of honey bees. However, persons with an urban background, especially the young, are not familiar with bees except for the negative aspect of the sting. Good public relations can help change this.

F. A Unique Bee Problem

Honey bees can cause an unusual and unique problem in urban areas as regards their scatologic behavior. After periods of confinement to a hive (a common occurrence in winter and early spring) fecal matter builds up in the gut of individual bees. On warm, sunny days these bees fly away from the hive to void their fecal matter. The excrement may land on vehicles, houses, or on laundry hanging outside to dry. The brown spots, especially on laundry can create a mess (Root, 1974).

Fecal spotting is not as serious a problem as it once was. Home use of clothes dryers has decreased the need to hang out the laundry. Bees used to be wintered in cellars or packed with wrapping paper for wintering more extensively than is now done. Spring flying to eliminate fecal matter was very heavy under such circumstances (Root, 1974). There is little that can be done about fecal spotting where it does occur, short of moving the winter locations.

VII. PRODUCTIVITY POTENTIAL

Three phases of a marked seasonal developmental cycle in bee colonies can be distinguished. The first phase is characterized by rapid population expansion. Early pollen sources are required to sustain this growth. The second phase, that of the nectar flow, is the time when the colony has a large population and sufficient flora available to enable bees to collect large surpluses of nectar to convert to honey. Geographical locations vary tremendously both in the number of nectar flows during spring and summer and in the quantitative intensity of nectar production during the flows.

The third phase is the fall storage period. This is different from the earlier nectar/honey flow phase in that nectar and pollen collected during the fall are required for the subsequent non-forage season, i.e., late fall and winter. The nectar will hopefully be of sufficient quantity for the winter, and it must be fully ripened to enable bee colonies to perform the necessary winter function of cluster thermoregulation without large accumulations of fecal matter in the gut of individual bees. Since pollen quickly loses nutritional value in storage, fall flowering plants ideally should have abundant pollen with an adequate protein content for the rearing of larval honey bees when brood rearing commences in early spring.

The urban environment differs from the rural or agricultural environment in all of these three phases. The buildup phase begins earlier because of temperature and moisture modifications in the urban environment. Urban areas are also characterized by more plant species, but numerically fewer individual plants. Large, unifloral nectar flows are practically non-existent in the urban environment because the concentration of flowering plants is smaller in urban settings. Finally, the fall storage flow is smaller because weed species, which are present in rural areas (e.g., *Solidago, Aster, Bidens*), are not as common in urban areas. Thus the urban beekeeper has fewer problems with spring buildup but greater problems in securing surplus honey or in obtaining adequate winter stores.

Trees are important pollen and nectar sources for the urban beekeeper. In the eastern U.S. maple, primarily sugar (*Acer saccharum* Marsh.) and red (*A. rubrum* L.), and elm (*Ulmus americana* L.), are valuable sources of pollen in early spring. Many other wind pollinated trees such as alder (*Alnus*), ash (*Fraxinus*), oak (*Quercus*), redbud (*Cercis canadensis* L.) and aspen (*Populus*) are secondary pollen sources.

Some urban areas do have a meaningful nectar flow and tree species are usually responsible. Along the eastern seaboard, for example, tulip poplar (*Liriodendron tulipifera* L.) is a major nectar source (Phillips and Demuth, 1922). Black locust (*Robinia pseudoacacia* L.) is another, although it is not a reliable nectar species. Basswood or linden (*Tilia americana* L.) is another major nectar source. In Florida and the southwest, citrus (*Citrus*) trees in urban areas provide a major nectar flow. There are other trees of more limited importance such as sourwood (*Oxydendrum arboreum* DC.), holly (*Ilex*), catalpa (*Catalpa*), sumac (*Rhus typhina* L.), buckeye (*Aesculus*), mimosa (*Albizzia julibrissin* Durazz.), and willows (*Salix*) (Cory, 1930). Fruit trees such as apple (*Malus*), pear (*Pyrus*) and cherry (*Prunus*), should also be mentioned as minor sources, but they usually bloom so early that the nectar is used by the colonies for population buildup rather than honey storage.

One other plant that can be described as a surplus nectar source in the urban environment is white Dutch clover (*Trifolium repens* L.). In northern U.S. cities, white Dutch clover thrives with constant cutting and watering. It blooms for a longer period than it would in an agricultural environment (Root, 1974), and it provides a generous surplus honey plant for many urban beekeepers.

A large number of ornamentals are useful to the bees as pollen and/or nectar sources. Although rarely found in large concentrations, a succession of ornamental blooming species often extends throughout the active foraging season when colonies are strong, and these plants can collectively lead to surplus honey production. Some that can be mentioned include barberry (*Berberis*), bush honeysuckle (*Diervilla lonicera* Mill.), firethorn (*Cotoneaster pyracantha* [L.] Spach), heather (*Calluna vulgaris* [L.] Hull), English ivy (*Hedera helix* L.), hawthorn (*Crataegus*), *Clethra* and thistle (*Cirsium*).

There has been some effort to identify ornamentals and other plants useful to the urban beekeeper. One seed supplier in Iowa, Pellett Gardens, specializes in honey plants, and their catalogue gives information on attractiveness of some plants to honey bees. Kroes (1951) listed the plantings by the beekeepers of Amsterdam, Netherlands, who established a bee park outside the city limits to keep bees (which were forbidden in the city). The colonies still had to be moved to other crops for surplus honey. Abts (1974) outlined efforts underway in Germany to improve bee pasture conditions in urban and industrial areas. Steche (1975) discussed the effects of industrial development on beekeeping in general terms.

One other potential honey source should be mentioned. In some sections of the world honeydew is collected by bees and stored as surplus honey. Bees collect the honeydew, actually the sugary excretions of aphids and certain other Homoptera, and ripen it into a product very similar to floral honey. It is a preferred product in many parts of Europe but is not widely accepted in the U.S. Some urban locations allow for surplus honeydew production. Most often it is left in the colonies to be used as food for winter.

In addition to securing a honey crop for beekeepers in urban areas, honey bees perform a valuable service by pollinating plants. Honey bees contribute to the attractiveness of ornamentals by enabling them to set fruit or produce seed that enhances their beauty and usefulness (Anon., 1970; Lecomte, 1975). Birds and other animals benefit from such unplanned pollination (Martin, 1973). McGregor (1976) lists a number of wild flowers and ornamentals dependent upon insect pollination.

With the growing interest in home vegetable gardening, honey bees are equally of importance in their pollination activities. The cucurbits must have insect pollinators to produce fruit. Most stone fruits except peaches are likewise insect dependent. Other foods commonly grown in home gardens produce higher yields with insect visitation. Berries, beans and many of the herbs fit this category. Many vegetables require bee pollination for seed production.

Fall nectar plants are lacking in some urban areas. The rural beekeeper relies upon a limited number of weed species that may or may not be numerous in the urban environment. Some late blooming ornamentals may help but there are seldom large concentrations of such plants. For proper wintering the urban beekeeper must leave more of the summer surplus or feed sugar water to his colonies.

VIII. HONEY BEES AS PESTS

Honey bees are not universally viewed as beneficial or desirable in the urban environment. Feral nests in a dwelling or human habitation area deserve pest designation and require control. For individuals with sting allergies, honey bees constitute a health hazard. Both aspects are considered below.

A. Honey Bees as a Health Hazard

Female honey bees possess a sting, a modified ovipositor, that is used in colony defense. Stinging behavior is evoked in colony defense or can be initiated when an individual bee is confronted by "surprise" outside of the hive context. As reported earlier, Parrish (1963) found an average annual U.S. mortality of 22.9 individuals from bee, wasp and ant stings for the period of 1950 to 1959. Slightly more than half of the total number of deaths during the period of study, 124, were attributed to bees and only four were caused by ants. Ebeling (1975) labels the honey bee as the worst offender among stinging insects.

On the other hand, Barr (1974) attempted to have his patients identify the insect that stung them. Of 249 patients, 35% could identify the offending insect with certainty; 34% could questionably identify the insect and 31% could not. Of the insects identified, yellowjackets (*Vespula* spp.) accounted for 47% of the stings, honey bees 27%, wasps (*Vespula* spp.) 14%, bumble bees (*Bombus* spp.) 6% and hornets (*Vespula* spp. or *Vespa* spp.) 6%. Barr points out that identification of the honey bee as the stinging insect is aided by the stinger remaining at the site of envenomation.

Death by envenomation by stinging Hymenoptera is usually very rapid. Of 208 deaths, 80% occurred in less than one hour (Parrish, 1963). Less than a dozen were the result of a massive stinging attack; most were allergic responses. Such responses have been shown to affect from four to eight percent of various population samples (Parrish, 1963; Chafe, 1970; Settipane *et al.*, 1972). Unfortunately, at present the diagnosis of a sting allergy is possible only by case history analysis (Lichtenstein *et al.*, 1974).

Once diagnosed, an insect sting allergy can be treated. A physician can prescribe an insect-sting kit consisting of a needle and syringe of epinepherine, two antihistamine tablets, a constriction bandage, a sterile alcohol pad, instructions and medic-alert information. In a few minutes individuals can be properly trained to treat themselves. Such "self-help" kits should be used cautiously and only if trained medical assistance is not readily available.

For permanent relief it is possible to treat sting sensitive patients with whole-body extracts of the various stinging insects (Loveless, 1976). Such treatments have not worked in all instances, and researchers now have federal permission to use pure venom extracts on an experimental basis for treatment of sting sensitivity. Results to date have been positive and most promising (Lichtenstein *et al.*, 1974, 1976).

Sting deaths from the "Africanized" bee deserve special mention. Taylor and Williamson (1975) label the bee as "not a general health hazard." Goncalves *et al.* (1974) point out that the number of deaths attributed to bee stings in Brazil is not any greater than the death rate due to bee stings in the U.S. on a per capita basis. As an aggressive bee, it does seem to attack *en mass* more frequently than bees which are kept in the U.S. But these attacks are associated with colony disturbance.

Mass envenomation for the non-allergic individual would likely produce more discomfort than one or a few stings, but would not likely be life threatening until 500 or more stings were received (Frazier, 1964). In one documented instance an individual received 2,443 stings and survived without any ill effects (Murray, 1964).

Although studies indicate that four to eight percent of the population can be expected to display an allergic response to stinging Hymenoptera, a much greater segment of the population believes that it is allergic. Confusion results from the unpredictable amount of swelling that may result from a sting. The degree of localized swelling depends upon such variables as location of sting, age of bee, physical condition of sting recipient, amount of venom injected and other factors. A large amount of swelling is frequently diagnosed by layman and physician alike as an allergy (Frazier, 1969).

In addition to allergies and discomfort from stings, honey bees are viewed by some individuals as life threatening *in extremis*. Such entomophobic reactions were recently reviewed by Olkowski and Olkowski (1976). These fears can be and are very real. Individuals will make tremendous alterations in their life styles because of them.

Recently, Crane (1976) illustrated with examples the range of human attitudes towards bees. The apiphilics are beekeepers; some go to extremes such as wearing a "bee beard" - a mass of thousands of bees clustered around their caged queen which is held below the apiphilic's chin. At the other extreme are the apiphobics. An extreme example of such a condition is a recent report of a 28 year old male who died within an hour of mistakenly believing he had been stung, but who in fact was "literally frightened to death" (Crane, 1976). There are all ranges of intermediate attitudes between the apiphilic and the apiphobic.

Honey bees and other stinging insects have been known to cause vehicle accidents (Ebeling, 1975). Once inside a vehicle, bees and most other insects will fly to the windows. The driver and/or other vehicle occupants that panic may cause an accident. Such mishaps are undoubtedly more common in the urban environment where vehicles would be starting, stopping and moving at a slower speed which would allow easier inadvertent vehicle access to bees. Stings are not common if the insect is left alone under such circumstances. The best technique for removing the insect is to bring the vehicle to a safe stop and open one or more windows to permit its escape.

B. Swarms

As mentioned earlier, a swarm of honey bees is often viewed with alarm by the urban homeowner or citizen. Although an integral part of bee biology, swarming is a phenomenon not understood by the layman. Swarms are usually very gentle as bees in a physiological swarm condition are little inclined to sting. If left alone, the swarm will nearly always find a new nest cavity and move from its temporary swarm cluster location. They may remain as a swarm cluster for a short time or for several days. Disturbance of the cluster, such as attempts to break it up or disband the swarm, or adolescent disturbances such as rocks and sticks, can and often will lead to stings.

The easiest way to handle a swarm is to contact a local beekeeper. Lacking a beekeeper, it is best to leave it undisturbed. In nearly all instances the swarm will relocate and move to a new home site.

Techniques to capture swarms are varied. Swarms are easily "hived" by a beekeeper who is familar with their biology. Any technique that insures capture of the queen will result in successful swarm collection. Under no circumstances should anyone attempt to capture one that has been clustered on a site for three days or longer. Such swarms may be low on food reserves, by having depleted the communal stomach common to swarms (Combs, 1972), and may have established normal colony defensive behaviors. Any disturbance can lead to multiple stings. Such swarms are often termed "dry."

When necessary, swarms can be destroyed using insecticides. Compounds labeled for bees and wasps can be delivered by aerosol, air pressure sprayers or dusted in some manner onto the bee cluster. Repeated applications may be necessary. Some individuals attempt to use fire or chemicals such as kerosene or gasoline to eliminate the bees. Such compounds are not recommended as they are not effective for rapid knockdown and can result in angry bees and multiple stings.

If a swarm is unable to locate a new home site they may build their nest at the cluster location. Such sites are usually somewhat darkened which enables the bees to efficiently produce beeswax for their comb (Morse, 1965). Such outside nests are frequently discovered in fall when the foliage begins to thin. These established nests cannot be captured in the same manner as swarms. Since the initial discovery is usually late in the season, it is frequently too late to transfer such nests into bee hives and successfully over-winter the resulting colony, because of the shortage of food stores. Nearly all outside nests will die out over the winter if left undisturbed.

Occasionally, a swarm of bees may be discovered just as it is moving into a tree hollow or into the wall void of a building. If the queen has not yet moved into the cavity, it may be possible to capture the swarm. In such instances, brushing bees and the queen into a hive box will result in the release of pheromones from the abdominal Nasanov gland by workers. This "assembly" pheromone will cause those bees which have previously moved into the cavity to abandon it and rejoin that portion of the swarm containing the queen. However, if the queen has already moved inside the cavity, there is nothing to be gained by brushing the remaining worker bees into a hive.

C. Removing Bees from Tree Hollows and Wall Voids

In the urban environment, honey bees readily select wall voids, attics, unused chimneys and similar protected locations as nest sites. They also utilize hollows in trees. Under most circumstances feral colonies in such locations should be considered pests. Colonies nesting in wall voids or other available cavities in houses may result in adult bees finding entrance to the interior rooms of the house. Additionally, the honey storage combs may result in honey stains on walls. Such bee nests also serve as an attractant to other insects.

Removal of feral colonies from such "unnatural" sites is not particularly easy. Most techniques for their removal or destruction involve a good knowledge of honey bee biology and a familiarity with colony management. Only competent beekeepers should attempt to handle such situations.

Several techniques are available for removal of bees in unwanted locations. The exact method will vary with individual situations. Caron (1974) and Youngs and Burgett (1975) have outlined several methods for removing bees from undesirable habitats. Most pest control operators possess the skill and necessary equipment to remove bees and their nest.

Often it is preferable to remove a feral colony without killing the bees. It is possible to do so under some circumstances, but the techniques are not recommended as a general practice. If a nest is in a situation where the siding can be removed and the entire nest exposed, a beekeeper can transfer the bees to hives without having to kill them. The exposed comb is carefully cut and placed in wooden frames. Sections of comb are held in place by string or rubber bands. This transfer technique is messy, may lead to stings and usually results in the death of a number of adult bees. Upon removing the total nest, the wall void should be thoroughly cleaned and filled with some type of insulation material and the siding replaced. This should prevent reestablishment of a new nest of bees at another time.

Another technique involves trapping adult bees outside the colony entrance. Using this method, a beekeeper closes all entrances to the nest except one. Over this entrance is placed a funnel constructed of wire screen. Funnel length should be slightly in excess of 25 cm and should end with a small opening approximately one centimeter in diameter. With the funnel's large end attached over the nest entrance, a dummy hive containing comb is positioned near the small opening of the funnel. As adult workers leave their nest through the funnel they have difficulty in reentering and will drift to the nearby dummy hive.

During the first few days after the funnel is in place, a large number of foraging bees are trapped outside. At this time the dummy hive should be given a queen and/or some frames of brood. The dummy hive now becomes a small, viable honey bee colony that will gradually increase in size as more bees are trapped out of the feral hive. After a period of one or more months, the feral colony will be considerably weakened and the newly established colony can then be moved.

Although the adult population of the feral nest is removed with this system, the queen will not leave and honey, wax combs and some brood will still be in the nest. It is often easiest to let the colony die out naturally over the winter, then remains of the nest can be removed in the manner described earlier.

Occasionally, feral colonies will be found in abandoned boxes, bird houses and other structures that are easily removed *in toto*. These should be taken at least five kilometers from where they were found. This will prevent older foragers from returning the the original location. In transferring bees from these nests the usual procedure is to create an opening in the top of the structure and place a standard hive body, with comb, on top of it. Over time the bees will move up into the standard hive. Once the queen has moved up, a queen excluder can be placed between the new hive and the old one. Eventually the brood will hatch out in the nest below. Bees will move up to the new brood area and the old nest and its structure can be removed.

Berto (1971) used the above technique to successfully remove bees from a bird house. Fisher and Caron (1974) discussed transferring of bees from nonstandard hives and included a discussion of the technique of drumming. When a colony is rhythmically pounded with a rubber hammer or some other instrument, the vibration will cause the bees to walk upwards in the hive. By placing a standard sized hive above a "wild-type" hive and then rhythmically drumming, one can transfer adult bees in a very short period of time. When adults have left the feral hive, it can be opened and the combs cut out and placed in frames as discussed earlier.

A feral nest of honey bees can be eliminated with an insecticide. Nearly any compound will kill the bees if applied directly on the bees or their comb. The best materials will supply a rapid knockdown, affect large numbers of the population and have a residual effect to eliminate newly emerging adults for a few days.

The insecticides chlordane, lindane and DDVP have been widely used in the past to control bees when they have assumed pest status. These materials are no longer available for this use. Replacement compounds such as Baygon, malathion and chlorpyrifos are as effective. The insecticide may be applied as an aerosol or with any type of spray equipment available. Commercial aerosols for bee and wasp control are available and are usually recommended for the urban homeowner with a bee, wasp or hornet problem.

Insecticide dusts may also be used. The compound, whether dust or spray, needs to be applied to the nest entrance and the interior of the colony for the most effective control. Repeated applications are often necessary. Pest control operators prefer to use a compound that eliminates the problem on a single visit. To insure that the insecticide reaches the nest interior, small holes must often be drilled through the interior or exterior walls for introduction of the insecticide.

After the adult population of a colony has been eliminated it is very important to remove the wax combs containing honey and brood. If bees are removed within a few weeks of their establishment of a feral nest, comb removal is usually not necessary because they haven't had sufficient time to build much comb or store honey. When the adult population of a nest is removed, honey cannot be properly cared for and the combs will begin to break up. The combination of dead brood and associated odor, and rancid dripping honey, are certainly viewed with disdain by the homeowner (Pence, 1955).

Additionally, other insects and pests may be attracted to

the unprotected nest. Ants, wax moths, carpet beetles, flies and cockroaches may all invade such unprotected colonies and result in a rapid destruction and elimination of the comb, brood and honey. Some of these scavengers may necessitate additional pest control action. For example, Happ (1966) observed wax moths pupating beneath wall paper, resulting in oval lumps.

Honey that stains interior walls presents a special problem. Since honey is hygroscopic it never dries. It is impossible to paint or wallpaper over honey soaked walls without the stain reappearing. Thus, it is always best to remove all the nest after the adults are eliminated.

D. Africanized Honey Bees

During the past decade a great deal of nonscientific reporting of the introduction and spread of African bees in South America has created a problem for apiculturists in both North and South America. A great deal of the publicity has been sensationalized, and the problem is neither as bad nor as good as some would label it (Jaycox, 1976).

This honey bee, often assuming the monikers "Killer," Brazilian" or "African," but most correctly Africanized, is now found in South America as the result of an importation which went awry. A geneticist, hoping to improve the productivity of honey bees in Brazil, imported 47 queens from Africa of the aggressive African bee subspecies *Apis mellifera adansonii*. Up to 26 of these queens, and accompanying swarms, accidentally escaped from an experimental apiary. This small genetic introduction has since hybridized with the honey bees already present in South America. The Africanized bees appear to have retained many of the undesirable characters of *A. m. adansonii* in its native state.

In South America the bee has been reported to spread northward at a rate in excess of 320 km per year. A team of North American scientists visited South America to assess the situation relative to aggressiveness, northward migration and potential impact on beekeeping in Mexico, the U.S. and Canada. Their report (Anon., 1972b) predicted the eventual spread of the Africanized bees to the U.S. Recently, Taylor and Williamson (1975) have reexamined the rate of dispersal, and they feel that it may take longer than originally predicted to reach North America.

Africanized honey bees are more difficult to manage than those subspecies with which North American beekeepers are familiar. The bee has a greater tendency to swarm and respond quickly to colony disturbance. However, beekeepers can manage the bee and as Goncalves *et al.* (1974) point out, the number of deaths attributed to the Africanized bee is not proportionally different than North American sting fatalities.

Africanized bees have been introduced previously into the U.S. and Europe. Morse *et al.* (1973) wrote of its introduction to Louisiana and Poland without adversely affecting bee populations in those areas. Stinging incidents and fatalities will undoubtedly continue to produce reports and confusion as to whether the bee has established itself in other areas, as for example Long Island (Peabody, 1976). If the Africanized bees from South America reach North America, there is enough time to enable beekeepers and scientists to evolve means of managing the bee as is apparently happening in South America (Michener, 1973; Taylor and Williamson, 1975).

IX. THE FUTURE OF URBAN BEEKEEPING

McGregor (1976) alludes to the changing beekeeping population of that from rural to urban. Each year the agricultural arable land base in the U.S. diminishes (Pimentel *et al.*, 1976). This influences beekeeping not only in the demographic transition of beekeepers, but also in the loss of forage pasturage for the bees. In a recent evaluation of the U.S. beekeeping industry by the U.S. International Trade Commission (U.S.I.T.C., 1976), it was shown that the average honey production per colony has begun to plateau after a 25 year period of increase. On the other hand, yearly decline in total number of honey bee colonies has stopped, and it is estimated that in 1976 there were five percent more colonies than in 1975. Some of this increase in colony numbers is attributed to the increased price of honey during the last six years, but also to the renewed interest in honey bees especially on the part of hobbyists (U.S.I.T.C., 1976).

One of the greatest challenges facing urban apiculturists is that of their inherent right to pursue beekeeping as an avocation. The beekeeper must remain aware of the possibility of statutory restrictions that could prohibit or severely limit beekeeping in the urban environment. Perhaps the greatest safeguard to such prohibitive legislation is for urban beekeepers to join together and organize active local beekeeper associations that will promote good bee management and an urban bee policy.

As commercial bee colonies have become more centralized, we are beginning to see a disruption of adequate pollinator densities in various agricultural areas (McGregor, 1976). To maintain satisfactorily high yields from those crops requiring insect pollinators, this situation will have to be ameliorated. The urban, hobby beekeeper may become involved in this insect-plant-man trinity.

The honey bee as an integrated biological entity performs an important ecological function in any environment, even in the vastly artificial urban one. The professional farmer and hopefully the home or backlot gardener are aware of their interdependence with the beekeeper.

REFERENCES

Abts, W. 1974. Bienenweideverbesserung in Grosstädten und Industriebereichen. *Rheinische Bienenzig 125*: 236-239.

Anonymous. 1970. Bees in towns and cities. *Glean. Bee Cult. 98*: 209, 251.

Anonymous. 1972a. Bees in the attic. *Am. Bee. J. 112*: 420.

Anonymous. 1972b. Final report, committee on the African honey bee. Nat. Acad. Sci., Washington, D. C.

American Honey Producers' League. 1924. A treatise on the law pertaining to the honeybee. Am. Honey Prod. League, Madison, Wisconsin.

Anderson, E. D. 1969. An appraisal of the beekeeping industry. USDA-ARS 42-150.

Avitabily, A., and J. Kasinskas. 1974. How to find, capture and hive a swarm of honey bees. *Am. Bee J. 114*: 181-183.

Bailey, N. E. 1975. A beekeeper's right to keep bees lawfully on property over a neighbor's objections. *Am. Bee J. 115*: 474-475, 490.

Barr, S. E. 1974. Allergy to Hymenoptera stings. *J. Am. Med. Assoc. 228*: 718-720.

Berto, V. J. 1971. Birdhouse bees. *Glean. Bee Cult. 99*: 296-297.

Blanchard, O. K. 1956. Rights and responsibilities of beekeepers. *Glean. Bee Cult. 84*: 658-664.

Burr, L. 1959. Laws relating to bees. *In* ABC and XYZ of bee culture. A. I. Root Co., Medina, Ohio, pp. 460-466.

Caron, D. M. 1970. Survey of Maryland beekeepers. Maryland Ext. Serv. Mimeo.

Caron, D. M. 1972. Swarming - its prevention and control. Maryland Entomol. Leafl. 77.

Caron, D. M. 1973. Ten tips for suburban beekeepers. Maryland Ext. Serv. Leafl. 75.

Caron, D. M. 1974. Removing bee swarms/bees from buildings. Maryland Ext. Serv. Leafl. 76.

Caron, D. M. 1975. Queen introduction. Maryland Ext. Serv. Mimeo.

Caron, D. M. 1976a. Locations for bee colonies. Maryland Ext. Serv. Mimeo.

Caron, D. M. 1976b. The pollen basket. Maryland Ext. Serv. Newsletter.

Chafee, F. H. 1970. The prevalence of bee sting allergy in an allergic population. *Acta Allergol. (Kbh) 25*: 292-293.

Combs, G. F. 1972. The engorgement of worker honeybees. *J. Apic. Res. 11*: 121-128.

Connor, L. J. 1974. Observation bee hives. Ohio State Univ. Ext. Serv. Beekeeping Information No. 10.

Connor, L. J. 1976. Promotion of honey and beekeeping. *Am. Bee J. 116*: 472-473.

Cory, E. N. 1930. Flowering plants, shrubs and trees for the beekeeper. *Rep. Maryland Agric. Soc. Maryland Farm Bureau Fed. 15*: 348-351.

Crane, E. 1975. History of honey. *In* Honey: A comprehensive survey (E. Crane, ed.). Heinemann, London, pp. 439-488.

Crane, E. 1976. The range of human attitudes to bees. *Bee Wld. 57*: 14-18.

Davidson, R. T. H. 1973. Personal communication to H. Wenzel, Islip, New York.

DeJong, D. 1975. Keeping bees in populated areas. *Glean Bee Cult. 103*: 319-320.

Ebeling, W. 1975. Urban entomology. Univ. Calif. Div. Agric. Sci., Berkeley, California.

Eckert, J. E., and F. R. Shaw. 1960. Beekeeping. Macmillan Co., New York.

Fisher, R., and D. M. Caron. 1974. Box hive transfer. *Glean. Bee Cult. 102*: 277, 290.

Frazier, C. A. 1964. Insect allergy. Warren Green, Inc., St. Louis, Missouri.

Frazier, C. A. 1976. Insect stings - a medical emergency. *J. Am. Med. Assoc. 235*: 2410-2411.

Gary, N. E. 1968. A glass-walled observation hive. *Am. Bee J. 108*: 92-94, 108, 143-144, 146.

Gary, N. E., and K. Lorenzen. 1976. How to construct and maintain an observation bee hive. Univ. Calif. Ext. Serv. Leafl. 2853.

Goncalves, L. S., W. E. Kerr, J. C. Netto, and A. C. Start. 1974. Some comments on the African honey bee. Nat. Res. Council, Nat. Acad. Sci. 1972 Mimeo.

Greve, C. 1974. Keeping bees in a city or suburb. *Glean. Bee Cult. 102*: 6-7, 26.

Haff, J. 1974. Metropolitan bees swarm too. *Glean. Bee Cult. 102*: 239-240.

Happ, J. 1966. Wax moth lumps wall paper. *Pest Contr. 34*: 108-110.

Haydak, M. H. 1968. Beekeeping in Minnesota. Ext. Bull. Univ. Minnesota Agric. Ext. Serv. No. 204.

Jaycox, E. R. 1976. Does entomology need a boogeyman? *Bull. Entomol. Soc. Am. 22*: 131-132.

Johanssen, T. S. K., and M. P. Johanssen. 1971. Queen introduction. *Am. Bee J. 111*: 146, 183-185, 226-227, 264-265, 306-307, 348-349, 384, 387,

Kauffeld, N. M., and H. Knutson. 1976. Bee culture. Kansas State Univ. Agric. Exp. Sta. Bull. 375 (Revised).

Kelley, W. T. 1975. How to keep bees and sell honey. W. T. Kelley Co., Clarkson, Kentucky.

Kerr, W. E. 1974. Advances in cytology and genetics of bees. *Ann. Rev. Entomol. 19*: 253-268.

Koover, C. J. 1971. Backlot beekeeping. *Glean. Bee Cult. 99*: 254-255.

Kroes, R. C. A. 1951. The Amsterdam bee park. *Bee Wld. 32*: 10-12.

Langstroth, L. L. 1853. A practical treatise on the hive and honey bee. A. O. Moore and Co., New York.

Lecomte, J. 1975. The bee, conservator of nature by pollination of wild plants. *Apiacta 10*: 107-110.

Lichtenstein, M. D., M. D. Valentine, and A. K. Sobotka. 1974. A case for venom treatment in anaphylactic sensitivity to Hymenoptera sting. *New England J. Med. 290*: 1223-1227.

Lichtenstein, M. D., M. D. Valentine, A. K. Sobotka, and K. J. Hunt. 1976. Diagnosis of allergy to stinging insects by skin testing with Hymenoptera venoms. *Ann. Internat. Med. 85*: 56-59.

Loveless, M. H. 1976. The sting: prophylactic venom prevents disaster. *Modern Med. 44*: 54-57.

Luver, B. 1973. Model city ordinances and how Seattle got them. *Am. Bee J. 113*: 336-337.

MacMillan, J. T. 1938. All-round Bronx bees. *Glean. Bee Cult. 66*: 145-146.

Martin, E. C. 1973. Bees and pollination in Michigan. *In* Res. report 188, fruit, vegetables, bees and pollination - now and in 1985. Michigan State Univ. Ext. Serv., pp. 23-28.

Martin, E. C. 1975. Basic beekeeping. Michigan State Ext. Serv. Bull. E 625.

McGregor, S. E. 1976. Insect pollination of cultivated crop plants. USDA Agric. Handbook No. 496.

Michener, C. D. 1973. The Brazilian honey bee. *BioScience 23*: 523-527.

Michener, C. D. 1974. The social behavior of bees. Harvard Univ. Press, Cambridge, Massachusetts.

Morse, R. A. 1965. The effect of light on comb construction by honeybees. *J. Apic. Res. 4*: 23-29.

Morse, R. A. 1972. The complete guide to beekeeping. E. P. Dutton and Co., Inc., New York.

Morse, R. A. 1975. Bees and beekeeping. Cornell Univ. Press, Ithaca, New York.

Morse, R. A., D. M. Burgett, J. T. Ambrose, W. E. Connor, and R. D. Fell. 1973. Early introductions of African bees into Europe and the new world. *Bee Wld. 54*: 57-60.

Muller, L. H. R. 1974. Long Island Beekeepers' Club guide for novice beekeepers. L. I. Beekeepers' Assoc. Leafl.

Murray, J. A. 1964. Case of multiple bee stings. *Central Afr. J. Med. 10*: 249-251.

Newton. D. C. 1968. Behavioral response of honeybees to colony disturbances by smoke. I. Engorging behavior. *J. Apic. Res. 7*: 3-9.

Newton, D. C. 1969. Behavioral response of honeybees to disturbances by smoke. II. Guards and foragers. *J. Apic. Res. 8*: 79-82.

Olkowski, H., and W. Olkowski. 1976. Entomophobia in the urban ecosystem, some observations and suggestions. *Bull. Entomol. Soc. Am. 22*: 313-317.

Owens, C. D., and S. E. McGregor. 1964. Shade and water for the honey bee colony. USDA Leafl. No. 530.

Parrish, H. M. 1963. Analysis of 460 fatalities from venomous animals in the United States. *Am. J. Med. Sci. 245*: 129-141.

Peabody, F. 1976. The survival of beekeeping in an urban environment. *Am. Bee J. 116*: 474-475, 494.

Peer, D. F. 1976. Selecting the apiary site. P.N.W. Beekeeping Shortcourse Mimeo, Washington State Univ., Pullman, Washington.

Pellet, F. C. 1938. History of American beekeeping. Collegiate Press, Inc., Ames, Iowa.

Pence, R. J. 1955. Bees and the PCO. *P.C.O. News 15*: 8-10.

Phillips, E. F. 1928. Beekeeping. Macmillan Co., New York.

Phillips, E. F., and G. S. Demuth. 1922. Beekeeping in the tulip tree region. USDA Farmers Bull. 1222.

Phipps, R. L. 1972. Watchdog bees. *Glean. Bee Cult. 100*: 267-268.

Pimentel, D., E. C. Terhune, R. Dyson-Hudson, S. Rochereau, R. Samis, E. A. Smith, D. Denman, D. Reifschneider, and M. Schepard. 1976. Land degradation: Effects on food and energy resources. *Science 194*: 149-155.

Root, A. I. 1974. The ABC and XYZ of bee culture. A. I. Root Co., Medina, Ohio.

Settipane, G. A., G. J. Newstead, and G. K. Boyd. 1972. Frequency of Hymenoptera allergy in an atopic and normal population. *J. Allergy Clin. Immunol. 50*: 146-150.

Steche, W. 1975. Industrial development and its effects on beekeeping. *Apiacta 10*: 119-124.

Taylor, O. R., and G. B. Williamson. 1975. Current status of the Africanized honey bee in northern South America. *Am. Bee J. 115*: 92-93, 98-99.

Thurber, P. F., and C. Johansen. 1974. Ten tips for suburban beekeeping hobbyists. Washington State Univ. Ext. Serv. EM 3083.

United States International Trade Commission. 1976. Honey. Report to the president on investigation no. TA-201-14 under section 201 of the Trade Act of 1974. U.S.I.T.C. Publ. No. 781.

Youngs, L., and M. Burgett. 1975. Removing bees from buildings. Oregon State Univ. Ext. Cir. 857.

APPENDIX

Appendix I. Seattle, Washington, Model Beekeeping Ordinance

"An ordinance relating to nuisances; and prescribing certain procedures for urban beekeeping.

BE IT ORDAINED BY THE CITY OF SEATTLE AS FOLLOWS:

Section 1. It shall be the duty of any person, firm or corporation having honey bees, *Apis mellifera*, on its property to maintain each colony in the following condition:

a) Colonies shall be maintained in moveable-frame hives:
b) Adequate space shall be maintained in the hive to prevent overcrowding and swarming:
c) Colonies shall be re-queened following any swarming or aggressive behavior.

All colonies shall be registered with the County Agricultural Extension Agent prior to April 1 of each year.

Activities or places not meeting these standards shall be deemed public nuisances. The Corporation Council shall maintain a civil action to abate and prevent such nuisances. Upon judgement and order of the court, such nuisances shall be condemned and destroyed in the manner directed by the court or released upon such conditions as the court in its discretion may impose to secure that the nuisance will be abated; the owner of such nuisance shall be liable for fine not to exceed one hundred dollars.

Section 2. Bees living in trees, buildings, or any other space except in moveable-frame hives; abandoned colonies; or diseased bees shall constitute a public nuisance and subject the owner to the penalties imposed by Section 1 of this ordinance.

ARTICLE 6. SINGLE FAMILY RESIDENCE LOW DENSITY ZONE

Section 6.3. Beekeeping, when registered with the State Department of Agriculture and subject to the following conditions:

1) Lots having less than ten thousand (10,000) square feet shall have not more than four (4) hives.
2) Hives shall not be located within twenty-five (25) feet of any property line except:
   a) When situated eight (8) feet or more above adjacent ground level, or
   b) When situated less than six (6) feet above adjacent ground level and behind a solid fence or hedge six (6) feet in height parallel to any property line within twenty-five (25) feet of the hive and extending at least twenty (20) feet beyond the hive in both directions."

PAST, PRESENT, AND FUTURE DIRECTIONS IN THE MANAGEMENT OF STRUCTURE-INFESTING INSECTS

Walter Ebeling

University of California
Los Angeles, California

I. INTRODUCTION

The use of the term "pest management" in place of "pest control" indicates an increased emphasis on measures other than the application of pesticides for the suppression of pests. Research has been directed primarily toward the management of agricultural pests. Pest management for crop protection provides many more options than does pest management of structure-infesting pests. Some measures that are theoretically possible in structural pest management are clearly not practicable. For example, it is unlikely that a homeowner would tolerate the "flooding" of his domicile with sterile male cockroaches as a control measure. The same may be said of the introduction of parasites or predators as natural enemies. The latter could prove to be more of a pest than the target species. A family

ISBN 0-12-265250-9

in Ohio was more annoyed by a species of hymenopterous parasite found throughout the home than by the host insect, the oriental cockroach, *Blatta orientalis* L., which occurred only in secluded areas of the basement (Edmunds, 1953). On the other hand, over an extended period a parasite might greatly reduce the severity of a structure-infesting pest in a region in which ecological conditions are favorable. According to E. C. Zimmerman, the encrytid parasite, *Comperia merceti* (Compere), practically eliminated the brownbanded cockroach, *Supella longipalpa* (F.) in Hawaii after the parasite was accidentally introduced (Roth and Willis, 1960). Parasites and predators might also be helpful against most cockroach species in subtropical and tropical areas where a significant proportion of the infestation is outdoors (Piper and Frankie, Chapt. 10, this volume).

The purpose of this paper is to discuss measures that are now employed to supplement or replace insecticides in the control of three of the most important groups of structure-infesting insects: subterranean termites, cockroaches, and domestic flies. These three groups represent, respectively, 1) insects that feed on wood and can maintain themselves indefinitely in buildings and in the soil beneath them, 2) insects that feed on human refuse and can maintain themselves indefinitely inside a building, and 3) insects that ordinarily breed outdoors and then invade buildings.

II. SUBTERRANEAN TERMITES

Subterranean termites (Isoptera: Rhinotermitidae, Termitidae, Mastotermitidae) are by far the most important insects attacking wood structures. This is a measure of the importance of their role in nature - breaking down and returning to the soil and atmosphere the enormous tonnage of dead and fallen trees and other cellulosic material that is continuously accumulating on the earth's surface. It has been man's attempt to prevent this generally desirable natural breakdown process, because of his need to preserve the wood used in his buildings, that has caused some species of termites to be his enemies rather than benefactors.

From the standpoint of prevention and management, the chief characteristic of subterranean termites is that they ordinarily must maintain contact with moist soil. When feeding on dry wood above ground they periodically crawl down to moist subterranean passages and nests, where they replenish water lost above ground. The nests of some species in some areas of the world may be in mounds that extend above ground level, but the galleries and nests remain moist. The mounds may be a considerable distance from buildings, in which case the termites reach structures via subterranean galleries. If the wood members of the infested building rest on a concrete foundation, the termites reach the structural wood through "shelter tubes" constructed from particles of earth, sand, and minute bits of wood coated with a gluey substance from their mouths and gullets (Pickens, 1934). The shelter tubes may be constructed directly from the ground to a wood member of the substructure or vice versa, but more often they are constructed against the surface of the concrete foundation. In either case the approximate location of their subterranean galleries and nests is revealed.

In countries with building codes, prevention of subterranean termite infestation depends on following these codes during site preparation and construction so as to adversely affect the subterranean environment for the termites and to deny their access to the wood members of the building substructure.

A. Site Preparation

Tree stumps, roots and wood or cellulose debris of any kind on or in the soil should be removed when the building site is graded, so as to remove potential termite colonies. The ground should have sufficient slope to enable surface water to drain away from the building and thereby avoid the damp soil that favors termites. Ideally, provision should be made for connecting eave gutters and downspouts to a storm sewer system. If the building is to have a basement, drainage tile around the outside is especially important.

B. Construction of Foundation and Substructure

Much can be done at the time of construction to retard infestation, to expose termite activity and to minimize damage and cost of control and repair. Poured concrete foundations or masonry unit foundations capped with reinforced concrete serve to expose termite activity, for termites must build tubes over the masonry surfaces to gain access to substructure wood. Before the foundation concrete sets, spreader sticks and grade stakes must be removed, and form boards and all lumber scraps should be removed from the construction sites, for buried wood scraps are an important source of termite infestation.

Door frames or jambs must not extend into or through concrete floors. The level of the bottom of the basement window well should be at least 15 cm below the nearest wood. For wooden porches or steps, any support, such as piers, should be separated from the building by five centimeters to prevent hidden access by termites. Wooden steps should rest upon a concrete base or apron extending at least 15 cm above grade.

Sills should be preservative-treated with a chemical that protects them for the life of the building, provided that cut ends are painted or soaked with a protective solution at the time of construction. Termites are then forced to build their tubes over the treated sills just as they are forced to build them over the concrete foundation. In addition, the sill itself is protected. This is a vital part of the substructure and is difficult to replace. Some may wish to use treated wood for other members of the substructure also.

Danger from termites is reduced further by providing proper clearance and ventilation between the ground and wood members of the substructure. Minimum clearance between the ground and bottoms of the joists in crawl spaces should be 46 cm, and the minimum clearance between ground and beams or girders should be 30 cm. Minimum clearance between outside finish grade and the tops of slab-on-ground foundations or tops of foundation walls in houses with a basement should be 20 cm with a least 15 cm exposed.

C. Correction of Environmental Conditions

Prevention and control of subterranean termites continues after completion of construction. All cellulose-bearing debris (forms, stakes, lumber, etc.) upon which termites could develop should be removed from under and around a building. Termites commonly infest such material when it rests on damp soil. Soil moisture for subterranean termites is optimum when the soil is continuously damp but is not saturated too long. Poor drainage and inadequate ventilation are the principal factors contributing to excessively damp soil. Remedial measures may involve surface or subsurface drainage, adjustment of soil grade in relation to the substructure of the building, correction of faulty guttering and downspouts, repair of leaky faucets, and repair of stopped sewers and drains. Excessive and careless irrigation of the area around a building should be avoided. Sprinklers should not be allowed to wet the side of a building or its foundation.

Building codes require that foundations contain vent openings that allow movement of air under the buildings. Vines and shrubs should not be allowed to cover ventilation openings. If drainage and ventilation do not reduce soil moisture sufficiently, it is helpful to cover exposed soil with heavy roofing paper. In slab construction a "vapor barrier," usually consisting of a sheet of some synthetic material impervious to water, may be placed on the ground previous to the pouring of the concrete. A gravel substrate is also helpful, providing a uniform structural base and breaking capillarity. In arid regions the advantage of low humidity is often nullified by excessive and careless irrigation of plants or lawns around the foundation. In localities of consistently high humidity, sufficient moisture may be present in various parts of the building to maintain colonies of subterranean termites that have no connection with the ground. Localized areas of excessive moisture in the superstructure may develop from leaf-filled gutters or water traps on flat roofs, leaky shower pans, plumbing leaks, or condensation from water pipes.

Concrete porches, terraces, patios, or steps are often supported on an earth fill. Such earth fills frequently harbor subterranean termites, often because carpenters have disposed of wood debris in these areas. In California, most subterranean termite infestations in dwellings originate in earth-filled extensions of the foundation. The concrete cap of the porch, terrace, etc., may break away from the foundation, allowing termites to gain access to the wood of the foundation. Remedial measures involve breaking out a strip of the concrete cap adjacent to the foundation, excavating the earth fill down to the top of the footing of the foundation, refilling with chemically-treated soil, and resealing the concrete cap to the foundation.

D. Soil Treatment

A highly effective method of isolating a building from subterranean termites is the application of insecticide to the soil. When done before and during the construction of the foundation, it is called "pretreatment." It is of particular importance when concrete slab-on-ground construction is employed, for treatment of the soil is difficult and expensive after the concrete has been poured. Concrete slab was once believed to be a termite barrier, but experience has proved otherwise. Termites can penetrate via cracks in the cap slab, expansion joints (in "floating slab" construction), and apertures around utility pipes.

The insecticides that are used for either pretreatment or remedial treatment of known infestations after construction and occupation of a building are organochlorines (aldrin, chlordane, dieldrin, and heptachlor). They are highly persistent in soil. Tests made by the United States Forest Service demonstrated that treated soil is an effective barrier for a quarter of a century or more (Johnston *et al.*, 1971; Smith *et al.*, 1972). There is neither downward nor lateral migration of these insecticides of any practical importance. Although insecticide residues in soil can be physically removed, as when soil is eroded by floodwater, this is rarely a problem in treated soil under buildings. In the United States, insecticides, dosages, and manner of application are standardized by the Environmental Protection Agency.

E. Injection of Insecticides into Shelter Tubes and Mounds

In some areas of the world it is possible to employ termite-control measures not applicable in the United States. Good results have been obtained in Australia by blowing white arsenic (arsenic trioxide 300 mesh, screened) into holes made in the large shelter tubes of *Coptotermes acinaciformis* (Frogatt). Insecticide-carrying termites return to the central nest in sufficient numbers to make this an efficient control method. Too much disturbance of a shelter tube may cause the termites to abandon it, and it is then difficult to determine whether the treatment has been successful. Generally, three to four grams of dust are blown in gently, and the hole is then sealed over (Casimir, 1957; Wilson, 1959; Gay, 1963). In areas where mound-building termites occur, they are sometimes controlled by boring a hole in the mound's central nest and blowing finely powdered arsenicals or organochlorine insecticides into it or pouring in dilute emulsions of organochlorine insecticides (Casimir, 1957; Das, 1958; Deoras, 1962).

F. Trail-marking Pheromones

Social insects maintain their social cohesiveness primarily through the utilization of pheromones. Termites secrete pheromones for caste regulation, attraction, communication, and trail-marking. These pheromones are produced in specialized tissues known as exocrine glands. In response to specific stimuli, these glands evacuate their contents into the environment (Blum, 1970). Of special interest from the standpoint of termite control are pheromones which termites use for marking trails. These pheromones are secreted by the sternal glands of the workers and soldiers of all termite families (Lüscher and Müller, 1960; Stuart, 1961, 1963, 1964, 1969; Noirot and Noirot-Timothée, 1965; Mosconi-Bernardini and Vecchi, 1966; Smythe and Coppel, 1966b; Stuart and Satir, 1968; Moore, 1969; Noirot, 1969; Howse, 1970; Mertins *et al.*, 1971). The pheromone-secreting gland of most termites is situated on the base of the fifth abdominal sternite. The trail-marking pheromone may at the same time be a food attractant (Smythe and Coppel, 1966a; Smythe *et al.*, 1967a, b; Ritter and Coenen-Saraber, 1969), and in this capacity it offers some potential as a possible means of termite control in the same way that other natural or synthetic attractants have been used, as described below.

There are many nonpheromone substances, some found in nature and some artificially produced, that have effects similar to those caused by pheromones. Wood rotted by the fungus *Lenzites trabea* (Pers. ex Fr.) produces an attractant for *Reticulitermes flavipes* (Kollar) as well as other species of *Reticulitermes* and *Coptotermes*. The fungus induces trail-following by termites simular tò that induced by the trail-marking pheromone secreted by the sternal glands of these insects (Esenther *et al.*, 1961; Esenther and Coppel, 1964; Allen *et al.*, 1964; Smythe *et al.*, 1965, 1967a, b; Esenther, 1969). At the United States Forest Service's Wood Products Insect Laboratory it was observed that *R. flavipes* followed marks made by a certain ballpoint pen with blue ink. Becker and Mannesmann (1968) found that nine glycol compounds, including some used in ink for ballpoint pens, proved to be trail-marking substances.

G. Future Possibilities in Termite Control

In southern Ontario, Canada, Esenther and Gray (1968), using a bait of wooden blocks infested with *L. trabea* and immersed for 10 seconds in a one percent solution of mirex in toluene, were able to attract subterranean termites, *R. flavipes*, to the wood and markedly suppress termite foraging activity and damage in treated areas, compared with nontreated ones. Later research revealed that even in southern Mississippi, in an area where termites are abundant and severely destructive, mirex-attractant blocks effectively suppressed termites for three years (Esenther and Beal, 1974). The use of insecticides combined with natural or synthetic attractants appears to be a promising means of greatly reducing both the quantity and area of distribution of insecticide used in termite control.

III. COCKROACHES

Cockroaches (Dictyoptera: Blattellidae, Blattidae) share with termites the reputation of being the most important structure-infesting pests in the United States. The domiciliary species have been distributed worldwide and are important pests wherever they occur. In the United States the German cockroach, *Blattella germanica* (L.), is by far the most important cockroach pest, and this is also true in most areas of the world where it occurs. The habits of all domiciliary species in buildings are so similar that a discussion of the German cockroach as an example will suffice for the group.

A. Distribution of German Cockroaches

The German cockroach, like other domiciliary species, is believed to have originated in north or tropical Africa. In temperate regions, it remains in a building throughout the year if food and water are present, although heavy population pressure can induce it to migrate from one building to another. Central heating and perhaps other changes in human life style have resulted in the German cockroach being able to continuously expand the areas in which it is commonly found in dwellings.

The German cockroach rarely if ever flies, but it is readily carried about on such items as sacks, cartons (particularly corrugated cartons), packages of food, in laundry or in kitchen appliances and furniture. Returnable beverage cartons are a particularly important means of distributing cockroaches. The cartons are often contaminated with spilled syrup or malts, which attract the insects. Empty and unrinsed soft drink or beer bottles in night clubs, restaurants, markets, and homes form a part of the infestation chain. Cockroaches may be found in large numbers in insulation in the walls of refrigerators and ranges, surviving for months, if necessary, without access to the usual foods and feeding only on cast skins and dead insects. If these appliances are moved from a cockroach-infested house or apartment to another, enormous numbers of cockroaches can be transferred with them. Cockroaches may become established in basements or crawl spaces, particularly if these places are dark and damp, and they may then enter the building around utility pipes, air ducts or ventilators, or under doors.

Cockroaches spend the daylight hours in dark, hidden areas such as under the refrigerator, stove, and sink and the back edges of shelves in pantries, cabinets, and closets. At night they search for food and water, but tend to follow intersections rather than crawl about indiscriminately. In a test made in an experimental kitchen, traps[1] were placed at intersections of the floor and the walls, and two in the middle of the kitchen floor. Trap yields were determined during alternate three-day periods when (a) food was removed from two food (dog chow) stations in the kitchen and (b) when the food was returned.

---

[1]*A cockroach trap as used for this purpose consists of a one liter jar with a film of sorptive clay on its interior surface and a teaspoonful on the bottom. Pieces of bread lure cockroaches into the jar, but they cannot escape because they slip on the dusty interior of the jar. The clay adsorbs the grease off the cuticles of the cockroaches, causing them to lose water at a lethal rate.*

The two traps placed at wall-floor intersections yielded an average of 11.7 and 15.3 cockroaches per trap when food was absent and one and 1.3 cockroaches when food was available in the kitchen. (Bread as a lure was always present in the traps.) At the same time, yields for the two traps near the middle of the room were 2.2 and 2.7 cockroaches when no food was available in the kitchen and zero when food was available. The tendency of cockroaches to follow intersections, as between the wall and the floor, rather than to crawl about over plane surfaces, was demonstrated. The experiment also demonstrated how exploratory activity can be greatly increased by a sudden removal of accustomed food sources. Also, it indicated that the efficacy of an insecticide treatment applied in the usual way, including treatment of all intersections, would be expected to be greatly increased if preceded by a thorough housecleaning, with special emphasis on removal of all food sources.

In the United States, houses and apartment buildings are commonly built with hollow walls. Cockroaches can move in and out of these hollow walls around utility pipes and conduits and imperfections in building construction. They can breed in the walls in enormous numbers and migrate via voids from one room to another and, in apartment houses, from one apartment to another. This type of distribution is accentuated by the application of insecticides with even a moderate degree of repellency and is stikingly increased by insecticides that are strongly repellent, such as propoxur and pyrethrins. After a heavily infested apartment is treated with very repellent insecticides, an increase in the numbers of cockroaches in adjoining apartments is soon noticeable and has been confirmed by trap data in such apartments before and after treatment of the adjoining infested apartment. A thorough housecleaning of an infested apartment, by removing food sources, may also cause a migration of cockroaches to adjoining apartments (Ebeling *et al.*, 1968b).

B. Exploratory Activity of Cockroaches

From her investigations of German cockroaches, Darchen (1952, 1955) concluded that the exploratory drive is the expression of a fundamental property of the nervous system and that, contrary to other stimuli such as hunger and thirst, it is never satisfied. Placed in an unfamiliar site, cockroaches crawled about actively, but their activity decreased during the first half hour to about 35% and eventually to 20% of the original. This minimal level of activity was retained indefinitely, being increased only when the insects were presented with new objects to explore. Activity was decreased if hunger and thirst were satisfied before exploration began. Mourier (1965) made similar observations regarding the exploratory activity of house flies, *Musca domestica* L. In experiments made in "choice boxes"[2] Ebeling and Reierson (1970) determined that within the population range investigated, the exploratory activity of German cockroaches decreased as population density was increased, in six increments, from one to 40.

It is fortunate that the daytime places of harborage and the nighttime movements of cockroaches are so well restricted by the physical features of a building that one can successfully "spot treat" a building without excessive use of insecticide. However, it is essential that *all* the harborages and typical paths of nighttime travel be treated. Cockroaches tend to return to an accustomed daytime harborage after their nocturnal exploratory activity. Such an area might be behind a calendar, a picture on the wall, or in an electric clock, radio or television set. If these areas are overlooked while treating, cockroaches using them as harborage may be able to crawl to and from sources of food and water without contacting lethal quantities of insecticide residue.

Spot treatment involves the application of insecticide to a very small percentage of the entire inside surface of a building. The greater the number of untreated places in or under which cockroaches can hide, the lower will be the percentage of insects that contact lethal quantities of insecticide residue. For example, in experimental closets, we found that in those filled with clutter the cockroach mortality from a given quantity applied to surfaces of the closet will be less than in a less cluttered closet or one in which contents are neatly arranged. In either case, exploratory activity of the cockroaches and mortality from a given quantity of insecticide residue is inversely related to the intensity of light, for increasing light intensity decreases exploratory activity (Ebeling *et al.*, 1967).

---

[2]*The choice box was 30 cm square, 10 cm high, and had a partition dividing the box into two equal compartments. One compartment had screens on three sides and was covered with a plexiglas cover and the other compartment was covered with Masonite to keep out light. When the two compartments were covered, a one centimeter hole drilled close to the top of the partition wall was the only means of entry from one compartment to the other. Normally, most of the cockroaches could be found in the dark half of the choice box in daylight hours.*

C. Influence of Repellency

Research in a mock-up kitchen had shown that the degree to which an insecticide repelled cockroaches had more influence than toxicity on the ultimate result of a treatment for cockroaches (Ebeling *et al.*, 1966). Experiments were then made with simulated wall voids, which consisted of plaster lath attached to studs by means of wing nuts. To the lower portion of each plaster lath was attached a 19 liter tin can with a 1.9 cm pipe at the bottom which led to the interstud void at a point 20.3 cm above the floor plate. The can contained a corrugated paper maze, food and water, and was equipped with an electric barrier to prevent escape of cockroaches. Fifty cc of German cockroaches (mostly adults) were place in each can (Ebeling *et al.*, 1966). Although conditions in the cans were ideal for the cockroaches, their continuous migratory activity resulted in large numbers in the wall voids at all times. If insecticide dusts were blown into the voids, a varying percentage of the cockroaches learned to stay out of them, and remained in the cans, before picking up a lethal dose of insecticide. In one experiment, after 30 days the numbers of cockroaches left alive in the cans and corresponding wall voids (few, if any, in the latter) with various insecticides were: boric acid, 29; borax, 219; Dri-die 67(R)[3], 418; and the untreated control, 1072. In another experiment the numbers left alive after 30 days were: boric acid, 92; diazinon, 205; Drione(R) (Dri-die + pyrethrins), 406; sodium fluoride, 452; propoxur, 791; and the untreated control, 1017. This is by no means a reflection of the relative toxicity; boric acid is the least toxic of the insecticides used, but is also the least repellent. Cockroaches entered boric acid-treated areas repeatedly until they contacted lethal quantities of the powder (Ebeling *et al.*, 1966).

Cockroaches remove dust from their labial and maxillary palps, antennae, and legs, by passing these appendages through their mouthparts. When the insects are placed on deposits of boric acid powder, large quantities of the powder are found in their crops in as brief a period as 30 minutes. Boric acid

---

[3]*Dri-die(R) is a very light, amorphous, nonabrasive silica aerogel powder with high sorptivity for oil (and wax) and low sorptivity for water. It contains 4.7% ammonium fluosilicate, present in less than a continuous monolayer, that causes the particles to have a positive electrostatic charge and increases their adherence to dusted surfaces. It kills insects by adsorbing a portion of their epicuticular wax, thereby causing a lethal rate of water loss.*

also enters their bodies by penetration through their cuticles (Ebeling *et al.*, 1975).

Choice boxes, described in III B, likewise simulated some structural features of a building such as voids under cabinets or appliances, in which cockroaches could find darkness and seclusion, and therefore were effective devices for testing repellency. Being small devices, inexpensive and requiring little space, they could be constructed in large numbers. Generally, five choice boxes were used for each insecticide tested, and as many as 80 were used in an experiment. Results followed the pattern of those reported for the experiments with mock-up wall voids. The more repellent insecticides tended to be the least effective, irrespective of their toxicity to cockroaches. All finely divided powders are repellent to cockroaches, even those that in another form can serve as food. In the choice-box tests, boric acid was shown to be the least repellent of all powders tested.

Powders that are highly sorptive, such as Dri-die 67 or highly toxic, such as sodium fluoride, are avoided by cockroaches to such an extent that they spend little time on the residues and this greatly delays mortality. In one experiment the average percentages of live German cockroaches in the dusted dark halves of choice boxes 15 hours after the insects were released were: sodium fluoride, 0.0; Dri-die 67, 8.3; boric acid, 56.7; untreated control, 66.4. As usual, boric acid was shown to be slightly repellent, i.e., the average percentages of live insects in the dark halves of the choice boxes were less than in uncontaminated boxes. But the degree of repellency was not enough to prevent their crawling about in the boric acid deposits sufficiently to pick up a lethal dose of this mild toxicant. The number of days required for 100% mortality in the choice boxes were, respectively, 7.7, >30, 3.7, and >30. The average KD_{50} values (in minutes) for the three insecticides when cockroaches were continuously confined with them in petri dishes were: sodium fluoride, 51; Dri-die 67, 127; and boric acid, 1,140 (Ebeling *et al.*, 1966).

The same tendency is manifested among organic insecticides applied as liquid formulations. The following insecticides are given in descending order of effectiveness (LT_{50}) against nonresistant German cockroaches confined in covered dishes containing residues: pyrethrins 1%, propoxur 1%, diazinon 1%, malathion 1%, ronnel 5%, and chlordane 2%. The descending order of effectiveness based on their performance in choice boxes, however, in which repellency was able to exert an influence, was: chlordane, diazinon, ronnel, propoxur, malathion, and pyrethrins. Pyrethrins and propoxur, which were first and second, respectively, in the dish test were sixth and fourth, respectively, in the choice-box test (Ebeling *et al.*, 1967).

In choice tests with boric acid, when percent mortality

was plotted against the period German cockroaches spent in treated boxes, sigmoid curves were obtained, the period for 100% mortality being six days. With propoxur the rate of mortality was much higher the first three days, but declined strikingly the fourth day and thereafter. Six percent of the insects were still alive after 30 days (Ebeling and Reierson, 1969). These curves followed the usual pattern that had been established with several highly toxic but repellent organic insecticides in previous investigations (Ebeling *et al.*, 1967).

The ability of a certain percentage of a cockroach population to survive for long periods in choice boxes with insecticide-tested dark areas is probably an indication of their ability to learn to avoid insecticide residues before picking up a lethal dose. Investigations of the learning capacity of cockroaches has traditionally involved statistical studies of their ability to utilize their negative phototaxis to guide them to a dark area where some "punishment" awaits them, generally an electric shock. The choice-box setup is evidently similar, the punishment being a repellent insecticide. The success of the associative learning (Thorpe, 1956) in choice boxes is enhanced by the ease with which cockroaches can habituate themselves to remain in a light area (Ebeling *et al.*, 1966).

In a building, avoidance of insecticide residues should be even easier because of the many secluded and dark areas in which they can seek harborage far removed from the typically treated areas. In field experiments, the nonrepellency of boric acid was found to be even more strikingly advantageous than in the choice-box tests and comparable in this respect to the experiments in our experimental kitchen, mock-up wall voids, and mock-up closets (Ebeling *et al.*, 1968a). The efficacy of boric acid in field tests was confirmed by Moore (1972, 1973), Reierson (1973), Gupta *et al.* (1973, 1975, 1976), and Wright and Hillman (1973).

Ebeling *et al.* (1968a, b) showed in field experiments that a cockroach population can be more rapidly eliminated by first treating certain heavily infested areas with a pyrethrin aerosol, then applying boric acid throughout the building. Pyrethrins kill cockroaches much more rapidly than boric acid. The efficacy of this treatment was confirmed in field experiments by Moore (1972, 1973), Reierson (1973), and, substituting resmethrin for pyrethrins, by Gupta *et al.* (1975, 1976).

D. Insect-proofing during Building Construction

In field experimentation two inorganic insecticides, the silica aerogel Dri-die 67 and boric acid, were depended upon for long-term residual control of cockroaches. Dri-die 67 possessed both repellency and insecticidal efficacy and boric acid possessed only the latter. (Dri-die 67 has the disadvantage in humid areas in that it tends to form aggregates and is then less effectively picked up by insects.) An experiment was made in a seven-story apartment building in Ventura, California, to determine to what extent these insecticides might aid in a long-term cockroach management program if they were applied in enclosed spaces, such as attics, drop ceilings, wall voids, and voids under cabinets, built-in applicances such as ranges and refrigerators, and other out-of-sight locations. Dri-die 67 was applied throughout the building at the time of construction and boric acid was applied under appliances, cabinets, and pallets in the community kitchen and food-storage room after these were installed. A modified water-type fire extinguisher was used for applying the dust (Ebeling *et al.*, 1969; Ebeling, 1975). By being practically nonrepellent, boric acid allows cockroaches to enter out-of-sight areas, and, being inorganic, acts as a permanent trap for cockroaches as long as it is left in place.

For nine years no cockroaches were seen in the kitchen and food storage room of the Ventura building. Then, through a change in management, the pallets were removed form the food-storage room, the floor was swept and washed, and the pallets were put back in place, but without retreatment with boric acid. Soon cockroaches were seen and the writer was consulted as to why cockroach infestation had followed the change in management. Boric acid was reapplied under the pallets and other suitable areas in the food-storage room and kitchen. About 18 months have passed since the retreatment without recurrence of cockroach infestation. The efficacy of cockroach-proofing at the time of construction was verified by controlled field experiments conducted by Moore (1973).

Insect proofing at the time of construction is no guarantee of freedom from cockroaches, for the insects are continually accidentally carried into buildings. There are many areas in the living space of a building as well as furniture, appliances, fixtures, and other furnishings that provide cockroach harborage. But cockroach control in a building containing an inorganic insecticide in out-of-sight areas is not likely to be as difficult as if the building had not been pest-proofed at the time of construction.

E. Future Possibilities for Cockroach Control

Cockroach control in the future will probably evolve increasingly toward the use of both dusts and liquids rather than liquids alone, as has been the case in recent decades. An inorganic dust such as boric acid is particularly effective because of its longevity and its low degree of repellency to cockroaches. The application of boric acid to enclosed spaces of a building such as attics, drop ceilings, wall voids, and voids under cabinets, at the time of construction, will probably come into increasing use.

Cockroach control follows the current trend toward reduction in insecticide dosage and area of distribution. Labels for residual insecticide formulations now bear the wording, "Applications of this product in the food areas of food-handling establishments, other than as a crack-and-crevice treatment, are not permitted." (A food-handling establishment is an area, other than a private residence, in which food is held, processed, prepared, and/or served.) This led to refinement of what has become known as crack-and-crevice (C & C) treatment, and the equipment to apply it. The good results that have been obtained will probably result in a greater use of the concept even in areas in which it is not legally required.

C & C treatment involves application of small amounts of insecticides into cracks and crevices in which insects hide or through which they may enter a building. Insecticides can be used in food areas if applied only into cracks and crevices, but not onto exposed surfaces.

Equipment for C & C treatment includes the traditional compressed air sprayer with the usual pin-stream nozzle and with low pressure. There are also some nozzles built especially for C & C treatment. The Whitmire Research Laboratories, Inc., have developed pressurized containers of insecticide in low-pressure compressed gas. The gas is directed by means of a narrow tube that can be readily inserted into cracks and crevices, where it expands approximately 240-fold at the tip and deposits the insecticide free of carriers such as oil, water, powder, solvents, and emulsifiers. The insecticides used are chlorpyrifos, diazinon, and propoxur.

Interest has been demonstrated in the Hercon(TM) Roach Tape or "Insectape." In these tapes, chlorpyrifos, diazinon, or propoxur are incorporated into a multilayered thin plastic strip. The insecticide is slowly released to the surface and an active insecticide residue is maintained there for long periods. The tapes are placed in cracks and around, in back of, and under dishwashers, refrigerators, sinks, and cabinets in the kitchen, bathroom, and other areas for the control of cockroaches. In one experiment the tapes were reported to be more effective than spray residues in counts made five months after treatment (Moore, 1976). In this experiment, after the tapes were installed, 3% pyrethrin fog was applied to flush roaches onto the tapes or areas sprayed with residuals.

In experiments made with Hercon Roach Tapes by Reierson (1976), none of the tapes alone produced a statistically significant reduction in the number of cockroaches trapped two, four, or eight weeks after the tapes were installed. (The tapes might be useful, however, for the protection of certain limited areas that are commonly infested.) Two applications of 0.25% pyrethrin aerosol following the installation of the tapes did not increase their efficacy and had no significant effect by themselves. However, when the installation of the tapes was followed with two applications of 3.34% pyrethrins, good results were obtained, but similar results were obtained with this high concentration of pyrethrins where no tapes were installed. Reierson suspects that a 3% concentration of pyrethrins as used by Moore would make it extremely difficult to evaluate the efficacy of laminated tapes. He also suggests that persons particularly sensitive to the insecticides used should avoid the tapes, for it is difficult or impossible to remove the pressure-sensitive backing of the tapes while wearing protective gloves.

IV. DOMESTIC FLIES

The domestic (synanthropic) flies (Diptera: Muscidae, Anthomyidae, Calliphoridae, Sarcophagidae) are those species that either require the human environment or are greatly benefitted by it. Best known and usually the most common among the domestic flies is the house fly, *Musca domestica* L., but in some areas and under some conditions other species may be more abundant and troublesome, including other muscids, as well as anthomyids, calliphorids, and sarcophagids.

A. Current Status of the Domestic Fly Problem

The domestic fly problem is continuously increasing in magnitude because of the development of high-density, low-area monocultures of beef and milk cattle and poultry on the fringes of large urban areas. The riding-horse population is also increasing in stables scattered throughout suburban areas. Before the current practice of high concentration of livestock in limited areas, the animals scattered their feces in small piles over large ranges or pastures, creating a favorable ecosystem for minimal fly development. Various pest arthropods that developed in the piles of feces could be consumed by poultry and hogs, but this is not possible where livestock are highly concentrated. In feedlots, all important predators associated with pasture feces have been eliminated, along with most of the coprophagous arthropods that efficiently convert pasture feces into arthropod life (Anderson, 1966).

B. Fly Control on Poultry Ranches

When a poultry ranch is situated near an urban area, flies breeding in the poultry manure can be a serious problem. Among these flies, the most important house pests are *M. domestica* and *Fannia canicularis* (L.), developing in manure, and *Phaenicia* spp., developing on fowl carcasses and broken-egg wastes. These species can fly for distances up to 32 km and may be attracted to food and shelter in homes and backyards (Peters, 1963).

Modern poultry production, being a monoculture, is at the same disadvantage as plant monoculture with regard to the limited opportunities of natural enemies to keep pests under control. The need for changes in cultural practices and modifications in the pest control program in such a way as to avoid the need for continually increased dosages of pesticides is equally urgent in plant and livestock monocultures. In either case the new approach to pest suppression may be referred to as "integrated control" or "pest management," the object being to integrate chemical control with the natural factors influencing populations (Smith and Allen, 1954; Anderson, 1965; Burton *et al.*, 1965; Axtell, 1968, 1970; UCAE, 1971-72; Legner *et al.*, 1973, 1975a).

A factor favoring success of pest management of domestic flies breeding in poultry manure is the many species and potentially large numbers of predators and parasites of the flies. The species of natural enemies of flies in accumulated excrement- or manure piles is quite different from the ones that are natural enemies in the undisturbed feces of livestock in pastures; the two groups are discussed separately by Legner and Poorbaugh (1972). Modern pest management involves the removal of manure in such a way as to cause minimal reduction of the influence of natural enemies of the pest fly population. Studies on the ecology and behavior of the pest species, and of the complexity of the ecosystem of which they are a part, must precede sound pest management programs. For example, the resting habits of pest flies during the day and night, indoors and outdoors, in urban and in rural areas, and when buildings are screened or not screened, are important elements of such studies.

A study was made on poultry ranches in northern California using a backpack vacuum sampling-machine and sticky fly tapes, that revealed which species of domestic flies and which species of natural enemies were involved, as well as much useful information on their habits, particularly their resting places (Anderson, 1964; Anderson and Poorbaugh, 1964). Unlike their widespread dispersion during the day, all species except *P. sericata* Meigen that remained outdoors at night rested predominantly on branches of trees and shrubs, and those inside the poultry houses rested predominantly in the general ceiling area. The flies remained at their overnight resting places for 12 to 16 hours, depending on temperature. Counts made of flies caught on sticky tapes revealed that 85% of resting *F. canicularis* and *M. domestica* were caught in houses and 91% of the beneficial predator *Ophyra leucostoma* (Wiedemann) and 95% of the ichneumonid parasites were caught in trees and shrubs. A knowledge of the nocturnal aggregation sites obviously is a useful guide in the application of insecticides.

The University of California Extension Service recommends that caged laying hens have a concrete base beneath them to catch droppings and prevent house fly larvae from pupating in the soil. A greater percentage of immature flies can then be destroyed by natural enemies. Only the top portion of the manure should be removed during the main fly season and a 15 to 20 cm pad of the manure should remain on the concrete base. Also, all manure may be removed on an alternative row basis. With either procedure, natural enemies from the manure that remains can quickly invade adjacent fresh droppings. A period of six to 12 months is required to reestablish natural enemies if all manure is removed at one time (UCAE, 1971-72).

Water systems in poultry houses (and cattle feedlots) should be continually checked. Water dripping from faulty watering systems results in wetter manure and increased fly problems.

On some poultry ranches in southern California on a supervised pest management program, an ecologically sound manure removal program was practiced, leaving a minimum residual deposit of at least 16 cm following every cleaning operation. This program resulted in a maximum fly predator and scavenger population and also hastened decomposition of the manure. The fly species involved were *F. canicularis, F. femoralis* (Stein), *M. domestica, Stomoxys calcitrans* (L.), and *Phaenicia* spp., the first two comprising over 70% of the total number. There were also two predatory fly species, *Muscina stabulans* (Fallen) and *O. leucostoma*.

On control ranches, poison baits were used on an average of about once every two months between April and September, compared with one such treatment on the supervised ranches. Tens of thousands of three native species of hymenopterous parasites (*Spalangia enduis* Walker, *Muscidifurax raptor* Girault & Sanders, and *M. zaraptor* Kogan & Legner, and an imported Australian species, *Tachinaephagus zealandicus* Ashmead) were released in 48 releases. The pest management resulted in a striking suppression of the fly population (Legner *et al.*, 1973, 1975a).

The entire community of arthropods that inhabit solid animal wastes, including predators and scavengers (Histeridae, Staphylinidae, Hydrophylidae, Dermaptera) is required for the management of synanthropic insects at their lowest densities.

Individual species of natural enemies may vary in importance according to the season of the year (Legner *et al.*, 1975b).

Reduced fly population was attributed to certain physio-chemical changes in the composition of the manure, making it less suitable for fly oviposition and development, as well as to increased populations of scavenger, predatory, and parasitic arthropods.

In Florida, in two field tests, sustained releases of *S. endius*, a pupal parasite, resulted in such large populations of this species that it could keep the flies at a low density level. This method of fly control was found to be both effective and relatively inexpensive (Goodin, 1976).

The importance of mites as well as insects as natural enemies of domestic flies, particularly the species *Macrocheles muscadomesticae* (Scopoli) and *Furscuropoda vegetans* (De Geer), has been emphasized by Axtell (1968, 1970). They are predacious on the immature stages of *M. domestica* and *F. canicularis* and are destroyed by conventional larviciding, but are unharmed by selective adulticiding. In North Carolina conditions the larviciding method required 16 to 18 applications of insecticide spray for satisfactory control while an integrated program required only five or six applications of spray to the resting sites of the flies, primarily the inside upper part of the poultry houses. Conventional larviciding would require five times more insecticide and 2.5 times more labor per season than would be needed for the integrated control program.

Insecticide sprays should be applied only to overhead and peripheral portions of poultry houses, provided insecticides are required for control. Droppings should not be sprayed, for natural enemies that can destroy more than 95% of the fly populations are killed by the insecticides. When applied to the ceiling and peripheral parts of the poultry houses, insecticides do not interfere with key natural enemies (Legner and Olton, 1968).

C. Sanitation in Domestic Fly Control

Probably no other order contains so many pestiferous building-infesting insects for which satisfactory control depends on effective community action to eliminate breeding places as does the order Diptera. Common examples of such insects are domestic flies, mosquitoes, midges and gnats. In the case of domestic flies, the major community contribution to control is garbage pickup and disposal. A city ordinance, for example, might require that garbage be wrapped or placed in bags and placed in covered containers to await pickup by the the municipal sanitation service once a week, along with prunings, litter, lawn clippings, or other organic materials that are potential sources of fly infestation. The trash and garbage may be dumped in areas which are continuously being covered with earth by means of bulldozers. This may be suitable when an earth fill is desired by the community in a particular locations, as for filling a ravine or canyon. In other areas, incineration of the refuse may be considered to be a suitable alternative, except where it may contribute substantially to air pollution. The wide use of automatic garbage-disposal units in the home has relieved city sanitation services of much of the burden of garbage removal, for much of the organic waste thus becomes a part of the city sewage.

Supplementing the community garbage removal service, or in case this service is not available, decaying organic material and animal excrement on residential properties should be deeply buried or otherwise disposed of by the homeowner, for house flies, as well as stable flies, flesh flies, bottle flies, blow flies, little house flies, and false stable flies, all breed in this type of material.

D. Future Possibilities for Fly Control

Fly control will continue to rely heavily on suppression of fly populations in outdoor breeding places such as poultry farms and cattle feedlots. Supervised pest management programs will involve ecologically sound manure removal procedures leaving minimum residual deposits of manure following every cleaning operation, so that natural enemies from the manure that remains can quickly invade adjacent fresh droppings. In addition, sustained releases of natural enemies will aid in keeping flies at a low density level.

One of the most promising approaches to chemical control of flies appears to be the use of insect growth regulators (IGRs). The chemical known as Dimilin$^{(R)}$, for example, needs only to be applied to surfaces on which species such as the house fly and stable fly land and rest, to be effective. Dimilin generally prevents the eggs from hatching but, if they do hatch, it interferes with the insect's ability to make chitin from glucose. The possibility of incorporating Dimilin in paint, as is now done with fungicides, has been suggested, thus eliminating application problems and the threat of contamination of people, pets, and food. Dimilin has been registered by the Environmental Protection Agency for use against the gypsy moth, but its use against other insects is still experimental (Carriere, 1976).

V. SUMMARY AND CONCLUSIONS

Pest management of structure-infesting insects provides fewer opportunities for utilization of biological and cultural control measures to reduce reliance on pesticides than does pest management for crop protection. Nevertheless, many such opportunities present themselves and have long been recognized by the pest control operator. Recent biological research has suggested interesting new procedures. Pest management of three important groups of structure-infesting insects - termites, cockroaches, and domestic flies - is discussed to illustrate current methods and future possibilities.

For termite control, time-honored insect managment measures begin with site preparation: removal of buried stumps or logs and provision for proper drainage. In construction, proper clearance in joist-type construction, proper ventilation, and removal of form boards and broken stakes, as specified in building codes, are a part of insect management. Post-construction alterations that favor termite infestation should be avoided.

During the last decade field tests have shown that wooden blocks infected with the fungus *Lenzites trabea* are able to attract subterranean termites, *Reticulitermes flavipes*. Such bait blocks, treated with mirex, have been found to satisfactorily suppress termite populations for a three year period. The ban on mirex for insecticidal purposes will result in a delay of at least five years for field testing and registration of a new environmentally acceptable insecticide. Bait blocks would presumably te renewed at two- or three-year intervals. A community-wide program of termite suppression with bait blocks would probably substantially increase the period between replacement of the bait blocks.

The principal feature of management of cockroaches is sanitation and good housekeeping. These insects cannot live in buildings without the food and water they sometimes provide. Cockroaches are also favored by a cluttered household, with piles of stored magazines, newspapers, paper bags, and corrugated-paper boxes. Such clutter not only provides harborage and breeding places but also provides areas into which cockroaches can escape from the generally repellent insecticide residues when control is attempted.

In food-handling establishments, insecticides must now be applied in limited amounts directed into cracks and crevices and not onto exposed surfaces. Special equipment is available for this type of treatment. This is called "C & C" treatment and will probably come into increasing use even in areas in which it is not now legally required.

In areas in which climatic conditions are such that a considerable proportion of the cockroach population occurs outdoors, such measures as outdoor trapping and liberation of parasites and predators might be helpful (Piper and Frankie, Chapt. 10, this volume).

Domestic flies breed outdoors. Often the flies affecting urban populations are those that develop in poultry houses and feedlots close to cities. Effective pest management programs have been developed to suppress such outdoor fly populations by avoiding complete removal of manure at any one time, thus avoiding complete removal of the many species of parasites and predators that suppress the fly population. Remaining parasites and predators can then invade fresh droppings. Introduction of insectary-reared parasites has been shown to be an effective adjunct to the general pest management program.

Progress has been made in the development of nonhazardous compounds when insecticide treatment is required. A promising compound is Dimilin, an insect growth regulator that needs only to be applied to surfaces on which flies rest.

REFERENCES

Allen, T. C., R. V. Smythe, and H. C. Coppel. 1964. Response of twenty-one termite species to aqueous extracts of wood invaded by the fungus *Lenzites trabea* Pers. ex Fr. *J. Econ. Entomol. 57*: 1009-1011.

Anderson, J. R. 1964. The behavior and ecology of various flies associated with poultry ranches in northern California. *Proc. Calif. Mosq. Contr. Assoc. 32*: 30-34.

Anderson, J. R. 1965. A preliminary study of integrated fly control on northern California poultry ranches. *Proc. Calif. Mosq. Contr. Assoc. 33*: 42-44.

Anderson, J. R. 1966. Recent developments in the control of some arthropods of public health and veterinary importance. *Bull. Entomol. Soc. Am. 12*: 342-348.

Anderson, J. R., and J. H. Poorbaugh. 1964. Biological control: Possibility for house flies. *Calif. Agric. 18(9)*: 2-5.

Axtell, R. C. 1968. Integrated house fly control: Populations of fly larvae and predaceous mites, *Macrocheles muscae domesticae*, in poultry manure after larvicide treatment. *J. Econ. Entomol. 61*: 245-249.

Axtell, R. C. 1970. Fly control in caged-poultry houses: Comparison of larviciding and integrated control programs. *J. Econ. Entomol. 63*: 1734-1737.

Becker, G., and R. Mannesmann. 1968. Untersuchungen über das Verhalten von Termiten gegenüber einigen spurbildenden Stoffen. *Z. angew. Entomol. 62*: 399-436.

Blum, M. S. 1970. The chemical basis of insect sociality. *In* Chemicals controlling insect behavior (M. Beroza, ed.).

Academic Press, New York and London, pp. 61-94.

Burton, V. E., J. R. Anderson, and W. Stanger. 1965. Fly control costs in northern California poultry ranches. *J. Econ. Entomol. 58*: 306-309.

Carriere, B. D. 1976. New era in insect control. *Agric. Res. 25(1)*: 11.

Casimir, M. 1957. Detection and control of termites. *Agric. Gaz. New South Wales 68*: 68-78.

Darchen, R. 1952. Sur l'activite exploratrice de *Blattella germanica*. *Z. Tierpsychol. 9*: 362-372.

Darchen, R. 1955. Stimuli nouveaux et tendance exploratrice chez *Blattella germanica*. *Z. Tierpsychol. 12*: 1-11.

Das, G. M. 1958. Observations on the termites affecting tea in north-east India and their control. *Indian J. Agric. Sci. 28*: 553-560.

Deoras, P. J. 1962. Some observations on the termites of Bombay. *In* Termites in the humid tropics. Paris: UNESCO, pp. 101-103.

Ebeling, W. 1975. Urban entomology. Div. Agric. Sci., Univ. Calif., Berkeley.

Ebeling, W., R. E. Wagner, and D. A. Reierson. 1966. Influence of repellency on the efficacy of blatticides. I. Learned modification of behavior of the German cockroach. *J. Econ. Entomol. 59*: 1374-1388.

Ebeling, W., D. A. Reierson, and R. E. Wagner. 1967. Influence of repellency on the efficacy of blatticides. II. Laboratory experiments with German cockroaches. *J. Econ. Entomol. 60*: 1375-1390.

Ebeling, W., D. A. Reierson, and R. E. Wagner. 1968a. Influence of repellency on the efficacy of blatticides. III. Field experiments with German cockroaches, with notes on three other species. *J. Econ. Entomol. 61*: 751-761.

Ebeling, W., D. A. Reierson, and R. E. Wagner. 1968b. Pyrethrin aerosol and boric acid - A combination treatment for cockroach control. *PCO News 28(5)*: 8, 10, 12-13, 22-25, 29.

Ebeling, W., and D. A. Reierson. 1969. The cockroach learns to avoid insecticides. *Calif. Agric. 23(2)*: 12-15.

Ebeling, W., R. E. Wagner, and D. A. Reierson. 1969. Insect-proofing during building construction. *Calif. Agric. 23(5)*: 4-7.

Ebeling, W., and D. A. Reierson. 1970. Effect of population density on exploratory activity and mortality rate of German cockroaches in choice boxes. *J. Econ. Entomol. 63*: 350-355.

Ebeling, W., D. A. Reierson, R. J. Pence, and M. S. Viray. 1975. Silica aerogel and boric acid against cockroaches: External and internal action. *Pest Biochem. Physiol. 5*: 81-89.

Edmunds, L. R. 1953. Some notes on the Evaniidae as household pests and as a factor in the control of cockroaches. *Ohio J. Sci. 53*: 121-122.
Esenther, G. R. 1969. Termites in Wisconsin. *Ann. Entomol. Soc. Am. 62*: 1274-1284.
Esenther, G. R., T. C. Allen, J. E. Casida, and R. D. Shenefelt. 1961. Termite attraction from fungus-infected wood. *Science 134*: 50.
Esenther, G. R., and H. C. Coppel. 1964. Current research on termite attractants. *Pest Contr. 32(2)*: 34-46.
Esenther, G. R., and D. E. Gray. 1968. Subterranean termite studies in southern Ontario. *Canad. Entomol. 100*: 827-834.
Esenther, G. R., and R. H. Beal. 1974. Attractant-mirex bait suppresses activity of *Reticulitermes* spp. *J. Econ. Entomol. 67*: 85-88.
Gay, F. J. 1963. Soil treatments for termite control in Australia. Building: Lighting: Engineering (Australia). No. 671: 52-55.
Goodin, P. L. 1976. Death knell for the house fly? *Agric. Res. 24(10)*: 8-9.
Gupta, A. R., Y. T. Das, J. R. Trout, W. R. Gusciora, D. S. Adam, and G. J. Bordach. 1973. Effectiveness of spray-dust-bait combination. *Pest Contr. 41(9)*: 20, 22, 24, 26, 58, 60-62.
Gupta, A. P., Y. T. Das, W. R. Gusciora, D. S. Adam, and L. Jargowsky. 1975. Effectiveness of 3 spray-dust-bait combinations. *Pest Contr. 43(7)*: 28, 30-33.
Gupta, A. P., Y. T. Das, W. R. Gusciora, and D. S. Adam. 1976. Inner-city control of German roaches with resmethrin/boric acid. *Pest Contr. 44(6)*: 43-44, 55.
Howse, P. E. 1970. Termites: A study in social behavior. Hutchinson, London.
Johnston, H. R., V. K. Smith, and R. A. Beal. 1971. Chemicals for subterranean termite control: Results of long-term tests. *J. Econ. Entomol. 64*: 745-748.
Legner, E. F., and G. S. Olton. 1968. The biological method and integrated control of house and stable flies in California. *Calif. Agric. 22(6)*: 2-4.
Legner, E. F., and J. H. Poorbaugh. 1972. Biological control of vector and noxious synanthropic flies: A review. *Calif. Vector Views 19*: 81-100.
Legner, E. F., W. R. Bowen, W. D. McKeen, W. F. Rooney, and R. F. Hobza. 1973. Inverse relationship between mass of breeding habitat and synanthropic fly emergence and the measurement of population densities with sticky tapes in California inland valleys. *Environ. Entomol. 2*: 199-205.
Legner, E. F., W. R. Bowen, W. F. Rooney, W. D. McKeen, and G. W. Johnson. 1975a. Integrated fly control. *Calif.*

Agric. 29(5): 8-10.
Legner, E. F., G. S. Olton, R. E. Eastwood, and E. J. Dietrick. 1975b. Seasonal density, distribution and interactions of predatory and scavenger arthropods in accumulating poultry wastes in coastal and interior southern California. *Entomophaga 20*: 269-283.
Lüscher, M., and B. Müller. 1960. Ein spurbildendes Sekret bei Termiten. *Naturw. 47*: 593.
Mertins, J. W., H. C. Coppel, and F. Matsumura. 1971. Sternal gland in *Coptotermes formosanus* (Isoptera: Rhinotermitidae). *Ann. Entomol. Soc. Am. 64*: 478-480.
Moore, B. P. 1969. Biochemical studies in termites. *In* Biology of termites (K. Krishna and F.M. Weesner, eds.). Academic Press, New York and London, Vol. I, pp. 407-432.
Moore, R. C. 1972. Boric acid-silica dusts for control of German cockroaches. *J. Econ. Entomol. 65*: 458-461.
Moore, R. C. 1973. Cockroach proofing: preventive treatments for control of cockroaches in urban housing and food service carts. Connecticut Agric. Exp. Sta. Bull. 740.
Moore, R. C. 1976. Efficacy of Hercon roach tape. *Pest Contr. 44(6)*: 37-38, 40, 42.
Mosconi-Bernardini, P., and M. L. Vecchi. 1966. Osservazioni istologiche e fluoromicroscopiche sulla ghiandola sternale di *Reticulitermes lucifugus* (Rhinotermitidae). *Symp. Genet. Biol. Ital. 13*: 169-177.
Mourier, H. 1965. The behavior of house flies (*Musca domestica* L.) toward "new objects." *Vidensk. Medd. Naturhist. Foren. Kjobenhavn 128*: 221-231.
Noirot, C. 1969. Glands and secretions. *In* Biology of termites (K. Krishna and F.M. Weesner, eds.). Academic Press, New York and London, Vol. 1, pp. 89-123.
Noirot, C., and C. Noirot-Timothee. 1965. Le gland sternal dans l'evolution des termites. *Ins. Soc. 12*: 265-272.
Peters, R. F. 1963. Urbanization's impact on the poultry industry. *Calif. Vector Views 10*: 69-72.
Pickens, A. L. 1934. The biology and economic significance of the western subterrancean termite, *Reticuliterms hesperus*. *In* Termites and termite control (C.A. Kofoid, ed.; partially revised and reprinted, 1965). Univ. Calif. Press, Berkeley, pp. 157-183.
Reierson, D. A. 1973. Field tests to control German cockroaches with ULV aerosol generators. *Pest Contr. 41(1)*: 26-32.
Reierson, D. A. 1976. Field tests with laminated tapes containing Dursban or Baygon to control German cockroaches in apartments in southern California. Univ. Calif. (mimeo).
Ritter, F. J., and C. M. A. Coenen-Saraber. 1969. Food attractants and a pheromone as trail-following substances

for the Saintonge termite. Multiplicity of the trail-following substances in *Lenzites trabea*-infected wood. *Entomol. Exp. Appl. 12*: 611-622.

Roth, L. M., and E. R. Willis. 1960. The biotic associations of cockroaches. Smithsonian Misc. Coll. 141.

Smith, R. F., and W. W. Allen. 1954. Insect control and the balance of nature. *Sci. Am. 190(6)*: 38-42.

Smith, V. K., R. H. Beal, and H. R. Johnston. 1972. Twenty-seven years of termite control tests. *Pest Contr. 40(6)*: 28, 42, 44.

Smythe, R. V., T. C. Allen, and H. C. Coppel. 1965. Response of the eastern subterranean termite to an attractive extract from *Lenzites trabea*-invaded wood. *J. Econ. Entomol. 58*: 420-423.

Smythe, R. V., and H. C. Coppel. 1966a. Some termites may secrete trail-blazing attractants to lead others to food sources. *Pest Contr. 34(10)*: 73-78.

Smythe, R. V., and H. C. Coppel. 1966b. A preliminary study of the sternal gland of *Reticulitermes flavipes* (Isoptera: Rhinotermitidae). *Ann. Entomol. Soc. Am. 59*: 1008-1010.

Smythe, R. V., H. C. Coppel, and T. C. Allen. 1967a. The response of *Reticulitermes* spp. and *Zootermopsis angusticollis* (Isoptera) to extracts from woods decayed by various fungi. *Ann. Entomol. Soc. Am. 60*: 8-9.

Smythe, R. V., H. C. Coppel, S. H. Lipton, and F. M. Strong. 1967b. Chemical studies of attractants associated with *Reticuliterms flavipes* and *R. virginicus*. *J. Econ. Entomol. 60*: 228-233.

Stuart, A. M. 1961. Mechanism of trail-laying in two species of termites. *Nature (London) 189*: 419.

Stuart, A. M. 1963. Origin of the trail in the termites *Nasutitermes corniger* (Motschulsky) and *Zootermopsis nevadensis* (Hagen) (Isoptera). *Physiol. Zool. 36*: 69-84.

Stuart, A. M. 1964. The structure and function of the sternal gland of *Zootermopsis nevadensis* (Isoptera). *Proc. Zool. Soc. London 143*: 43-52.

Stuart, A. M. 1969. Social behavior and communication. *In* Biology of termites (K. Krishna and F.M. Weesner, eds.). Academic Press, New York and London, Vol. 1, pp. 193-232.

Stuart, A. M., and P. Satir. 1968. Morphological and functional aspects of an insect epidermal gland. *J. Cell Biol. 36*: 527-549.

Thorpe, W. H. 1956. Learning and instinct in animals. Methuen and Co., London.

UCAE. 1971-72. Univ. Calif. Agric. Ext. Serv. AXT-72.

Wilson, H. B. 1959. Termites (white ants) and their control. *J. Dept. Agric. Victoria (Australia) 57(2)*: 35-73; *(3)*: 89-95.

Wright, C. G., and R. C. Hillman. 1973. German cockroaches:

Efficacy of chlorpyrifos spray and dust and boric acid powder. *J. Econ. Entomol. 66*: 1075-1076.

INTEGRATED MANAGEMENT OF URBAN COCKROACH POPULATIONS

Gary L. Piper
Gordon W. Frankie[1]

Department of Entomology
Texas A&M University
College Station, Texas

I. INTRODUCTION

The cockroach is among the most adaptable of all insects and has been able to survive many changing environments as evidenced by the more than 3,500 species in the world today. Fortunately, few cockroach species are domiciliary pests (Rehn, 1945). Of the 57 species found in the United States, only 10%

---

[1]*Present address: Department of Entomological Sciences, University of California, Berkeley, California.*

ISBN 0-12-265250-9

of these have a predilection for infesting human habitations. The most important and frequently encountered are the German cockroach, *Blattella germanica* (L.); the oriental cockroach, *Blatta orientalis* L.; the brown-banded cockroach, *Supella longipalpa* (F.); the American cockroach, *Periplaneta americana* (L.); and the smokybrown cockroach, *P. fuliginosa* (Serville). The success of household-infesting species probably can be attributed to a combination of characteristics which include omnivorous habits; high reproductive potential, permitting rapid development of insecticide resistant strains; and secretive habits, which protect them from detection and destruction.

Problems associated with domiciliary cockroaches in urban environments are well documented (Roth and Willis, 1957, 1960). Cockroaches are among the most important and disagreeable of all household insects. To many individuals, roaches are an aesthetic abhorrence and their presence in an urban structure may be psychologically disturbing and cause considerable mental distress. An infestation implies a condition of uncleanliness and this tends to be a major embarrassment for urban dwellers regardless of their socioeconomic status. Cockroaches either consume or contaminate human foodstuffs with salivary secretions and excrement and produce secretions which impart a persistent, fetid odor to materials they contact. They are known to serve as natural vectors for pathogenic bacteria of the *Salmonella* group, causative agents of dysentery, enteric fever, food poisoning and gastroenteritis in man (Truman, 1961; Cornwell, 1968). In addition to bacteria, other disease producing organisms can be transmitted by cockroaches (Roth and Willis, 1957), but in general they are not particularly associated with widespread contagion or disease outbreaks (Guthrie and Tindall, 1968). Roaches are also medically significant in that they may induce allergic reactions in humans (Bernton and Brown, 1964, 1970a, b).

A. Problems with Current Control Methods

Insecticides, when properly applied, generally provide adequate short-term control of roaches found within the immediate confines of urban structures (Moore, 1971). However, several species (e.g., American, oriental, smokybrown) invade residential and commercial structures from outside reservoir sources (Beatson and Dripps, 1972), particularly in areas of the world with warm climates. When insecticides are applied around the perimeter of such structures, their killing effect on roaches often diminishes very quickly. High temperatures and associated high humidities, UV radiation, water pH and type of surface to which the material is applied all contribute to insecticide degradation (National Pest Control Association,

1966). Thus, the premises are subject to reinvasion which in turn requires further chemical treatment. As a result, roach-plagued individuals are placed on an "insecticidal treadmill" (van den Bosch and Messenger, 1973) not unlike that of the grower who seeks to continually chemically suppress the insects feeding on his crops.

To date, programs aimed at suppressing populations of the aforementioned roach species in the southern United States have met with less than satisfactory results and traditional control is still dependent upon the establishment of an insecticidal barrier between the household or other structure and outdoor cockroach population reservoirs. This "rote" approach reflects an attempt to achieve abatement by routine application of insecticides without regard to habits or abundance of the species involved. Periodic treatments to outdoor situations by most applicators (i.e., commercial companies, public agencies and homeowners) result in 1) wasteful and pollutive overuse of chemicals, 2) low percentage kill of roach populations in affected areas, and 3) maintenance and/or increase in resistance levels to roach-oriented insecticides.

Most of the above difficulties with insecticides can ultimately be traced to an inadequate understanding and appreciation of the general habits and ecology of cockroaches in urban environments. This situation is further aggravated where man's role in the ecology of cockroaches is overlooked. In this paper, we explore the components of a cockroach management program that has at its base a strong emphasis on the behavioral ecology of the cockroach. The program stresses an integration of methods which results in an environmentally sound management scheme that is safe and efficient in terms of resources expended.

Times are propitious for acceptance and implementation of a cockroach management program. Population growth at an unprecedented rate and expanding urbanization have spawned conditions conducive to development of intolerable roach populations in metropolitan areas. Individuals have become increasingly aware of the problems associated with insecticide use and their national and local implications relative to personal health and environmental quality. Pest control industry personnel are learning to contend with more stringent federal pesticide regulations and restrictions, increased costs of petroleum-based chemicals and an environmentally conscious and concerned public.

II. COCKROACH MANAGEMENT FRAMEWORK

The management program described in this paper, which is currently in its fourth year of operation, has been developed in residential areas in urban Texas environments. Although many parts of Texas may be considered subtropical, it is our contention that much of the basic information and most of the underlying concepts are applicable to other residential areas of the United States and possibly to residential situations in other parts of the world.

The goal of integrated roach management is to incorporate all principles of population regulation or limitation, natural and artificial, into a multidisciplinary and closely coordinated program. In theory, this approach should suppress roach numbers, reduce or eliminate outdoor population reservoirs and promote a more discriminatory usage of roach-oriented insecticides.

The basic elements of the generalized urban roach management program in Texas are illustrated in Figure 1. Individual management techniques are discussed in the sections which follow. Examples of operational integrated control programs for several domiciliary roaches are also presented.

III. COMPONENT TACTICS OF A COCKROACH MANAGEMENT PROGRAM

A. Education

Our experience with attitudes of urbanites toward cockroaches has convinced us that public education is an essential component of any management program. Integrated roach management is people-oriented, and implementation of successful control strategy is dependent upon a highly effective information

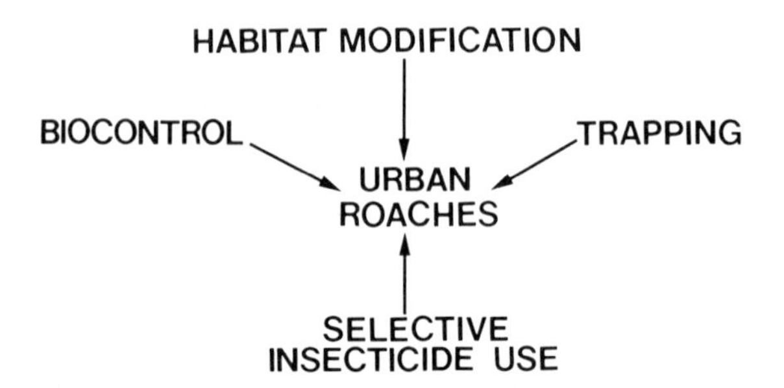

Fig. 1. Generalized component tactics of an integrated cockroach management program.

delivery system, which in turn is based on cooperative efforts of research and extension entomologists, public agencies, pest control operators and individuals with roach problems. Homeowner attitudes toward, understanding of, and participation in a roach management effort are extremely important relative to whether a program will attain maximum effectiveness. Nontechnical newspaper and magazine articles, TV appearances, talks before civic organizations and pest control associations have been used with success to promote an understanding of the objectives and operational mechanics of integrated cockroach management among the general populace (Piper and Frankie, 1978).

People have a misguided faith that every pest problem has a chemical cure (Plapp, Chapt. 16, this volume). Pest control operators and homeowners must come to realize that a unilateral chemical control approach will never solve all roach problems. This approach merely provides temporary suppression of roach populations which eventually resurge and must be treated again. The philosophy of integrated roach control is to manage populations at unobjectionable levels rather than to eradicate them. Obviously, this philosophy is incompatible with the public's current preconception and expectation of pest control.

Determination of realistic levels of tolerability or aesthetic injury levels, A.I.L., (i.e., levels of insect abundance or damage which offend the aesthetic values of people) (Olkowski *et al.*, 1976) are important to the development of an effective cockroach management system. Human attitude toward these domiciliary insects ranges from indifference to intolerance, but in most instances only very low numbers (zero to five cockroaches observed per week) inside a residence are acceptable. Tolerance levels are strongly influenced by an individuals's age and socioeconomic background, and such variability must be considered by a pest manager when planning control strategy (Piper and Frankie, 1978).

The A.I.L. is also influenced by the homeowner's understanding of the specific problem. In the case of homeowners who are willing to learn about their specific roach problem, we have observed that tolerance for roaches increases slightly (Piper and Frankie, 1978). For example, when a homeowner learns that an occasional roach in a clean household is probably a transient, a higher degree of tolerance can be expected. The success of certain subsequent control practices administered indoors (see later sections) also contributes to this pattern.

B. Habitat Modification

Habitat modification must be at the forefront of an integrated cockroach management program. The objective is to alter the immediate outdoor and indoor environments of urban structures in order to effect roach population decline. This can be accomplished by denying these pests resources such as moisture, food, warmth and shelter. Acceptance and adoption of recommended modification measures by homeowners are essential prerequisites for any effort to initiate and conduct a roach management program.

Success in prevention of American, oriental and smokybrown cockroach infestations inside buildings depends primarily on suppression or elimination of outdoor populations. Trash accumulations, firewood and lumber piles and other debris near dwellings provide excellent hiding and breeding areas for roaches. These aggregation sites serve as reservoirs for the production of individuals, many of which ultimately invade nearby structures (Piper and Frankie, in prep.).

To demonstrate the importance of outdoor reservoirs as infestation foci, a 14-month study of the outdoor movement patterns of the smokybrown cockroach was conducted in College Station, Texas. This species is a pest of major importance in urban areas of the gulf coastal United States. Roach activity was influenced by a variety of factors, including temperature, rainfall, food availability and population size and structure (Fleet *et al.*, 1978). The majority of cockroach movements were localized, with activity being confined to areas adjacent to structures. Once established, outdoor populations exhibited little tendency to leave as long as conditions for their survival remained suitable. The findings of this study suggest that any long-term program to eliminate cockroaches from urban structures will be difficult, if not impossible, unless reservoir populations in the vicinity are greatly suppressed or eliminated.

Roaches can be discouraged from entering urban structures from outside sources by sealing cracks in foundations and exterior walls. The seal or caulking around windows, doors, air conditioning units and other openings into the premises should be checked to insure there are no areas which facilitate roach entry. Similar corrective measures should be applied around water and steam pipes or other utility service lines to prevent intra- and/or interstructural roach movement.

If roach infestations are to be prevented or controlled, conditions within the structure must be made unsuitable for their development. Potential hiding and breeding areas should be eliminated. Cracks and holes in floors, walls and ceilings should be repaired, and openings around plumbing fixtures, furnace flues, electrical outlets and along baseboards or ceiling molding should be sealed with appropriate materials.

Good housekeeping and thorough cleaning are essential in controlling cockroaches, especially the German and brown-banded. Unwashed dishes and kitchen utensils and uncovered food should not be left overnight. Areas beneath refrigerators, stoves and sinks should be cleaned periodically as should cupboards, shelves and bins where minute amounts of food can accumulate. Dry pet food should be kept in lidded containers. Unconsumed pet food should not be allowed to remain in the feeding dish overnight. Garbage must be confined to containers with tight lids and removed at frequent intervals.

Cockroach infestation is favored by a cluttered household (Ebeling, 1975). Accumulations of stored papers, boxes or other items should be disposed of. These accumulations serve as indoor reservoirs and also provide areas into which cockroaches can escape from insecticide residues when chemical control is attempted (Gupta *et al.*, 1973).

The importance of minimizing roach harborages, preventing roach ingress into dwellings, and good housekeeping cannot be overemphasized. A pest control operator, being in the best position to observe sanitation defects conducive to cockroach infestations, should bring these to the attention of the customer, together with his recommendations for corrective action. The operator should also provide explanations for any recommended corrective measures.

Sanitary measures are often recommended by pest control operators; however, they are seldom practiced by their clients. As a result, many pest control firms have begun to charge for outdoor and indoor habitat modifications that lead to roach-proofing (Piper and Frankie, 1978).

C. Trapping

Over the years, a great variety of devices or techniques have been developed for catching or trapping insects. However, to our knowledge, the feasibility or actual use of traps as a cockroach population suppression technique have never been documented. The only use of traps has been to capture specimens for experimental research purposes and to aid in surveying or monitoring roach population movements in the urban environment (Washburn, 1913; Schoof and Siverly, 1954; Jackson and Maier, 1955; Dold, 1964; Lord *et al.*, 1964; Whitlaw and Smith, 1964; Ebeling *et al.*, 1966). Recently, a number of Texas pest control operators have expressed interest in utilizing roach traps as a control technique (Piper and Frankie, 1978). This can be attributed to increasing U.S. Environmental Protection Agency insecticide restrictions, awareness of an environmentally conscious public and failure of present methods to provide suitable control.

Experiments have shown that trapping is an effective and environmentally safe way to reduce roach populations, especially when used in conjunction with other suppressive measures (Piper and Frankie, 1978). Cockroaches can be trapped both outside and inside residential dwellings and commercial establishments. Outdoor trapping can reduce the size of roach reservoir populations and therefore limit the number of cockroaches entering a premise. Trapping inside a structure can suppress indoor-breeding roaches such as the German and brown-banded and may substantially reduce populations of primarily outdoor-dwelling species such as the American and smokybrown (Piper *et al.*, 1975; Piper and Frankie, 1978).

Knowledge of domiciliary roach habits, habitat preferences and the physical features of a habitat which influence roach movement is essential for the proper placement of traps in outdoor and indoor environments (Fleet and Frankie, 1974; Ebeling, 1975). Traps should be positioned to intercept roaches as they travel from harborages to feeding areas. Properly deployed, a trap can catch numerous nymphal and/or adult roaches daily, although the numbers trapped may vary with total population size, weather and season of year (Fleet *et al.*, 1978).

Depending upon the cockroach species involved, the severity and location of the infestation within or around a structure, both the number and trap type used will vary. Two types of trapping devices, each with a different operational format, have been utilized to reduce roach populations in urban Texas situations (Piper and Frankie, 1978). The first type of food-baited lure trap, designed for either outdoor or indoor use against larger domiciliary roach species, is fabricated from a lightweight, durable, rectangular plastic utility box with removeable lid. Trap sides have wire screen-covered ventilation and bait odor dispersion ports and several inwardly directed cylindrical entry corridors. A delicately balanced hinged door, opening into the trap interior, is attached to the end of each corridor. Upon detection of the bait odor, a cockroach crawls through the entry corridor, encounters and pushes against the one-way door and falls into the trap.

Various sized black metal and glass cylindrical containers have proven to be excellent indoor traps for domiciliary roaches, especially the German and brown-banded. A band of petrolatum three to four centimeters wide is applied to the container's upper interior surface (about one centimeter of the uppermost interior surface is left ungreased). The receptacle, with bait inside, is placed upright such that the outer surface is in contact with some substrate in an area frequented by roaches. Roaches are attracted by the bait odor and enter the container. Once inside, they are unable to exit due to the petrolatum barrier.

The most effective baits employed to date have been fresh or dehydrated apple segments and/or pieces of sponge saturated with a commercially manufactured oil-based, apple flavoring compound (Piper and Frankie, 1978). However, there is a pronounced tendency for these baits to lose their attractiveness after several weeks of usage. Inexpensive, potent and residual synthetic attractants must be developed if trapping is to achieve widespread recognition, acceptance and utilization by pest control practitioners. The feasibility of utilizing synthesized behavior-modifying chemicals such as cockroach sex and/or aggregation pheromones to lure individuals to traps also should be explored and evaluated.

The homeowner is encouraged to become indirectly involved with any trapping effort. In instances where this technique is employed, residents are familiarized with the operational format and advantages of trapping. Further, they are briefed on how this particular method is integrated with others in a management scheme. Information on the types and numbers of roaches caught over an elapsed time interval and the relevance of the trapping results is presented to these individuals. We believe that time and effort expended in such activities generates homeowner interest which, in turn, contributes to an enhanced appreciation of the integrated management concept.

D. Natural Enemies

According to Clausen (1940), every insect species probably is attacked to some degree by one or more natural enemies. Cockroaches are no exception. They have many and varied associations with a wide range of organisms including bacteria, helminths, arthropods (mites, spiders, scorpions, centipedes and other insects) and vertebrates (amphibians, reptiles, birds and mammals) (Roth and Willis, 1960). Of these associates, insects, especially hymenopteran parasites of cockroach eggs, appear to offer the greatest potential for short- and long-term population regulation because they possess certain intrinsic characteristics most often associated with effective natural enemies (Flanders, 1947; Doutt and DeBach, 1964). Domiciliary cockroach eggs are destroyed in different areas of the world by representatives of six hymenopteran families (Roth and Willis, 1960). Accounts of the bionomics and/or host affinities of some of these parasitic species are given by Flock (1941), Edmunds (1954, 1955), Lawson (1954), Roth and Willis (1954, 1960), Cameron (1955, 1957), Gordh (1973), Vargas and Fallas (1974) and Fleet and Frankie (1975). However, it should be noted that many of the parasite-host relationships are still imperfectly known.

Relatively few biological control attempts have been made against pests of public importance and exceptional opportunities unquestionably exist in this field (Legner *et al.*, 1974). The potential for using natural enemies for the management of urban cockroaches has received virtually no serious consideration to date (Roth and Willis, 1954, 1960).

One of the major components of an integrated control program is the utilization of biological control techniques. Applied biological control techniques or procedures are of three types: importation, conservation (enhancement) and augmentation of natural enemies.

When insects invade new environments, either fortuitously or through the activities of man, they often leave their natural enemies behind; once freed from these regulatory agents, they often attain pest status. The objective of importation is to locate a pest's native habitat, obtain there its natural enemies, transport them to the invaded area and colonize them in the hope that they will become established, thrive and subjugate their host. The domiciliary cockroaches of the United States are, for the most part, exotic invaders, their areas of origin being the Ethiopian and Oriental biogeographical regions of the world (Rehn, 1945; Cornwell, 1968; Guthrie and Tindall, 1968). Cockroach natural enemy importation from these areas into the continental United States has never been attempted. Exploration for and multiple-species introductions of control agents is deemed a most advisable and promising endeavor.

Effective conservation and enhancement of established natural enemies is often essential to the success of a biological control effort. Conservation involves manipulation of the environment to favor biotic mortality agents, either through removal or mitigation of adverse factors. Enhancement entails usage of measures which increase the attractiveness of an area to natural enemies.

Insecticide overuse and/or indiscriminate application is representative of an adverse environmental factor (DeBach, 1974). Insecticide interference can render normally effective resident natural enemies ineffective. Improper usage can also inhibit the effectiveness or even prevent the establishment of newly-imported exotic enemies. Decreased parasite efficacy and survival resulting from excessive insecticide usage has been observed for *Tetrastichus hagenowii* (Ratzeburg) (Hymenoptera: Eulophidae), a naturally-occurring egg parasite of American and smokybrown cockroaches in Texas (Piper, Frankie and Loehr, in prep.). It was observed that parasite activity was very low or nonexistent where preventive insecticide treatments had been applied frequently and indiscriminately to the exterior of urban structures. On the other hand, parasitization rates were much higher around structures where insecticide usage had been limited to spot treatments or where no toxicants had been used. The findings suggest that parasite populations can be conserved and increased through either restricted application or elimination of roach-oriented chemicals in the outdoor environment.

Artificial manipulations designed to enhance the efficacy of cockroach natural enemies in the urban environment should be considered when planning control strategies. McKittrick (1964) found that female American and smokybrown cockroaches usually attach their egg cases to suitable substrates with a mucilaginous secretion produced by the salivary glands. Experiments indicate that this oral secretion may be the sign stimulus for the induction of searching behavior leading to egg case location and subsequent parasitization by *T. hagenowii* (Piper, in prep.). If synthesis of the attractant component is realized, the chemical could ultimately be used in a management scheme aimed at artificially increasing parasitization rates, through applications of the material to cockroach harborage or aggregation sites where large numbers of egg cases are usually present. Parasite activity would be concentrated at these sites, resulting in increased parasite efficacy.

Augmentation of naturally-occurring biological control agents is a technique that offers great promise; however, to date it has received little testing in urban areas. This approach involves propagation and periodic release of large numbers of biotic mortality agents. When cockroaches are effectively controlled by natural enemies in certain situations, or during certain seasons, periodic colonization is warranted. In urban residential areas, releases of this type would be made in small restricted habitats such as selected indoor/outdoor environments or in adjacent wooded areas.

DeBach (1974) and Stehr (1975) consider periodic releases to be of three major types: inoculative, supplementary and inundative. Before any type of release is contemplated, the parasite population density in the target area must be ascertained. This can be achieved by either collecting all visible cockroach egg cases to determine parasitization rate or, if egg cases cannot be found, by distributing a variable number of laboratory cultured, non-parasitized egg cases, each attached to an identifiable cardboard slip, throughout the area. After a specified time interval, the egg cases are removed and dissected to determine the presence or absence of parasites. These techniques have been used to assess the relative abundance of the cockroach egg parasite, *T. hagenowii*, in urban Texas situations (Piper and Frankie, 1978).

Outdoor inoculative releases may be made once or several times a year to reestablish a natural enemy whose populations are periodically or accidentally decimated in an area as a result of adverse conditions (e.g., unfavorable weather, excessive insecticide usage, etc.). Cockroach control results from the progeny of released parasites and from subsequent generations. Outdoor releases should be made in areas where cockroach activity and egg cases are regularly observed.

Supplementary releases of natural enemies can be made when sampling indicates a cockroach population is about to escape control by its natural enemies. Reestablishment of control is expected from the released individuals or their progeny (Stehr, 1975).

The objective of an inundative release is to completely overwhelm the pest with the release, with little or no reliance put on subsequent generations of the natural enemy (DeBach and Hagen, 1964). This type of release has enormous potential as both an outdoor and indoor urban cockroach control measure. A study of smokybrown cockroach seasonal activity in Texas revealed that outdoor populations increase rapidly during June and reach peak abundance in July and August (Frankie, Piper and Fleet, in prep.). It is suggested that inundative parasitoid releases be made early in the year to damp the annual population surge of this cockroach.

Experimentation with *T. hagenowii* has shown that inundation can be an effective control technique in closed systems such as residences, commercial buildings and storehouses (Piper, in prep.). American cockroach egg cases were distributed from floor to ceiling in kitchens, bathrooms and living rooms of several residences prior to wasp release. Subsequent parasitization data indicated that wasps found and oviposited into cases located in all parts of the rooms. Wasp:egg case ratios varied from 2:1 to 16:1, with maximum parasitization (57-62%) occurring at ratios of 8:1 to 12:1.

It should be mentioned that not all urbanites will be amenable to indoor parasite releases to control cockroaches just as many people will never use insecticides in their residences because of their reservations about the safety of the compounds. It is evident that public understanding and acceptance of the use of biological control agents will be a major contributory factor to the success of an urban roach mangement scheme.

E. Insecticides

Insecticides are important components of roach management programs. They are effective, convenient to use, adaptable to many situations and rapid in curative action. However, problems of mismanagement, overuse and resistance, which have arisen from habitual or prophylactic application of natural or synthetic killing agents, suggest that changes must be made relative to their employment. Reductions in frequency of insecticide applications and dosage rates are at the base of a cockroach management program. To accomplish these reductions, a pest control specialist must move toward more selective means of insecticide application and to replace routine fixed-schedule treatments with treat-when-necessary schedules. This approach is based on a sound knowledge of both cockroach and natural enemy behavior and ecology. Insecticide selectivity can be enhanced substantially and application rates reduced by specific timing or placement of the chemical in relation to the cockroach's behavioral characteristics. For example, Frankie, Piper and Fleet (in prep.) presented evidence suggesting that insecticide applications to outdoor areas during the cooler months of the year in Texas represent a poor practice since most roaches are apparently inactive at this time. If exterior spraying becomes necessary, applications should be made just prior to and during the June cockroach population increase and periodically thereafter when warranted. Treatments should be restricted to outdoor cockroach harborage sites and potential entry areas of a structure.

Organophosphate, organochlorine and carbamate insecticide spray applications are most frequently used to combat roaches. Extensive field experiments by Ebeling *et al.* (1968) and Gupta *et al.* (1973) have shown that the use of insecticidal dusts in conjunction with insecticidal sprays or pyrethrins is superior to using either dusts, sprays or pyrethrins alone. An inorganic dust such as boric acid is a very effective roach control material (Ebeling, Chapt. 9, this volume) and should be included in a management program (Piper and Frankie, 1978).

Boric acid is only occasionally employed by most pest control operators. Operators rationalize disuse by stressing that dust applications are often too time-consuming and messy. What they fail to mention is that the dust provides effective, long-term roach control thereby reducing or eliminating the necessity for monthly or quarterly maintenance schedules. To most pest control practitioners this represents a loss of income. It is most unfortunate when such profit-motivated considerations prevent the utilization of an effective roach control practice.

IV. OPERATIONAL MANAGEMENT PROGRAMS

Figures 2 and 3 diagramatically illustrate the containment strategies available for integration into management programs for the five most prevalent and offensive urban cockroaches in the United States. It is most important to realize that every cockroach infestation is, to a certain extent, a unique situation. Objective decisions for optimal management in a particcular instance depend upon a thorough bioecological understanding of the species involved and application of such knowledge within a framework tempered by educational, social and economic considerations. Successful cockroach management will result only when pest managers make a serious and conscientious effort to treat each infestation as an individual case, diagnosing relevant details and establishing controls accordingly.

The authors have followed these guidelines in the develop-

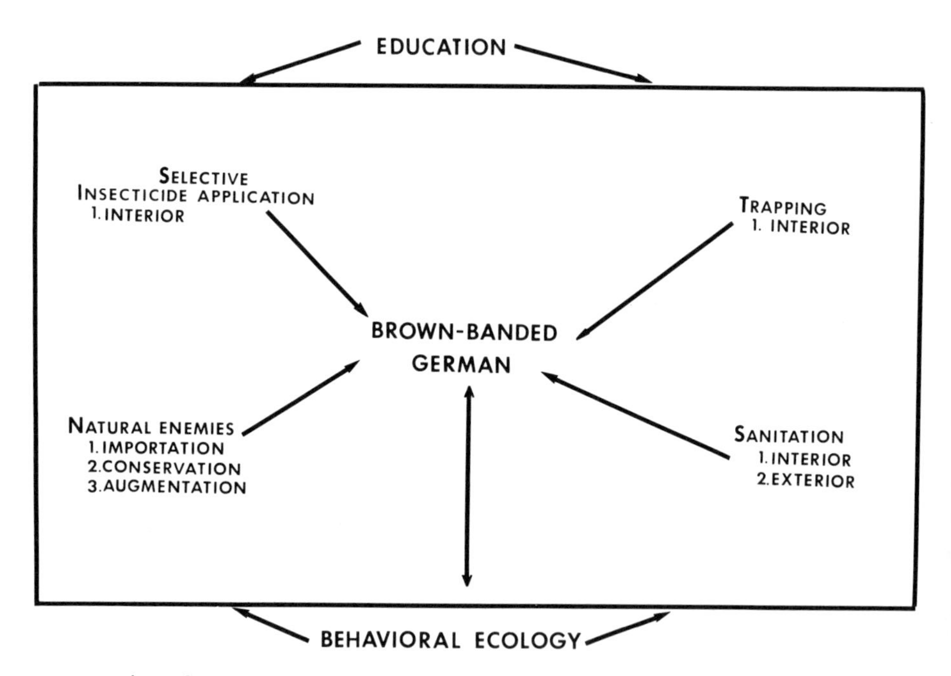

Fig. 2. Component tactics of a management program designed for utilization against German and brown-banded cockroaches.

ment and implementation of experimental integrated control programs against several cockroach species in urban Texas environments. A full account of several case history studies is provided in Piper and Frankie (1978).

V. PROSPECTUS

To bring the promise of cockroach management to fruition, all persons connected with a program, either directly or indirectly, must be cognizant of the philosophies, goals and tactics of roach management. Particularly essential to the implementation and success of a managed system is an informed citizenry and pest control industry. An integrated program incorporates multiple pest suppression techniques and, therefore, demands a greater knowledge of pest behavior, biology and ecology than does contemporary pest control. Additionally, these programs at the onset require greater time involvement and decision-making on the part of the pest control operator.

Presently, we are uncertain as to how many contemporary

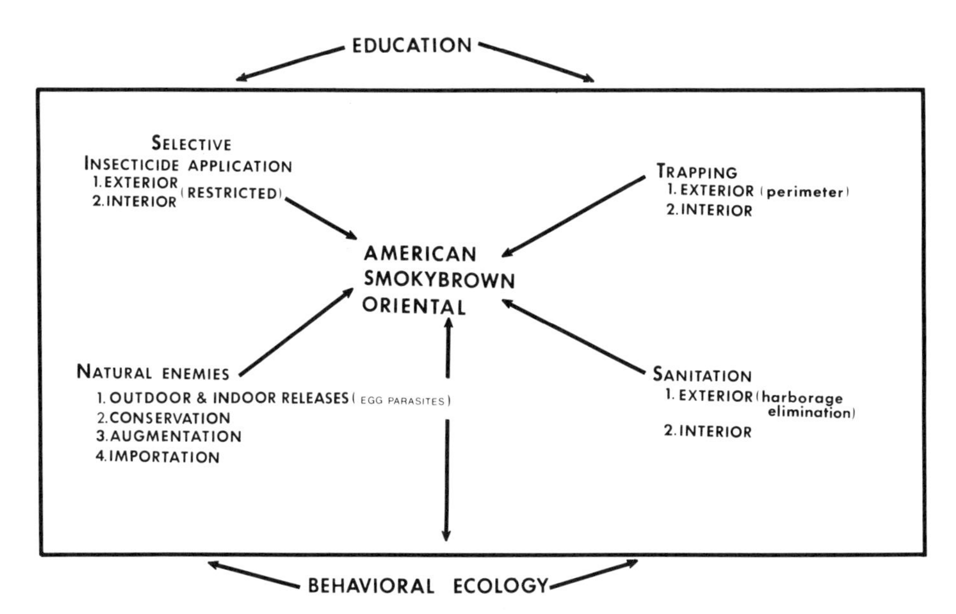

Fig. 3. Component tactics of a management program designed for utilization against American, oriental and smokybrown cockroaches.

pest control operators in Texas or in other states could actually set up and maintain a program as proposed in this paper. Our uncertainty stems from a lack of information on the general knowledge and level of expertise in the pest control industry. Added to this problem is the realization that some appreciation for conceptional thinking must underlie any integrated management program, and we have limited evidence to suggest that most pest control operators lack the exposure necessary to grasp theoretical concepts (Piper and Frankie, 1978). During the next five years, we expect that only the most knowledgeable pest control practitioners may attempt to use the integrated approach. However, we also believe that certain trends in pesticide use (as described earlier) may attract more highly trained individuals into this industry. We envision that many of these people will hold college degrees in entomology, ecology or environmental sciences and will likely practice a more sophisticated brand of pest control. It is highly conceivable that some of these trained individuals will become consultants to other pest control operators and thereby provide the latter group with newer techniques and approaches, one of which will be the integrated approach. Such services, termed supervised control, are presently available in agroecosystems.

ACKNOWLEDGMENTS

We sincerely thank Drs. W. Ebeling, University of California, Los Angeles; R. R. Fleet, Stephen F. Austin University, Nacogdoches, Texas; D. A. Reierson, University of California, Riverside; and L. M. Roth, U.S. Army Laboratories, for reviewing the manuscript and for providing pertinent and appreciated suggestions.

REFERENCES

Beatson, S. H., and J. S. Dripps. 1972. Long-term survival of cockroaches out-of-doors. *Environ. Health. Oct.*

Bernton, H. S., and H. Brown. 1964. Insect allergy: Preliminary studies of the cockroach. *J. Allergy 35*: 506-513.

Bernton, H. S., and H. Brown. 1970a. Cockroach allergy: Age of onset of skin reactivity. *Ann. Allergy 28*: 420-422.

Bernton, H. S., and H. Brown. 1970b. Insect allergy: The allergenicity of the excrement of the cockroach *Blattella germanica*. *Ann. Allergy 28*: 543-547.

Cameron, E. 1955. On the parasites and predators of the cockroach. I. *Tetrastichus hagenowii* (Ratz.). *Bull. Entomol. Res. 46*: 137-147.

Cameron, E. 1957. On the parasites and predators of the cockroach. II. *Evania appendigaster* (L.). *Bull. Entomol. Res. 48*: 199-209.

Clausen, C. P. 1940. Entomophagous insects. McGraw-Hill, New York.

Cornwell, P. B. 1968. The cockroach, Vol. I. Hutchinson, London.

DeBach, P. 1974. Biological control by natural enemies. Cambridge Univ. Press, London.

DeBach, P., and K. S. Hagen. 1964. Manipulation of entomophagous species. *In* Biological control of insect pests and weeds (P. DeBach, ed.). Reinhold, New York, pp. 429-458.

Dold, J. 1964. How to trap and rear roaches for display and resistance testing. *Pest Contr. 32*: 18-20.

Doutt, R. L., and P. DeBach. 1964. Some biological control concepts and questions. *In* Biological control of insect pests and weeds (P. DeBach, ed.). Reinhold, New York, pp. 118-142.

Ebeling, W. 1975. Urban entomology. Univ. Calif. Div. Agric. Sci., Berkeley, California.

Ebeling, W., R. E. Wagner, and D. A. Reierson. 1966. Influence of repellency on the efficacy of blatticides. I. Learned modification of behavior of the German cockroach. *J. Econ. Entomol. 59*: 1374-1388.

Ebeling, W., D. A. Reierson, and R. E. Wagner. 1968. The influence of repellency on the efficacy of blatticides. III. Field experiments with German cockroaches with notes on three other species. *J. Econ. Entomol. 61*: 751-761.

Edmunds, L. R. 1954. A study of the biology and life history of *Prosevania punctata* (Brulle) with notes on additional species (Hymenoptera: Evaniidae). *Ann. Entomol. Soc. Am. 47*: 575-592.

Edmunds, L. R. 1955. Biological notes on *Tetrastichus hagenowii* (Ratzeburg), a chalcidoid parasite of cockroach eggs (Hymenoptera: Eulophidae; Orthoptera: Blattidae). *Ann. Entomol. Soc. Am. 48*: 210-213.

Flanders, S. E. 1947. Elements of host discovery exemplified by parasitic Hymenoptera. *Ecol. 28*: 299-309.

Fleet, R. R., and G. W. Frankie. 1974. Habits of two household cockroaches in outdoor environments. Texas Agric. Exp. Sta. Misc. Publ. 1153.

Fleet, R. R., and G. W. Frankie. 1975. Behavioral and ecological characteristics of a eulophid egg parasite of two species of domiciliary cockroaches. *Environ. Entomol. 4*: 282-284.

Fleet, R. R., G. L. Piper, and G. W. Frankie. 1978. Movement of the smokybrown cockroach, *Periplaneta fuliginosa*, in an urban environment. (in manuscript for submission to *Environ. Entomol.*)

Flock, R. A. 1941. Biological control of the brown-banded roach. *Bull. Brooklyn Entomol. Soc. 36*: 178-181.

Gordh, G. 1973. Biological investigations on *Comperia merceti* (Compere), an encyrtid parasite of the cockroach *Supella longipalpa* (Serville). *J. Entomol. (A) 47*: 115-123.

Gupta, A. P., Y. T. Das, J. R. Trout, W. R. Gusciora, D. S. Adam, and G. J. Bordash. 1973. Effectiveness of spray-dust-bait combinations and the importance of sanitation in the control of German cockroaches in an inner-city area. *Pest Contr. 41*: 20-26, 58-62.

Guthrie, D. M., and A. R. Tindall. 1968. The biology of the cockroach. St. Martin's Press, New York.

Jackson, W. B., and P. P. Maier. 1955. Dispersion of marked American cockroaches from sewer manholes in Phoenix, Arizona. *Am. J. Trop. Med. Hyg. 4*: 141-146.

Lawson, F. 1954. Observations on the biology of *Comperia merceti* (Compere) (Hymenoptera: Encyrtidae). *J. Kansas Entomol. Soc. 27*: 128-142.

Legner, E. F., R. D. Sjogren, and I. M. Hall. 1974. The biological control of medically important arthropods. *CRC Crit. Rev. Environ. Contr. 4*: 85-113.

Lord, T. H., V. D. Foltz, and R. van Sickle. 1964. Natural incidence of *Staphylococcus aureus* Rosenbach in the brown-banded cockroach. *J. Insect. Pathol. 6*: 21-25.

McKittrick, F. A. 1964. Evolutionary studies of cockroaches.

Cornell Univ. Agric. Exp. Sta. Mem. 389.

Moore, R. C. 1971. Chemical control of German cockroaches in urban apartments. Conn. Agric. Exp. Sta. Bull. 717.

Nat. Pest Contr. Assoc. 1966. Cockroaches and their control. Tech. Release No. 9-66.

Olkowski, W., H. Olkowski, R. van den Bosch, and R. Hom. 1976. Ecosystem management: A framework for urban pest control. *BioScience 26*: 384-389.

Piper, G. L., R. R. Fleet, G. W. Frankie, and R. E. Frisbie. 1975. Controlling cockroaches without synthetic organic insecticides. Texas Agric. Exp. Sta. & Ext. Serv. Leafl. 1373.

Piper, G. L., and G. W. Frankie. 1978. Integrated management of urban cockroach populations: Final report. U.S. Environmental Protection Agency, Washington. D. C.

Rehn, J. A. G. 1945. Man's uninvited fellow traveler - the cockroach. *Sci. Month. 61*: 265-276.

Roth, L. M., and E. R. Willis. 1954. The biology of the cockroach egg parasite, *Tetrastichus hagenowii* (Hymenoptera: Eulophidae). *Trans. Am. Entomol. Soc. 80*: 53-72.

Roth, L. M., and E. R. Willis. 1957. The medical and veterinary importance of cockroaches. Smithson. Misc. Collect. 134.

Roth, L. M., and E. R. Willis. 1960. The biotic associations of cockroaches. Smithson. Misc. Collect. 141.

Schoof, H. F., and R. E. Siverly. 1954. The occurrence and movement of *Periplaneta americana* (L.) within an urban sewerage system. *Am. J. Trop. Med. Hyg. 3*: 367-371.

Stehr, F. W. 1975. Parasitoids and predators in pest management. *In* Introduction to insect pest management (R.L. Metcalf and W.H. Luckmann, eds.). John Wiley and Sons, Inc., New York, pp. 147-188.

Truman, L. C. 1961. Lesson no. 6. Cockroaches. *Pest. Contr. 29*: 21-28.

van den Bosch, R., and P. S. Messenger. 1973. Biological control. Intext Educational Publishers, New York.

Vargas, M., and F. Fallas. 1974. Notes on the biology of *Tetrastichus hagenowii* (Hymenoptera: Eulophidae): A parasite of cockroach oothecae. *Entomol. News 85*: 23-26.

Washburn, F. L. 1913. A successful trap for cockroaches. *J. Econ. Entomol. 6*: 327-329.

Withlaw, J. T., Jr., and L. W. Smith, Jr. 1964. Equipment for trapping and rearing the American cockroach, *Periplaneta americana*. *J. Econ. Entomol. 57*: 164-165.

POTENTIAL FOR DEVELOPING INSECT-RESISTANT PLANT MATERIALS FOR USE IN URBAN ENVIRONMENTS

David L. Morgan

Texas Agricultural Experiment Station
Texas A&M University Research & Extension Center at Dallas
Dallas, Texas

Gordon W. Frankie[1]

Department of Entomology
Texas A&M University
College Station, Texas

Michael J. Gaylor[2]

Texas Agricultural Experiment Station
Texas A&M University Research & Extension Center at Dallas
Dallas, Texas

---

[1] *Present address: Department of Entomological Sciences, University of California, Berkeley, California.*

[2] *Present address: Department of Entomology and Zoology, Auburn University, Auburn, Alabama.*

ISBN 0-12-265250-9

I. INTRODUCTION

Ecological and evolutionary aspects of insect-plant relationships have recently received considerable attention from biologists in basic and applied fields (de Wilde and Schoonhoven, 1969; Chamber, 1970; Sondheimer and Simeone, 1970; Watson, 1970; van Emden, 1973; Chew, 1974). This interest may in part be attributed to evolving ecological theory and to the availability of newer methods[3] of analysis and testing of natural plant and animal products. Theoretically, at least, many recent studies on insect-plant relationships also stress the need to view these interactions from the standpoint of both the insect and the plant. It follows, therefore, that expertise of entomologists and plant scientists must often be combined to adequately investigate and evaluate given systems. It is in this integrated context that we present an account of recent research developments and new considerations in the field of insect-host plant resistance as they pertain to urban environments. We also interface information on attitudes and practices of man as they actually or potentially relate to the development and ultimate usage of resistant plant materials in these environments.

The text consists of five major sections. Parts II and III are concerned with a historical perspective on insect-host plant resistance in agricultural, forest and urban environments. Two current programs on host plant resistance in urban areas are included. Factors affecting development and acceptance of resistant plant materials for urban use are considered in Part IV. Finally, the prospects of utilizing this approach in future insect management programs are explored in Parts V and VI.

---

[3] *These methods include several kinds of chromatography and mass spectrometry for analyses of natural products, and electrophysiological methods for recording insect responses to plant chemicals.*

II. INSECT-HOST PLANT RESISTANCE: AN OVERVIEW

The use of insect-resistant plant materials is a valuable tool in the protection of many field and orchard crops. Painter (1951) argued that because insects frequently can be a limiting factor in crop yields, insect resistance should be a normal part of a plant breeding program, receiving equal consideration with plant disease, susceptibility to drought and cold, and other environmental factors. He further suggested that there are two general sources of resistance to be considered: those provided by the variability within crop species and those in plant species closely enough related to the crop species to cross with it. The former, he said, provides greater opportunity for finding resistance. Painter delineated research efforts in corn, wheat, cotton, potato, sorghum and other crops in which the planting of insect-resistant varieties was a major reason the crop could be economically produced. Two outstanding cases given were phylloxera-resistant grapes in France and Hessian fly-resistant wheat in the United States.

Host plant resistance research also has been conducted on various tree species. Many of them are conifers that have been developed for use in a forest environment or for eventual use as Christmas trees (Hall, 1937; Austin *et al.*, 1945; Kawahata, 1955; Beier-Petersen and Soegaard, 1958; Harris, 1960; Holst, 1963; Beck, 1965; Plank and Gerhold, 1965; Coutts and Dolezal, 1966; Gerhold, 1966; Ewert, 1967; Mitchell and Nagel, 1969; Thielges and Campbell, 1972; van Buijtenen and Santamour, 1972). Gerhold (1966) lists 39 species of forest trees worldwide, including deciduous representatives, in which insect resistance has been found. Most of these are *Pinus* spp., but also included are *Larix, Abies, Alnus, Castanea, Quercus, Robinia, Populus, Salix,* and *Juglans* spp.

In the 26 years since Painter's book was published, much research has been designed by American experiment station and U.S. Department of Agriculture workers to identify and propagate insect-resistant plant materials and to understand the basic insect and plant mechanisms involved in resistance (Beck, 1965).

Painter (1951) defined resistance as "the relative amount of heritable qualities possessed by the plant which influence the ultimate degree of damage done by the insect." According to this definition, which is widely accepted by entomologists, resistance cannot be changed by environmental factors. Painter included under the term pseudoresistance factors such as host evasion and induced resistance which may be affected by environmental factors. However, Painter (1966) recognized that pseudoresistance can be useful in controlling insects.

In contrast, van Emden (1966) defined resistance as any means by which the plant may reduce insect attack, including escape. The study of resistance is not limited to insects; indeed, some plant pathologists recognize the possibility that the expression of resistance to disease may be modified by environmental factors (Walker, 1965; Strobel and Mathre, 1970). Because of the possible usefulness in a diverse urban environment of plant characteristics that may be modified by environmental conditions, the term "resistance" is used in the broadest sense in this discussion.

In urban areas, levels of resistance expressed by ornamental plants may be affected by many management practices which vary more widely than those used in field, orchard, or forest plantings. For example, well adapted varieties of fruit trees usually are planted in commercial orchards, and within a particular geographical area most growers follow similar irrigation, fertilization and cultural practices. In contrast, poorly adapted ornamentals are often knowingly or unknowingly planted in urban areas. Once planted in urban environments, some of these plants receive almost no fertilizer or supplemental water, while others receive proper amounts, and still others receive excessive amounts. Johnson (1968) offers a pertinent review and discussion of insect responses to trees that have been conditioned artificially with water and fertilizer.

Some modification of resistance has been recognized where plants have been moved from native environments to nearby urban areas. For example, larvae of the cypress bark moth, *Laspeyresia cupressana* (Kearf.) (fam. Tortricidae), feed almost exclusively on cones of their primary host, Monterey cypress, *Cupressus macrocarpa* Hartw., in endemic stands in California (Frankie and Koehler, 1971). In this feeding site, *Laspeyresia* is considered innocuous. However, when the cypress is planted away from native stands, e.g., urban areas, the lepidopteran also infests the inner bark of trunks and primary branch nodes where it may cause severe resin exudation. Change in insect-plant interaction is related to better soil conditions and various cultural activities practiced in these new habitats. Trees in adventive situations grow faster and in the process produce thinner bark, which ultimately leads to reduced resistance by facilitating entry of *Laspeyresia* into these "new" host feeding sites (Frankie and Koehler, in prep.).

A similar case, involving Texas mountain laurel, *Sophora secundiflora* Lag., and a lepidopteran, *Uresiphita reversalis* (Gn.) (fam. Pyralidae), has also been recognized in Texas (Gaylor, personal observation). In its native dry habitats in southwestern United States and Mexico, *Sophora* is not known to be noticeably damaged by any insect. However, when planted in urban environments, *Sophora* is often damaged severely by *Uresiphita* (Simpson, 1977).

Additionally, the perception of what constitutes damage to

a tree in an urban environment compared to a tree in a forest could determine whether the tree is considered resistant. In a forest, a tree is damaged by a pest when an infestation results in a loss of wood production. A tree in an urban environment can be "damaged" when its appearance is not pleasing to a human observer even if no loss of wood production occurs. For example, some aphids that do not measurably weaken a tree, but which produce honeydew on which sooty mold grows, would not damage a tree in a forest, but would in an urban environment. Further, usage patterns within urban areas (as described later) will also determine, to a large extent, plant damage.

Great varietal differences exist in the susceptibility of ornamental plant species to pests (Hamilton, 1924; Wyatt, 1965; Markkula *et al.*, 1970; Anonymous, 1971; Poe and Wilfret, 1972; Fischer and Shanks, 1974; Whitcomb, 1975; Johnson and Lyon, 1976). However, few host plant resistance studies have been conducted on ornamentals (Helgesen and Tauber, 1974), even though the need for such research has been evident for many years.

Around the turn of the century, entomologists in New York began to recommend the use of insect-resistant species, and they prepared ratings representing comparative resistance of certain shade trees (Felt, 1905). More recently, selection and vegetative propagation of trees exhibiting natural insect resistance has been encouraged (Callaham, 1966; Painter, 1966). Tree selection has been recommended for controlling insects on trees, including midges on Douglas fir (Mitchell and Nagel, 1969), *Psylla* on *Acacia* and *Albizia* (Munro, 1965), and a stem gall moth, *Periploca ceanothiella* (Cosens), on *Ceanothus* (Munro, 1963).

Some research has been conducted on selecting shrubs resistant to insect attack. In the state of Washington, screening tests among thousands of rhododendrons in hobbyists' gardens have indicated the existence of predictable interspecific patterns of susceptibility and resistance to weevils (Campbell, 1977). Several species of weevils infest rhododendrons in Washington. The most important are the woods weevil, *Nemocestes incomptus* (Horn), the obscure weevil, *Sciopithes obscurus* Horn, and the black vine weevil, *Otiorhynchus sulcatus* F. Although weevil larvae feed on root tissue, the greatest concern is due to foliage damage by adults. Campbell reports that differences in weevil response have been consistent enough to allow for categorization of the rhododendrons into susceptible and resistant species and hybrids. This information is now being made available to homeowners and growers. In cooperation with the USDA, an attempt at finding the basis for the observed resistance is in progress. It is hoped that the results will be useful to breeders interested in incorporating weevil resistance into new hybrids.

A few studies designed to determine the basis for resistance have been conducted. Lines of white spruce, *Picea glauca* Voss., and Norway spruce, *P. abies* Karst., have been found resistant to the eastern spruce gall aphid, *Adelges abietis* L., a serious pest in the midwestern and northeastern United States (Thielges and Campbell, 1971). Asexual propagation of selected trees was recommended so that the resulting clones could be used for further resistance testing. Using thin-layer chromatography, Tjia and Houston (1975) found a higher content of phenols in *P. abies* resistant to the adelgid than in susceptible trees.

Limited research also has been conducted to locate turfgrass cultivars resistant to insects. Leuck *et al.* (1968) evaluated 441 clones of Bermuda grass, *Cynadon dactylon* Pers., in Georgia for resistance to first instar fall armyworms, *Spodoptera frugiperda* (J. E. Smith), and found 11 accessions resistant or intermediate in resistance to larval attack. Leuck and Skinner (1970) determined that different Bermuda grasses as diets of fall armyworms may be important in the natural control of these insects, because larval feeding on a nonpreferred host produced few adults.

Bermuda grass accessions have been evaluated in Georgia for resistance to adult two-lined spittlebug, *Prosapia bicinata* (Say), (Stimmann and Taliaferro, 1969; Taliaferro *et al.*, 1969). Resistance was shown to be due partly to preferential feeding and partly to tolerance to the toxin injected with the insect's saliva. One accession, P.I.-289931, shown to be a nonpreferred host for feeding by first instar fall armyworm larvae in other investigations, was reported to be a nonpreferred host for adult spittlebugs.

Pass *et al.* (1965) reported variability among strains of Kentucky bluegrass, *Poa pratensis* L., and suggested the possibility of resistance to sod webworm, *Crambus* spp., damage. Resistant strains had heavier weights of rhizomes during May than susceptible strains, indicating the possibility of a relationship between stored food reserves prior to the stress period and resistance to sod webworm injury during the summer in Kentucky (Buckner *et al.*, 1969). Kindler and Kinbacher (1975) evaluated 15 cultivars of Kentucky bluegrass under field conditions in Nebraska for resistance to bluegrass billbug, *Sphenophorus parvulus* Gyllenhal, and none escaped infestation or damage. A range of resistance was shown, with Nebraska Common, Park, and South Dakota Certified showing the lowest infestation rates.

Tolerance to the Rhodes grass scale, *Antonina graminis* (Maskell), which is a serious pest of St. Augustine grass, *Stenotaphrum secundatum* (Walt.) Kuntze, and Bermuda grass, has been illustrated in Rhodes grass, *Chloris gayana* Kunth, and a resistant cultivar, Bell, was released in Texas (Schuster and

Dean, 1973).

Accessions of St. Augustine grass in Florida have also been located which show a high degree of resistance to the southern chinch bug, *Blissus insularis* Barber. Chinch bug resistance will be discussed in greater depth later in the chapter.

There are several reasons why so little research is being conducted on breeding insect-resistant ornamental plants for use in urban environments. In the past, most ornamentals breeding has been conducted by private individuals and companies. Few individuals working alone have the expertise or the resources to implement successful breeding programs for insect resistance. Private companies have felt no demand, expressed through the marketplace, for developing pest-resistant plants. The demands on individuals and breeders have instead been oriented toward selecting and propagating desired morphological characteristics. For example, Burpee Seed Company, after 21 years of offering $10,000 for development of a white marigold, finally paid the bounty in 1975 (Neary, 1976).

In the United States, federal, state and commercial research organizations have been concerned historically with the production of food, feed and fiber crops. Although the needs of essentially urban populations have grown, agricultural lobbies have contributed toward maintaining primary emphasis on traditional crop production. Recently, there has been more emphasis placed on city-oriented research by some land grant universities; for example, from 1973 to 1977 Texas A&M University established five urban research positions at the Texas A&M University Research and Extension Center at Dallas. Much more national effort is needed, and professional grounds maintenance associations, the pest control industry, nurserymen's organizations, arborists' organizations, and garden clubs can help influence appropriate bodies through lobbying activities and personal contacts with legislators and civic leaders.

The present lack of emphasis on urban research in most states may be due to several factors. Scientists who conduct research in urban settings are inundated by problems and challenged daily by a public that demands immediate solutions. This situation often forces scientists to concentrate most, if not all, of their efforts toward developing short-term solutions. As a result, many scientists simply do not have time to devote to relatively long-term, pest-resistance breeding programs. As more scientists begin conducting research on problems affecting urbanites, they should be able to devote more time to projects designed to provide more lasting solutions.

Another reason so little research is being conducted on developing insect-resistant ornamentals may be due to the belief by some entomologists that the public will not tolerate insects on ornamental plants. The probability of finding levels of resistance high enough to satisfy a public which has a strong aversion to insects (Olkowski and Olkowski, 1976) may seem small. However, entomologists often forget that the public usually is not as aware of insect populations as are professional entomologists. Few people notice pests until populations are very large, or until after damage would have been obvious to an entomologist. Helgesen and Tauber (1974) found that a density of 0.3 to 0.7 whiteflies/cm^2 was acceptable on commercially grown poinsettias. Much higher aesthetic damage levels would be acceptable on woody ornamentals grown as landscape plants, particularly where the distance between the urbanite and the plant is great and/or the degree of contact between urbanite and plant is low. Freeway, cemetery and certain park plantings provide examples of this kind of situation. Unless insect populations are very large, complaints about insects on ornamentals may not be related to the size of the population. Olkowski and Olkowski (1976) stated that most citizen complaints involve cases where little damage, or few or no insects, occur.

Although all these reasons are important, the primary reason so little research is being conducted on breeding pest-resistant ornamental plants probably is that a large proportion of urban populations is unaware of the possibilities of developing and using plant materials less susceptible to pests. Many urbanites believe that the only way, or at least the best way, to control pests is by chemicals. Consequently, nurserymen, seed companies and universities have felt little public pressure to develop pest-resistant ornamentals. Urbanites must be informed of the possibilities and benefits to be gained before they can be expected to begin to influence administrators and legislators

III. INSECT HOST PLANT RESISTANCE STUDIES IN PROGRESS

The case histories discussed below represent examples of diverse research on grasses and trees. They are research projects whose terminations are not planned; each is a continuing effort. It is the purpose here to present them as such, and to share with the reader the significance of the data collected to this point.

The first program is concerned with the development of a St. Augustine grass resistant to southern chinch bug feeding in Texas and Florida. Most of the work to date has been conducted in Florida. The second research effort is concerned with the development of gall-resistant oak trees in Texas.

A. St. Augustine Grass Resistance to Chinch Bugs

The improved St. Augustine grass, *Stenotaphrum secundatum* (Walt) Kuntze 'Floratam', is being used increasingly in Florida and Texas where St. Augustine Decline (SAD) and/or southern chinch bug, *Blissus insularis* Barber, are major problems. 'Floratam' originated as a SAD-resistant cultivar of St. Augustine grass and was a joint release of the University of Florida and the Texas Agricultural Experiment Station. Subsequently, 78 accessions of St. Augustine grass were evaluated in the field for resistance to the southern chinch bug in Florida (Reinert, 1972). Nine of the accessions had seasonal mean populations of 1.5 or fewer chinch bugs/0.09 m^2 of turf. In the laboratory three of the selections exhibited 49 to 58% antibiosis to the adult chinch bug populations confined on them for four days and 71 to 91% antibiosis to fifth instars confined for seven days (Reinert and Dudeck, 1974). Resistance to the southern chinch bug in one of the accessions, 'Floratam', was confirmed by Carter and Duble (1976). In further testing, Reinert (1977a) has found 10 additional St. Augustine grass accessions having varied levels of resistance to chinch bugs.

'Floratam' potentially has great impact in Florida and Texas. In Florida, St. Augustine grass is the most common turf used in ornamental plantings, comprising 46% of the home lawns (McGregor, 1976). It is estimated that St. Augustine grass accounts for 56% of the lawns in Texas (Carter and Duble, 1976) and 96% of the lawns throughout the Gulf Coast areas. The southern chinch bug is the most serious insect pest of St. Augustine grass in Florida and other southern coastal states (Reinert, 1972; Reinert and Dudeck, 1974). Stroble (1971) estimated that more than $25 million was spent annually to control this insect in Florida. The replacement value of turf in Florida exceeds $800 million (Horn *et al.*, 1973).

'Floratam' compares favorably with other cultivars of St. Augustine grass in color, soil adaption, and growth characteristics, though it is coarser than the common cultivar. It is also a more vigorous cultivar. In Texas it shows critical winter damage in the northern and central areas of the state. During the severe winter of 1976 to 1977, 'Floratam' was severely damaged in North Texas. Evaluations are continuing on this new cultivar.

B. Live Oak Resistance to Mealy Oak Gall

The mealy oak gall wasp, *Disholcaspis cinerosa* Bass. (Hymenoptera: Cynipidae), induces gall formation on live oaks[4] in Texas, western Louisiana and Mexico (Frankie *et al.*, 1977; Frankie, in prep.). The *Disholcaspis*-oak system was selected as a model for study in host plant resistance for several reasons. First, the host plants are widely used in landscape plantings. Second, the presence of large numbers of *Disholcaspis* galls causes concern to some people because they consider the gall unsightly, or because they fear plant damage will result. However, it is known that even the most severe infestations cause no measurable harm to the tree (Frankie *et al.*, 1977; Frankie, in prep.). Third, some people have indicated that they find the gall attractive. This appreciation is based on the gall's unique character and an awareness of its innocuous effect on the oak. It is noteworthy that a few people have expressed interest in the gall's potential use in artistic displays (Fig. 1).[5] Fourth, genetic variation in live oak is high, with great differences in gall-forming capacity among individual trees. Finally, the live oak tree can be propagated asexually, allowing for cloning of resistant and susceptible genotypes.

1. The Insect: Disholcaspis cinerosa *Bass.*

D. cinerosa has two generations per year on live oak, each of which induces a different gall type on affected host trees (Frankie *et al.*, 1977; Morgan, Frankie and Gaylor, in prep.). The more obvious growth induced by *D. cinerosa*, and the one that causes concern to homeowners, is roughly spherical in shape (10-30 mm in diam.) and is called the mealy oak gall (Fig. 2). This structure is found on branches and occasionally

---

[4]*Live oak in Texas is known to consist of four species:* Quercus virginiana *Mill.,* Q. fusiformis *Small.,* Q. oleoides *Cham. & Schlecht., and* Q. minima *(Sarg.) Small.* Q. virginiana *is coastal in natural distribution and extends eastward along the southeastern seaboard; it is the "live oak" known to the Deep South.* Q. fusiformis *is the live oak of central and southwestern Texas.* Q. minima *is a shrubby tree growing in deep sands of the east and central Texas Gulf Coast, and* Q. oleoides *is primarily a Mexican oak, but is found as far north as the sand of Aransas County (Muller, 1970).*

[5]*There are other galls on oak species that are considered attractive. Homeowners in California occasionally have expressed positive interest in two gall types provided by* Antron douglasi *(Ashm.) and* Andricus kingi *Bass. on* Q. lobata *Nee (Koehler, 1977) suggesting that propagating for susceptibility as well as resistance may be desirable.*

Fig. 1. Decorative arrangement of mealy oak galls in a home in College Station, Texas.

on trunks. The parthenogenetic adults of this generation emerge from the galls in early winter and oviposit in swollen leaf buds. These eggs give rise to the sexual generation which forms within the second gall type (3-4 mm long) as new leaves appear in early spring (Fig. 3). Adults of this generation emerge, mate and oviposit primarily into branches shortly before the new spring leaves are fully expanded. Thus, the yearly cycle is completed.

2. Experimental Studies

D. cinerosa gall-carrying capacity of live oak was investigated in Dallas from 1974 to 1976. The study was designed

Fig. 2. Asexual generation galls of Disholcaspis cinerosa *on a live oak branch.*

Fig. 3. Sexual generation gall of Disholcaspis cinerosa *arising from leaf shoot on a live oak branch. Note emergence hole.*

to gain insight on patterns of gall formation on selected trees through time. A complete account of this study is presented in Morgan, Frankie and Gaylor (in prep.). For the purpose of this paper, the following summary is provided.

In 1973, a large field planting of live oaks, *Q. virginiana* (about 1,300 trees), was located within the city limits of Dallas, Texas. The seven-year-old trees were owned by one of the city cemeteries and were maintained as stock material for new plantings on nearby cemetery grounds. All trees were surveyed for past heavy infestations of *D. cinerosa,* as evidenced by the persistent asexual generation galls. These trees were classified as apparently-susceptible (A-SUS) individuals. Trees directly adjacent to the A-SUS individuals having very few or no *D. cinerosa* galls were classified as apparently-resistant (A-RES). A total of 13 trees, seven A-SUS and six A-RES, was selected for study. During 1974 to 1976, sexual and asexual generation *D. cinerosa* were reared from galls collected throughout Dallas. A standard number of wasps from each generation were caged on each of several branches of each tree, and the resulting number of galls was subsequently recorded.

Results of the caging study indicated that A-RES trees have little capacity to form galls of either generation. Gall-carrying capacity remained relatively low on all A-RES trees during the three year period, averaging 2.0 galls per caged branch. In the case of A-SUS trees, however, an average of 14.3 galls per caged branch was recorded.

Resistance may persist for the life of the tree since extensive surveys of live oaks in several Texas cities and natural areas revealed that most trees, about 80%, had little or no tendency to form *D. cinerosa* galls (Frankie *et al.*, 1977). At the opposite extreme, the A-SUS trees varied greatly in their capacity to form *D. cinerosa* galls. Further, this capacity appeared to change through time on a given tree. It is tempting to suggest that annual differences in weather accounted for this phenomenon. However, during the same year some trees were forming fewer galls, other nearby A-SUS trees formed more galls, indicating that weather probably is not the only determining factor. If climatic factors are important, they apparently affected the trees differently. It also seems conceivable that aging processes and/or other physiological changes, acting independently of weather, may have altered a tree's gall-forming capacity (Frankie, in prep.).

3. *Propagating Gall Resistant Live Oaks*

Because live oaks are wind pollinated, their progeny vary greatly in genetic composition. Texas nurserymen puzzle over trees which differ in drought hardiness, salt tolerance, and appearance in height, shape, branching habit, growth rate, and even leaf color. Recently, however, new techniques have been developed enabling horticulturists to propagate live oak asexually, thereby establishing clones of genetically identical plant materials (Morgan and McWilliams, 1976). Through these practices, oak trees resistant or susceptible to the gall maker's activities can be produced. The use of woody cuttings from the basal section of the tree, and root cuttings, have been the most successful methods. Because rooting success decreases with tree age, a great number of cuttings is required from older trees, so that a sufficient number of plants for testing can be propagated. Once established, these young rooted plants then are used for cutting wood. Interestingly, these propagules are more easily rooted than are the cuttings taken from the original mature tree, so that once a difficult-to-root tree has been established by a few clones, subsequent rooting may not be so difficult (Morgan, 1976).

Present efforts center around caging insects on rooted propagules of resistant and susceptible trees to learn about comparative gall-forming tendencies on very young individuals. At least two more years of data will be collected from the caging studies, and higher numbers of insects, including insects from Mexico and other possible biotypes, will be caged on the trees. We also want to learn if trees of suspected resistance will remain gall-free over a long period of time, and whether greater numbers of insects could influence a tree to become susceptible, even after it had shown resistance for a period of several years. As we have demonstrated, the patterns of susceptibility varies; perhaps resistance also varies. During these same studies we might be able to gain insight into the mechanisms of resistance. Does competition for ovipositing sites exist? If so, what is the relationship between gall production and numbers of ovipositing females? Is host attraction related to the chemistry of the individual plant? In this regard, biochemical and plant physiological experiments must be initiated to determine the nature of insect attraction, if indeed it is of chemical origin. Finally, the propagation of resistant and susceptible genotypes should be a continuing effort, so that these individuals may be used for further screening, propagation, and, in the final stage of the work, distribution for public use.

IV. FACTORS AFFECTING THE DEVELOPMENT AND ACCEPTANCE OF INSECT-RESISTANT PLANTS IN THE URBAN ENVIRONMENT

In a research program to develop insect-resistant urban plants, an effort should be made to answer several socioeconomic questions such as: What are some of the relevant, homeowner attitudes and practices toward insects? What constitutes insect "damage" to a plant? Can homeowner attitudes and practices be changed? Will insect-resistant plants be produced by the nursery industry? If made available, will these plants be used by the public?

A. Attitudes and Practices

Efforts to document urban attitudes and practices towards insects are in their infancy. However, in a questionnaire survey conducted in Bryan-College Station and Dallas, Texas, from 1974 to 1976 (Frankie and Levenson, Chapt. 15, this volume), an attempt was made to explore public opinion and behavior toward insects. Some of the findings are germane to this paper. Overall, the Texas study revealed that basic attitudes and practices vary greatly. For example, at one extreme were those people who claimed that insects were aesthetically pleasing. At the opposite extreme were the entomophobic people. The survey also indicated that people use a wide variety of methods to control insects. Although most people use insecticides, a substantial number, 21 to 54% (depending on the city and the year), also use nonchemical control means.

B. Insect Damage

Some target insect species (and the respective damage) may be found even on reportedly resistant plants. The degree to which insects infest plants, and the degree to which this is accepted, must be considered from the standpoint of the homeowner.

The public's perception of insect damage on ornamentals is highly subjective. Olkowski (1974) coined the term, "aesthetic injury level" (AIL), in his attempt to describe this phenomenon. He points out that perception of damage on ornamentals is an individual matter that is subject to wide variation. Many factors may affect the AIL on a given ornamental plant. For example, the species of insects infesting the plant may be important to an individual perception of "damage." Mealybugs (Pseudococcidae) are more obvious to most people than are spider mites (Tetranychidae); damage by chewing insects is more obvious than that by insects with piercing-sucking mouthparts. Differences may even exist in the public's appreciation of damage by closely related arthropod species. *Tetranychus cinnabarinus* (Boisd.), a red mite, can be seen more easily on most plants than *T. urticae* Koch, usually a green mite. This is especially true in areas having low light, such as shopping center malls (Gaylor, personal observation). Unless the amount of damage to an ornamental plant is obvious, the public will tolerate smaller species of insects in relatively high numbers and will usually complain about the presence of a few larger, more visible insects.

Host plant morphological characteristics may affect the AIL. Cultivars with compact or cupped leaves may hide insects. Mite damage is not as obvious on plants with variegated leaves, such as *Dieffenbachia* 'Exotica', as on more uniformly dark green leaves such as *D. amoena* Bull. (Gaylor, personal observation). Most people untrained in entomology do not carefully examine plants for insects. Therefore, unless the insects or the damage they cause is obvious, populations may go undetected.

Location and use of an ornamental plant are important factors affecting AILs. Insects on plants grown in homes have lower AILs than the same plant species in enclosed shopping center malls. Shrubs and trees grown in parks will have different AILs than the same plants in home yards. Pests on plants in parks in affluent neighborhoods may have different AILs than those in less affluent neighborhoods. Pests on a tree planted near a parking lot or picnic table probably have lower AILs than those in a more inaccessible area of the park.

Finally, the size and value of the plant affect insect AILs. People will tolerate more insects and damage on an inexpensive plant than on an expensive one, but insects or damage on a large plant may not be noticed as soon as on a smaller, more accessible plant.

C. Attitude Change

In many cases, if not most, AILs are based on misconceptions of actual or potential insect "damage" to ornamental plants. Through appropriate educational efforts, AILs may be altered in the direction of tolerating more insects per plant, thereby increasing the AIL. Limited experience in Texas with this kind of educational effort offers support for this notion. For example, when homeowners have received appropriate information about such organisms as bagworms, webworms, and galls on their ornamentals, the tendency to chemically treat low and even moderate infestations is greatly reduced (Frankie, personal observation). Inferred evidence from a public opinion survey (Frankie and Levenson, Chapt. 15, this volume) also indicates that people's attitudes toward insects and insect control can be changed. Relevant findings of this study will be considered in some detail in the section on future prospects.

Researchers must assume a substantial share of the responsibility for developing and distributing appropriate information concerning resistant plants. They should be prepared to write lay-oriented articles. This should be a relatively easy task when both entomologists and plant scientists are involved in the research program. In the case of U.S. land grant institutions, associated science writers should be kept informed as to the research progress in developing new plant materials. Wherever and whenever possible, researchers should seek the assistance of the state extension service in distributing new information to grower groups, environmental groups and other relevant organizations that may be interested and able to help in disseminating this information.

D. Production of Insect-resistant Ornamental Plants by Nurseries

Written and oral presentations of new information represent one general form of communication. To insure the acceptance and commercial propagation of "new" plant varieties, researchers must be willing to work directly with growers during developmental research phases to demonstrate specific propagation procedures and also to discuss potential uses (and possible limitations) of new plant materials. These kinds of contacts should also provide insight and useful feedback to the researchers with regard to the practical application of the research. One of the positive points that should be stressed is the fact that insect-resistant plants are environmentally desirable since they should require few or no pesticide applications for target organisms. Finally, researchers should follow growers' progress and practical experience in developing and selling resistant plants.

Several contacts were made with Texas nurserymen concerning the development of gall-resistant live oaks. In a survey of eight individuals who represent some of the largest tree growers in the state, all reported that they were familiar with *Disholcaspis* and other galls. In the case of *Disholcaspis* galls, none of the respondents believed that gall growth caused any damage to the tree. When questioned about the attitudes of their customers toward galls in general, the responses varied. Four said their customers occasionally complained or mentioned dissatisfaction with the galls on their trees. One nurseryman suggested that certain leaf galls might be advertised as attractive ornamental features. Another respondent said he preferred gall-free trees and would ask higher prices for them if they were made available. One nurseryman indicated he was not concerned about the galls because he knew they were caused by insects, and, "I can spray for them anyway, can't I?" All respondents agreed, however, that they would prefer a tree "more attractive" to the customer and would charge a higher price for it - whether gall-free or galled. Finally, seven of the eight nurserymen indicated they would prefer to propagate the live oak asexually to maintain genetic uniformity.

E. Public Acceptance of Resistant Plants

In contrast to agronomic or orchard crops, many insect-resistant plant materials developed for urban areas probably would receive relatively limited usage. This is, to a large extent, related to geographical restrictions of insect pests, respective host plants, or both. Further, even within areas where resistant plant materials become available we should expect usage to be restricted because of the diversity of values, wants, needs and monetary resources of the people who populate these areas. For example, the use of 'Floratam' St. Augustine grass is on the increase throughout Florida and Texas. However, within given cities usage varies considerably from one residence to the next. When it was first released, it was an "upper middle class grass" due to its cost (more than 30% higher than the common cultivar of St. Augustine grass), and primarily people in the higher income levels purchased it. Presumably, its cost will decrease as it is more widely propagated. In Texas, many homeowners now recognize a need to plant 'Floratam' to replace turfgrass infected with SAD to which the new cultivar is resistant. In Florida, few yards are being stripped to replace healthy St. Augustine grass with 'Floratam', but many yards are being resodded with it when chinch bugs damage the original turf. A high percentage of new lawns, cemeteries, school grounds and athletic fields are, however, being sodded to 'Floratam' with the promise of reduced maintenance problems. In Florida, sod growers are able to sell as much 'Floratam' as they can produce.

With regard to plants with unique morphological characteristics, for example gall-resistant or gall-producing trees, we should also expect restricted use. Although use may be low, it probably will be maintained consistently since there are always people in urban areas who search for unique plants. Even among live oaks, which are now grown exclusively from seed, sale of superior individuals is an effective practice, as demonstrated by Storm Nursery, Premont, Texas. Storm's "Heritage Live Oak," a selection based on phenotype, is preferred by its customers. We consider this restricted use pattern to be a positive part of any effort to produce resistant plant materials for urban environments. This view is based on the fact that selective pressures that might produce new insect biotypes are unlikely to develop where a diverse mosaic of plant material is maintained in given urban areas.

Scientists who wish for quick answers to satisfy homeowners or university administrators should shy away from these types of projects. Only those willing to devote time, funding and patience will be able to collect enough data to successfully maintain a host plant resistance program. A commitment of this nature is required due in great part to the interactions that exist between plant and animal in seasonal weather changes; but it also reflects a vacuum in man's knowledge of plant physiology in relation to insect activities.

V. THE NEED FOR INTERDISCIPLINARY COOPERATION

Interdisciplinary cooperation is the heart of any program of this nature (Hanover, 1975). Benefits of an interdisciplinary team research effort include the following: 1) enhanced creativity - the potential for generating new stimuli; 2) improved understanding of the problem; 3) more efficient use of resources; 4) better use of the data collected; 5) improved communication of results; and 6) an explanation of our disciplines to others, including administrators, scientists in other fields, legislators and the general public (Mitchell, 1977). Painter, in his classic work on insect resistance in crop plants (Painter, 1951), gave serious consideration to interdisciplinary effort yet warned of narrow-sightedness. Suggesting that a "lack of genetic viewpoint on the part of entomologists" was one reason for slow acceptance of resistance as a means in insect control, Painter urged interdisciplinary cooperation among "the plant breeder, agronomist or horticulturist, and entomologist" as basic members of the research team. "But for an understanding of the mechanisms of resistance, the plant and insect physiologists and biochemists should also be members of the team," he stressed.

The nature of our present activities demands a certain broadness in specific areas of scientific interest, particularly entomology and horticulture. Yet, a plant physiologist or biochemist might add even more strength to our studies in exploring, for example, the behavior of the live oak tree in response to the insect's ovipositing. Herein may lie answers to some crucial questions: Can resistance be detected through chemical or metabolic determinations? The discovery of phenols in the foliage of *Picea abies* genotypes resistant to the eastern spruce gall aphid, *Adelges abietes* L., (Tjia and Houston, 1975), and the absence of crystallization of cortical oleoresins of white pines correlating with resistance to attack by the white pine weevil (Santamour and Zinkel, 1975) suggest the presence of chemical compounds may be a factor of resistance. Why do patterns of resistance change? Is climate a factor in susceptibility? Will cultural factors (i.e., watering, fertilization and pruning) influence the plant's resistance? A lawn fertilization test by Horn (1962) demonstrated that heavy fertilization programs utilizing fast-release inorganic sources of nitrogen resulted in greater damage from chinch bugs. Damage was much less on grass receiving an organic source of slow-release nitrogen. It has also been observed (Reinert, 1977b) that the higher the nitrogen fertilizer rate on turfgrass, the higher the sod webworm populations and damage levels.

Due to the unique nature of "damage" on ornamentals, a human behavioral scientist would strengthen an ornamentals program by assessing consumer attitudes.

In future studies we can forsee the need for a geneticist to give us greater research dimensions during which we might attempt to alter genetically a plant of known susceptibility.

Whether we determine that the leadership and direction of the team effort be through entomology, horticulture or physiology, we might reexamine the work of Painter (1951) who recommended a "broad biological background" by all team members, and who said that the "real success" of an insect-resistance study "depends on sharing work as well as credit, and on the true meeting of minds in the field plot, the laboratory, or greenhouse, as well as about the conference table."

VI. FUTURE PROSPECTS

In terms of case histories, a determination of future use of insect-resistant plants in urban environments may be difficult to assess, since little literature is available. While encouraging more research into host plant relationships, Weidhaas (1976) concluded that the use of insect-resistant ornamental trees in the near future was not feasible. His pessimism was based on problems associated with locating and utilizing pest-free plants, and with maintaining such plants in a pest-free condition. Weidhaas argues that the genetic potential of any shade tree may not be expressed because of great "horticultural diversity" in landscaping, resulting in a low likelihood of discovering an individual with genetic resistance, generally considered a recessive characteristic. He predicts the great numbers of insect pests on elms and oaks may represent a latent disaster, because plague-like conditions could develop among tree biotypes resistant to one type of insect as they become susceptible to another insect. Weidhaas fears that, due to fluctuations in tree cultural practices and new site introductions, pest problems will appear faster than the slower process of developing insect resistance, which may require time-consuming and expensive research.

Some researchers look favorably on insect resistance in ornamental (Santamour, 1977) and forest (Painter, 1966) tree species. Our research with the live oak tree appears promising. It has been grown as an ornamental tree in Dallas, where it is not native, for 25 years and elsewhere in Texas where it is native since the turn of the century (Griffing, 1977) with few reported serious problems, despite the presence of about 350 species of arthropods inhabiting its limbs and foliage (Frankie, unpub. data). Similarly, the Shumard red oak, *Quercus shumardii* Buckl., and the cedar elm, *Ulmus crassifolia* Nutt., are regularly transplanted from native sites into Texas cities with negligible insect damage reported. It thus appears unlikely that secondary pests will increase to epidemic levels.

The great numbers of these three species of forest trees, open-pollinated and propagated from seeds, would suggest that their genetic potentials are, indeed, expressed in the urban landscapes, horticultural diversity notwithstanding. One could hardly imagine a monoculture of live oak trees developing in Dallas, Texas, a city already noted for its oak-lined streets, parks and shaded homes. In contrast, most agricultural field crops are products of centuries of selection in which the genetic base has been dangerously narrowed (Miller, 1973; Wilkes, 1977). In the case of live oak resistance to *D. cinerosa*, since only approximately 20% urban live oak trees may be susceptible (only 6% severely) to the gall (Frankie *et al.*, 1977), resistance appears to be the rule rather than the exception, and many gall-free cultivars could be propagated for public use. Finally, the fact that vegetative propagation can be applied to the live oak provides the opportunity to circumvent lengthy and costly breeding work through selection of resistant biotypes.

Support for insect-resistance work in ornamental plants appears to be increasing. Yet, most present-day insect control practices in urban environments rely to a great extent on chemical insecticides (von Rumker *et al.*, 1972; Frankie and Levenson, Chapt. 15, this volume). Although these materials have proven effective in many cases, their continued use at the same levels and frequency is doubtful.

Since 1945 we have experienced resistance to insecticides in more than 200 arthropods throughout the world (Georghiou, 1972). Based on this history, we should expect more resistance cases in the future coupled with increasing difficulty in chemically controlling those species in which resistance has already developed. Reinert and Niemczyk (1977, unpub. data) have shown high levels of organophosphate insecticide resistance in southern chinch bug populations in areas of Florida. Disturbing side effects of insecticides have been described by numerous workers (von Rumker *et al.*, 1972; Glass, 1975; Plapp, Chapt. 16 this volume). In the case of ornamentals, one of the commonly observed side effects is that of phytotoxicity associated with some chemicals. Reinert and Neel (1974, 1976a, 1976b), Neel and Reinert (1975), and Short and McConnell (1973 a, 1973b) have shown phytotoxicity symptoms for most of the presently used insecticides on a wide range of ornamental plants. Because of more serious side effects which are related to human health and environmental contamination, the U.S. Environmental Protection Agency has severely restricted use of certain insecticides such as DDT, aldrin, dieldrin and chlordane (Plapp, 1977). Tighter restrictions on the registration of new compounds by chemical companies have resulted in reluctance on the part of industry to develop new insecticides for use on low-volume, high unit value crops like ornamentals. Finally, the cost of petroleum-based insecticides has increased significantly in recent years, and we should expect this trend to continue in the future.

There presently are indications that homeowner attitudes toward the use of chemical insecticides are changing. In Texas, the public opinion poll referred to earlier (Frankie and Levenson, Chapt. 15, this volume) disclosed that 29 to 46% of the people (depending on the city and year) had experienced a change in attitudes toward insecticides in recent years. Most said they had become more cautious in their use of chemicals.

It seems likely that when selecting for certain resistance characteristics, other desirable attributes will be sought at the same time. For example, combinations such as insect-disease resistance and drought tolerance may be possible. Selections for combinations would be especially desirable in areas where resources such as water might be seasonally or otherwise limited.

It seems clear that a combination of biological, economic and social factors will also enter into the development and acceptance of insect-resistant plants in urban areas. These kinds of factors must be recognized by researchers and given appropriate considerations and attention.

ACKNOWLEDGMENTS

The authors express their appreciation to NorthPark Center and Restland Cemetery of Dallas, Texas, and Wash Storm, Jr., Premont, Texas, for the use of live oak trees in this study, and to J. M. Tucker and C. H. Muller for tree species identification. They also wish to thank the following persons for their contributions and reviews of the manuscript: R. L. Campbell, M. K. Harris, James A. Reinert, F. S. Santamour, Jr., R. W. Toler, and J. A. Weidhaas, Jr. For their technical assistance, we appreciate the contributions of Larry Schaapveld and B. J. Simpson.

REFERENCES

Anonymous. 1971. Growing the Bradford ornamental pear. USDA Home and Garden Bull. 154.

Austin, L., J. S. Yuill, and K. G. Brecheen. 1945. Use of shoot characters in selecting ponderosa pines resistant to resin midge. *Ecology 26*: 288-296.

Beck, S. D. 1965. Resistance of plants to insects. *Ann. Rev. Entomol. 10*: 207-232.

Beier-Petersen, B., and B. Soegaard. 1958. Studies on resistance to attacks of *Chermes cooleyi* (Gill.) on *Pseudotsuga taxifolia* (Poir.) Britt. *Det forstl. Forsgsv. Danmark 25*: 37-45.

Buckner, R. C., B. C. Pass, P. B. Burrus, II, and J. R. Todd. 1969. Reaction of Kentucky bluegrass strains to feeding by the sod webworm. *Crop Sci. 9*: 744-746.

Callaham, R. Z. (Chm.). 1966. General guidelines for practical programs toward pest-resistant trees. *In* Breeding pest-resistant trees (H.D. Gerhold *et al.*, eds.). Pergamon Press, London, pp. 489-493.

Campbell, R. L. 1977. Unpublished data. Western Washington Research & Extension Center, Puyallup, Washington.

Carter, R. P., and R. L. Duble. 1976. Variety evaluation in St. Augustinegrass for resistance to the southern lawn chinch bug. Texas Agric. Exp. Sta. Prog. Rep. PR-3374C.

Chamber, K. L. 1970. Biochemical coevolution. Proc. 29th Ann. Biol. Colloq. Oregon State Univ. Press, Corvallis, Oregon.

Chew, R. W. 1974. Consumers as regulators of ecosystems: An alternative to energetics. *Ohio J. Sci. 74*: 359-370.

Coutts, M. P., and J. E. Dolezal. 1966. Polyphenols and resin in the resistance mechanism of *Pinus radiata* attacked by the wood wasp, *Sirex noctilio*, and its associated fungus (*Amylostereum* sp.). Leafl. For. Timber. Bur. Austral. No. 101.

de Wilde, J., and L. M. Schoonhoven. (Eds.). 1969. Insect and host plant. Proc. Second Internation. Symp. Insect and host plant, Wageningen, The Netherlands. *Entomol. Exp. Appl. 12*: 471-810.

Ewert, J. P. 1967. Untersuchungen über die Dispersion der Fichtengallenlaus *Sacchiphantes* (*Chermes*) *abietis* (L.) auf gewöhnlichen Kulturen, Einzelstammabsaaten und Klonen ihrer Wirtspflanze. *Z. Angew. Entomol. 59*: 272-291.

Felt, E. P. 1905. Insects affecting park and woodland trees. N. Y. State Educ. Dept. N. Y. State Museum Mem. 8:46-48.

Fischer, S. J., and J. B. Shanks. 1974. Host preference of greenhouse whitefly: effect of poinsettia cultivar. *J. Am. Soc. Hort. Sci. 99*: 261-262.

Frankie, G. W., and C. S. Koehler. 1971. Studies on the biology and seasonal history of the cypress bark morth, *Laspeyresia cupressana* (Lepidoptera: Olethreutidae). *Canad. Entomol. 103*: 947-961.

Frankie, G. W., D. L. Morgan, M. J. Gaylor, J. G. Benskin, W. E. Clark, H. C. Reed, and P. J. Hamman. 1977. The mealy oak gall on ornamental live oak in Texas. College Station: Texas Agric. Exp. Sta. MP-1315.

Georghiou, G. P. 1972. The evolution of resistance to pesticides. *Ann. Rev. Ecol. Syst. 3*: 133-168.

Gerhold, H. D. 1966. In quest of insect-resistant forest trees. *In* Breeding pest-resistant trees (H.D. Gerhold *et al.*, eds.). Pergamon Press, London, pp. 305-318.

Glass, E. H. (Coordinator). 1975. Integrated pest management: Rationale, potential, needs and implementation. Entomol. Soc. Am. Spec. Publ. 75-2.

Griffing, R. C. 1977. Personal communication. Griffing Nursery, Beaumont, Texas.

Hall, R. C. 1937. Growth and yield in shipmast locust in Long Island and its relative resistance to locust borer injury. *J. For. 35*: 721-727.

Hamilton, C. C. 1924. The biology and control of the chrysanthemum midge. Maryland Agric. Exp. Sta. Bull. 269.

Hanover, J. W. 1975. Physiology of tree resistance to insects. *Ann. Rev. Entomol. 20*: 75-95.

Harris, P. 1960. Production of pine resin and its effect on survival of *Rhyacionia buoliana* (Schiff.). *Canad. J. Zool. 38*: 121-130.

Helgesen, R. G., and M. J. Tauber. 1974. Biological control of greenhouse whitefly, *Trialeurodes vaporariorum* (Aleyrodidae, Homoptera), on short-term crops by manipulating biotic and abiotic factors. *Canad. Entomol. 106*: 1175-1188.

Holst, M. J. 1963. Breeding resistance in pines to *Rhyacionia* moths. World Consultation on Forest Genetics and Tree Improvement, Stockholm, 1963. FAO/FORGEN 63-6b/3.

Horn, G. C. 1962. Chinch bugs and fertilizer, is there a relationship? Florida Turf-Grass Assoc. Bull. 9.

Horn, G. C., A. E. Dudeck, and R. W. Toler. 1973. 'Floratam' St. Augustinegrass: A fast-growing new variety for ornamental turf resistant to St. Augustine decline and chinch bugs. Florida Agric. Exp. Sta. Cir. S-224.

Johnson, N. E. 1968. Insect attack in relation to the physiological condition of the host tree. Entomology and Limnology Mimeograph Review No. 1. Cornell University, Ithaca, New York.

Johnson, W. T., and H. H. Lyon. 1976. Insects that feed on trees and shrubs. Cornell Univ. Press, Ithaca, New York.

Kawahata, K. 1955. Selection tests of varieties of Sugi (*Cryptomeria japonica*) resistant to gall midge (*Contarinia inouyei*). (in Japanese) Kagoshima Prefec. For. Exp. Sta. Bull. No. 5.

Kindler, S. D., and E. J. Kinbacher. 1975. Differential reaction of Kentucky bluegrass cultivars to the bluegrass billbug, *Sphenophorus parvulus* Gyllenhal. *Crop Sci.* *15*: 873-874.

Koehler, C. S. 1977. Personal communication. Cooperative Extension, Univ. of California, Berkeley, California.

Leuck, D. B., C. M. Taliaferro, G. W. Burton, R. L. Burton, and M. C. Bowman. 1968. Resistance in bermudagrass to the fall armyworm. *J. Econ. Entomol.* *61*: 1321-1322.

Leuck, D. B., and J. L. Skinner. 1970. Resistance in bermudagrass affecting control of the fall armyworm. *J. Econ. Entomol.* *63*: 1981-1982.

Markkula, M., K. Roukka, and K. Tiittanen. 1970. Reproduction of *Myzus persicae* (Sulz.) and *Tetranychus telarius* (L.) on different chrysanthemum cultivars. *Ann. Agric. Fenniae* *8*: 175-183.

McGregor, R. A. (Chief). 1976. Florida turfgrass survey, 1974. Florida Crop and Livestock Rep. Ser. Orlando, Florida.

Miller, J. 1973. Genetic erosion: crop plants threatened by government neglect. *Science* *182*: 1231-1233.

Mitchell, R. G., and W. P. Nagel. 1969. Tree selection for controlling midges on Douglas-fir. *Am. Christmas Tree J.* *13*: 11-13.

Mitchell, R. L. 1977. The maintenance of professional integrety in the interdisciplinary team research effort. *HortScience* *12*: 36-37.

Morgan, D. L. 1976. Factors influencing propagation of *Quercus virginiana* Mill., and some aspects of the bionomics of *Disholcaspis cinerosa* Bass. (Hymenoptera: Cynipidae). Ph.D. dissertation. Texas A&M University, College Station, Texas.

Morgan, D. L., and E. L. McWilliams. 1976. Juvenility as a factor in propagating *Quercus virginiana* Mill. *Acta Hort.* *56*: 263-268.

Muller, C. H. 1970. *Quercus*. *In* Manual of the vascular plants of Texas (D.S. Correll and M.C. Johnston, eds.). Texas Res. Found., pp. 467-492.

Munro, J. A. 1963. Biology of the ceanothus stem-gall moth, *Periploca ceanothiella* (Cosens). *J. Res. Lepidoptera* *1*: 183-190.

Munro, J. A. 1965. Occurrence of *Psylla uncatoides* on *Acacia* and *Albizia*, with notes on control. *J. Econ. Entomol.* *58*: 1171-1172.

Neary, J. 1976. The search for a white marigold. *Hort. 54*: 20-28.
Neel, P. L., and J. A. Reinert. 1975. An evaluation of the phytotoxicity of five insecticides on selected ornamental plants. *Proc. So. Nurs. Res. Conf. 20*: 76-78.
Olkowski, H., and W. Olkowski. 1976. Entomophobia in the urban ecosystem, some observations and suggestions. *Bull. Entomol. Soc. Am. 22*: 313-317.
Olkowski, W. 1974. A model ecosystem management program. *Proc. Tall Timbers Conf. Ecol. Anim. Control Hab. Manage. 5*: 103-117.
Painter, R. H. 1951. Insect resistance in crop plants. Univ. Press of Kansas, Lawrence, Kansas.
Painter, R. H. 1966. Lessons to be learned from past experience in breeding plants for insect resistance. *In* Breeding pest-resistant trees (H.D. Gerhold *et al.*, eds.). Pergamon Press, London, pp. 349-355.
Pass, B. C., R. C. Buckner, and P. B. Burris, II. 1965. Differential reaction of Kentucky bluegrass strains to sod webworms. *Agron. J. 57*: 510-511.
Plank, G. H., and H. D. Gerhold. 1965. Evaluating host resistance to the white pine weevil, *Pissodes strobi* (Coleoptera: Curculionidae), using feeding preference tests. *Ann. Entomol. Soc. Am. 58*: 527-532.
Plapp, F. W. 1977. Personal communication. Dept. Entomology, Texas A&M University, College Station, Texas.
Poe, S. L., and G. J. Wilfret. 1972. Factors affecting spidermite (*Tetranychus urticae* Koch) population development on carnation: relative cultivar susceptibility and physical characteristics. *Proc. Florida St. Hort. Soc. 85*: 384-387.
Reinert, J. A. 1972. Turf-grass insect research. *Florida Turf-Grass Manag. Conf. Proc. 20*: 79-84.
Reinert, J. A. 1977a. Antibiosis of the southern chinch bug by St. Augustinegrass. (unpub. manuscript).
Reinert, J. A. 1977b. Personal communication. Agric. Res. Center, Ft. Lauderdale, Florida.
Reinert, J. A., and A. E. Dudeck. 1974. Southern chinch bug resistance in St. Augustinegrass. *J. Econ. Entomol. 67*: 275-277.
Reinert, J. A., and P. L. Neel. 1974. Insecticide phytotoxicity studies on ornamental plants. *Proc. So. Nurs. Res. Conf. 19*: 36-38.
Reinert, J. A., and P. L. Neel. 1976a. Phytotoxicity of miticides on selected environmental plants. *Proc. So. Nurs. Res. Conf. 21*: 44-47.
Reinert, J. A., and P. L. Neel. 1976b. Evaluation of phytotoxicity of malathion, ethion, and combinations of FC-435 spray oil with each on twenty-eight species of environmental plants under slat shade. *Proc. Florida State*

Hort. Soc. 81: 368-370.

Santamour, F. S., Jr., and D. F. Zinkel. 1975. Weevil-induced resin crystallization related to resin acids in eastern white pine. *Proc. N. E. For. Tree Impr. Conf. 23*: 52-56.

Santamour, F. S., Jr. 1977. The selection and breeding of pest-resistant landscape trees. *J. Arbor. 3*: 146-152.

Schuster, M. F., and H. A. Dean. 1973. Rhodesgrass scale resistance studies in rhodesgrass. *J. Econ. Entomol. 66*: 467-469.

Short, D. E., and D. B. McConnell. 1973a. Phytotoxicity of pesticides to selected foliage plants. *Proc. So. Nurs. Res. Conf. 18*: 56-59.

Short, D. E., and D. B. McConnell. 1973b. Pesticide phyto-toxicity to ornamental plants. *Proc. Florida State Hort. Soc. 86*: 439-443.

Simpson, B. J. 1977. Personal communication. Texas Agric. Exp. Sta., Dallas, Texas.

Sondheimer, E., and J. B. Simeone. (Eds.) 1970. Chemical ecology. Academic Press, New York.

Stimmann, M. W., and G. M. Taliaferro. 1969. Resistance of selected accessions of bermudagrass to phytotoxemia caused by adult two-lined spittlebugs. *J. Econ. Entomol. 62*: 1189-1190.

Strobel, G. A., and D. E. Mathre. 1970. Outlines of plant pathology. Van Nostrand Reinhold Co., New York.

Stroble, J. 1971. Turfgrass. *Florida Turf-Grass Manag. Conf. Proc. 19*: 19-29.

Taliaferro, C. M., D. B. Leuck, and M. W. Stimmann. 1969. Tolerance of *Cynadon* clones to phytotoxemia caused by the two-lined spittlebug. *Crop Sci. 9*: 765-766.

Thielges, B. A., and R. L. Campbell. 1971. Genetic resistance: Alternative to chemical control. *Ohio Rep. 56*: 55-56.

Thielges, B. A., and R. L. Campbell. 1972. Selection and breeding to avoid the eastern spruce gall aphid. *Am. Christmas Tree J. 16*: 3-6.

Tjia, B., and D. B. Houston. 1975. Phenolic constituents of Norway spruce resistant or susceptible to the eastern spruce gall aphid. *For. Sci. 21*: 180-184.

van Buijtenen, J. P., and F. S. Santamour, Jr. 1972. Resin crystallization related to weevil resistance in white pine (*Pinus strobus*). *Canad. Entomol. 104*: 215-219.

van Emden, H. F. 1966. Plant resistance to insects induced by environment. *Scient. Hort. 18*: 91-101.

van Emden, H. F. (Ed.) 1973. Insect/plant relationships. Symp. Roy. Entomol. Soc. Lond. No. 6. Blackwell, Oxford.

von Rumker, R., R. M. Matter, D. P. Clement, and F. K. Erickson. 1972. The use of pesticides in suburban homes and gardens and their impact on the aquatic cnvironment.

Final report by Ryckman, Edgerle, Tomlinson, and Associates, Inc. on Contract No. 68-01-0119 for the Environmental Protection Agency.

Walker, J. C. 1965. Use of environmental factors in screening for disease resistance. *Ann. Rev. Phytopath. 3*:197-208.

Watson, A. (Ed.) 1970. Animal populations in relation to their food resources. Blackwell, Oxford.

Weidhaas, J. A. 1976. Is host resistance a practical goal for control of shade-tree insects? *In* Better trees for metropolitan landscapes (F.S. Santamour, Jr. *et al.*, eds.). USDA For. Serv. Gen. Tech. Rep. NE-22, pp. 127-133.

Whitcomb, C. E. 1975. Know it and grow it. Whitcomb, Stillwater, Oklahoma.

Wilkes, G. 1977. Breeding crisis for our crops: is the gene pool drying up? *Hort. 55*: 53-59.

Wyatt, I. J. 1965. The distribution of *Myzus persicae* (Sulz.) on year round chrysanthemums. I. Summer season. *Ann. Appl. Biol. 56*: 439-459.

BEHAVIOR-MODIFYING CHEMICALS AS A BASIS FOR MANAGING BARK BEETLES OF URBAN IMPORTANCE

Gerald N. Lanier

State University College of Environmental Science and Forestry
Syracuse, New York

I. BARK BEETLES

A. Impact

Since the beginning of commercial timber production in North America, bark beetles (Scolytidae) have been considered a major forestry problem, and in northern Europe and Asia, *Ips typographus* (L.) has for centuries been infamous as the harvester of spruce. With attention on the enormous economic and ecological impact of bark beetles on forests, the depredation of shade and ornamental trees by these insects is underrated. Only the assault on the elms in our parks and shady lanes by Dutch elm disease (DED)-elm bark beetle complex is

ISBN 0-12-265250-9

widely known. Losses to bark beetles occur wherever conifers, especially pines, are a major component of the shade tree mixture. Broad leafed trees are also attacked by bark beetles, but usually only when they are extremely stressed or moribund. For identification of specific bark beetles and details on their biology one can refer to works by Baker (1972) for eastern species and Bright and Stark (1973) for western species.

The potential for bark beetle problems in the urban environment appears to be increasing for these reasons: 1) increased use of conifers in landscaping, 2) expansion of suburbs and second homes into existing conifer stands, and 3) the advancing age of trees which replaced the original forest ("second growth") renders them less resistant to bark beetle attack.

Air pollution, off-site planting, tree abuse, abrupt exposure of trees due to clearing and inadvertent augmentation of breeding material are factors which continually cause or exacerbate bark beetle damage to urban trees. Photochemical oxidants (smog) generated in the Los Angeles megalopolis chronically injure highly valued ponderosa pine, *Pinus ponderosa* Laws., in communities within the San Bernardino Mountains; the pines decline in vigor until they are killed by bark beetles (Stark and Cobb, 1969). In Ft. Collins, Colorado, mountain pine beetles, *Dendroctonus ponderosae* Hopkins, bypassed numerous ponderosa pine to search out and attack 15 of 17 suitably large ornamental Scotch pines, *Pinus sylvestris* L., within that city (McCambridge, 1975). In Marin County, California, I observed an area in which the developer of an expensive residential tract carefully preserved selected Bishop pines, *Pinus muricata* D. Don., only to have them killed by hoards of *Ips plastographus* (Leconte) generated in pines which had been cut and neglectfully left near the building sites. Bark beetles capitalizing on pruned limbs for breeding material frequently cause localized damage. Translocation of infested firewood has undoubtedly accelerated the spread of certain pests, especially the European elm bark beetle, *Scolytus multistriatus* (Marsh.), and DED.

B. Biology

Bark beetle adults bore into and breed in the phloem-cambium region of the root collar, bole and branches. Their tunneling and feeding activities, together with the microorganisms introduced, are incompatible with the life of the tree; therefore, the beetles must infest dying wood or kill the tree, or at least the parts of the tree they colonize.

Odors emitted by potential host trees, especially weakened or damaged individuals, stimulate initial attacks. Attacks by the more aggressive species, e.g., southern pine beetle, *Den-*

droctonus frontalis Zimm., sometime occur on any tree of a suitable host species, regardless of vigor. Pioneer attackers release an aggregating pheromone that, in combination with the host odors, attracts both sexes and insures mass attack, quick colonization and the death of the tree. This aggregating mechanism makes it possible for bark beetles to overcome tree defenses and accounts for the extreme clustering observed in bark beetle populations. When breeding material is fully utilized, the pheromone signal is modified so that attack is terminated and the attention of the population is directed elsewhere.

C. Control

Owing to seclusive habits, bark beetles are difficult and expensive to control with insecticides. Most students of population dynamics of these insects have concluded that parasites and predators may be instrumental in keeping an endemic population of bark beetles in check, but natural enemies are generally not responsible for the collapse of bark beetle outbreaks (Beaver, 1967; Berryman, 1973).

The outcome of attacks by bark beetles is a function of the number and duration of attacks versus the vigor of the tree. Most trees are capable of destroying by resinous reaction the number of bark beetles that are likely to attack them when population levels are normal; only trees in poor condition will usually be killed at this time. However, simultaneous attacks by large numbers of beetles can overwhelm virtually any tree. Outbreaks generated by the availability of large amounts of susceptible host material may be perpetuated by the ability of the burgeoning population to generate massive attacks that kill vigorous trees. Long term control, therefore, is based upon maintaining tree vigor above the threshold for successful bark beetle colonization and preventing population buildup. This entails, at the least, sanitization of moribund and recently infested trees. In the short term, trees which are temporarily stressed by drought, etc., might be protected and periodic outbreaks suppressed with the aid of behavior-modifying chemicals.

II. BEHAVIOR-MODIFYING CHEMICALS

A. History and Terminology

Since the chemical characterization of the silk moth sex pheromone by Butenandt and co-workers in 1961 (Jacobson, 1972), this burgeoning field has produced identifications of behavior-modifying chemicals for more than 100 insect species (mostly Lepidoptera, Coleoptera and Hymenoptera). Those interested in keeping up with recent advances in this field might refer to books by Beroza (1970), Jacobson (1972), Birch (1974) and Shorey (1976).

Natural behavior-modifying chemicals are usually classified as pheromones, kairomones or allomones. Pheromones are those chemicals which affect communication within a species. Examples are the morphological phase-governing pheromone of locust, the alarm pheromone of the honey bee, sex attractants of moths, aggregating pheromones of bark beetles and trial pheromone of ants (see Birch, 1974, for many specific examples). The other terms apply to chemical communication between species. If the chemical acts to the advantage of the receiver (e.g., the odor of a prey to a predator), it is kairomone; if it works to the advantage of the producing species (i.e., repellents produced by stink bugs [Pentatominae]) it is called an allomone.

B. Rationale for Use

Behavior-modifying chemicals are potentially useful in the survey and control of bark beetle populations. This may be especially true in urban environments where access is excellent, tree values high and the use of conventional insecticides restricted. Factors favoring success are as follows:

1. The action is directed against the damaging stage. The bark beetle adult is responsible for tree killing and disease dissemination. Interfering with, or killing adults will have immediate impact on the level of damage. This is in contrast to Lepidoptera for which the strategy is to reduce mating by adults, but it is the late instar larva which does most of the damage. The numbers of breeding adults and the eggs may be only loosely correlated with number of larvae surviving to late instars.

2. Both sexes are affected, thus reproductive potential of the population is directly reduced. If only males were affected, survivors could compensate by mating more frequently.

3. Successful breeding depends upon mustering enough individuals over a short period of time to overcome the resistance of the host tree. Destroying part of the population by trapping or disrupting its aggregation will result in many beetles being lost in flight or absorbed in fruitless scattered attacks on trees that cannot be killed.

C. Strategies for Use

Approaches for the employment of behavior-modifying chemicals can be categorized as attraction, disruption and repellency.

Attraction entails the use of pheromones or kairomones or a combination of these to concentrate the target species where they can be sampled, destroyed, or led away from areas where they are most likely to be damaging.

Disruption implies interference with normal chemical communication, i.e., for mate-finding, aggregation or trail-following. This may be accomplished by confusing the target species with an abundance of artificial pheromone sources, by adapting or blocking its olfactory receptors, or by interrupting essential activities by releasing anti-attractants, alarm pheromones or mating suppressors.

Repellents, such as allomones produced by some plants, might be transferred or augmented by topical or systemic application or by selective breeding for plants with high allomone concentrations.

Each of these basic techniques has been used successfully to reduce damage by a variety of insect pests. This paper will concentrate on experiments designed to manipulate bark beetle populations by using pheromones. Particular emphasis will be placed on recent work with the European elm bark beetle.

III. CONTROL OF BARK BEETLES BY USE OF PHEROMONES

A. Trap Trees

For the last three centuries European foresters have utilized the aggregating behavior of bark beetles as a means to control them. The strategy for this control was based upon providing felled or girdled trees which bark beetles would attack in preference to standing timber. Once infested, trap trees were debarked or burned to destroy beetle broods (Schwerdtfager, 1973). The prevailing thought of the foresters was that odors of logs, *per se*, attracted all of the beetles; the phenomenon of aggregation by bark beetles in response to pheromones was apparently not understood until the middle of the twentieth century (Anderson, 1948).

B. Trap Trees Augmented by Pheromones and Cacodylic Acid Treatment

Synthetic pheromone has been used to attract beetles to trap logs, living trees and trees killed by treatment with the silvicide, cacodylic acid (Pitman, 1971, 1973; Copony and Morris, 1972; Knopf and Pitman, 1972; Coulson *et al.*, 1975). The rationale for attracting beetles to living trees is that populations of a tree-killing species, such as the southern pine beetle, can be concentrated in a predetermined area so that brood trees are easily located for prompt harvesting or insecticidal treatment. Injecting cacodylic acid into pheromone-baited trees further increases their attractiveness to some species, yet brood production in these trees is drastically reduced because bark loosens and becomes unsuitable for maturation of beetle brood. This method is cheap because further action is not necessary. Buffam and Yasinski (1971) found that cacodylic acid treatments of trees to be felled for road and trail construction prevented buildup of the spruce beetle, *Dendroctonus rufipennis* (Kirby), at a cost of $3.80 per tree, as opposed to a cost of $20 per tree for conventional insecticide treatment.

Although the practices outlined above are effective and economically feasible in the forest where trees are being harvested and/or can be sacrificed for the good of the stand, their application to the control of bark beetles in urban environments will be limited to special situations in which some trees are scheduled for removal for expansion projects or because they are senile and diseased. In municipalities where DED is rampaging, the best way to regain control of the situation might be to kill diseased and low value elms with cacodylic acid and bait them with pheromone so that they act as super trap trees. Because injected trees would not yield beetle brood (Rexrode, 1974), they could be removed in an orderly manner rather than on an urgent schedule. Cacodylic acid applied to axe frills on elms did not prevent brood development (Hostetler and Brewer, 1976); injection may be essential. Wild elms at the periphery of urban areas treated with cacodylic acid and pheromone could make a positive contribution to DED control programs rather than serving as reservoirs of beetles and the disease-causing fungus.

C. Pheromone-baiting Trees Unsuitable for Colonization

Bark beetles responding to high concentrations of pheromone, such as occurs during outbreaks, often attack trees of a species or size class unsuitable for successful breeding. Rasmussen (1972) baited (*trans*-verbenol + α-pinene) lodgepole pines, *Pinus contorta* Dougl., too small (less than 22.6 cm dbh) to serve as brood trees for the mountain pine beetle. Baited trees were quickly attacked but few were killed. However, adjacent larger trees were killed, and there was no evidence that the absorption of early attacks on small trees affected the final number of trees killed or the number of beetles in the next generation.

In trapping elm bark beetles, we have placed pheromone traps on a number of tree species and have found evidence that this practice induced attack only when the tree was an elm. Beetles land on any object in the vicinity of a pheromone source, but they are able to distinguish potential host trees by specific feeding stimulants (Doscotch *et al.*, 1970) and non-host trees by the presence of repellents (Gilbert and Norris, 1968; Norris, 1970).

D. Disruption by Aggregation-suppressing Compounds

Rudinsky *et al.* (1972) field tested three compounds believed, on the basis of laboratory bioassays, to be components of the aggregating pheromone of the Douglas fir beetle, *Dendroctonus pseudotsugae* Hopkins. Surprisingly, one of these, 3-methyl-2-cyclohexane-1-one (3,2-MCH), drastically decreased the numbers of beetles responding to the other compounds. Subsequent tests (Furniss *et al.*, 1972, 1974) proved that 3.2-MCH acted to suppress aggregation and could prevent attack on Douglas fir, *Pseudotsugae menziesii* (Mirb.), even if pheromone-releasing females were already present. This technique appears to hold promise for the protection of shade trees. For example, verbenone, which is released by males of the southern and western pine beetles and is known to depress the attractiveness of their aggregating pheromones (Renwick and Vite, 1969), might be used to prevent attacks on individual or groups of ornamental pines. This technique must be used very cautiously because an inhibitor for one species may be an attractant for a closely related species; for example, ipsenol, which inhibits aggregation of *Ips pini* (Say), is a component of the attractant pheromone of the more dangerous *Ips paraconfusus* Lanier (Birch and Wood, 1975). *Endo*- and *exo*-brevicomin, which are produced by the western pine beetle, *Dendroctonus brevicomis* Leconte, disrupted aggregation by preventing landing on host trees by southern pine beetles (Payne *et al.*, 1977).

E. Disruption by Attractant Pheromones

Superabundance of attractant pheromones may prevent tree killing by disrupting the aggregation process. Pheromone receptors might become adapted and render the beetles unresponsive, or the beetles might be unable to concentrate in the vicinity of a particular pheromone source. This approach has been successful in preventing mate-finding in certain Lepidoptera species (Gaston and Shorey, 1974). Aggregation and tree-killing by the southern pine beetle was not interrupted by aerially distributed rice grains soaked with frontalure (frontalin + α-pinene) (Vité *et al.*, 1976). However, catches on traps baited with frontalure decreased markedly during the period of treatment. Vité *et al.* (1976) suggested that beetles were utilizing cues other than frontalin to locate trees under colonization.

That disruption of attack by bark beetles can result from super concentrations of attractant pheromone is suggested by the observation that western pine beetles attacked trees around traps releasing pheromone at 10 mg/day whereas traps releasing one mg/day caught almost as many beetles while stimulating many fewer attacks on adjacent trees (Bedard and Wood, 1974).

F. Mass-trapping Western Pine Beetles

The first attempt at controlling bark beetles with synthetic pheromone was directed against the western pine beetle in the Bass Lake basin of the central Sierra Nevada in California (Bedard and Wood, 1974). The site was 65 km^2 within a single drainage bordered on most sides by forest types which contained few *D. brevicomis*. Within the area two tests and two check plots, each 2.56 km^2, were selected. Large sticky traps were positioned in a 161 m grid in each suppression plot. In addition, small "survey" traps were placed at 0.8 km intervals throughout the basin. Tree mortality was monitored by aerial photographs and the within-tree population of bark beetles was followed by a previously developed sampling system.

Results were highly encouraging. The suppression and survey traps caught an aggregate of 894,000 beetles and beetle-killed trees within the basin dropped from 227 immediately before the test to 73 trees killed by the generation flying during the suppression period. The survey traps also apparently played an important role in suppression and consequently influenced the check area. Bedard and Wood (1974) concluded that "pheromone-induced mortality contributed substantially to the reduction of populations of *D. brevicomis* and subsequent tree mortality..."

Attempts to mass trap mountain pine beetle (Pitman, 1971) and Douglas fir beetle (Pitman, 1973) apparently were not effective because the attractants used were not competitive with the natural pheromones.

G. Mass Trapping of the European Elm Bark Beetle

Since 1970 I have had the good fortune to work with J.B.

Simeone and R. M. Silverstein of the College of Environmental Science and Forestry at Syracuse and J. W. Peacock, A. C. Lincoln and R. A. Cuthbert of the U.S. Forest Service's Northeastern Forest Research Laboratory.[1] As a team we have isolated, identified and synthesized the aggregating pheromone of the European elm bark beetle. These are: 1) an alcohol, 4-methyl-3-heptanol (H); 2) a bicyclic ketal, α-multistriatin (M); 3) a sesquiterpene, α-cubebene (C) (Pearce *et al.*, 1975). Two components, H and M, are released by tunneling virgin females while the third, C, is produced by moribund elm wood (Gore *et al.*, 1977). We have investigated the effects of varying ratios of the pheromone components, dosages, trap designs and trap placement. In addition, the biology, dispersal and pheromone response of the beetle have been studied (Cuthbert and Peacock, 1976; Lanier *et al.*, 1976). We are currently in the third year of an investigation on the effectiveness of mass-trapping elm bark beetles as a means of reducing rates of DED. The trapping technology described below is the synthesis of these studies.

Traps used in beetle suppression studies consist of sheets of 46 x 66 cm white plastic-coated cardboard covered on the outer surface with Stikem Special®. These were fixed at three meters above the ground on utility poles and trees. During 1975 we used one of these sheets per trapping site; in 1976, two contiguous sheets were used for each trap. Tests of various sizes, styles, and colors of traps showed that visibility, size and height were important in trapping efficiency (Lanier *et al.*, 1976). Color and hue were important only as they maximized contrast to the background. It appears that beetles attracted by the pheromone tend to land on the most conspicuous object. Against a dark background, white traps caught three times more beetles than dark traps. The reverse was true if traps were suspended so that they were viewed against the sky.

Pheromone dispensers are affixed on one corner or in the center of the traps. Each bait contains the equivalent of 5,000,000 virgin females and releases the pheromone at a rate equivalent to 2,000 females/hour for about 100 days. We tested five controlled-release dispensers of three basic types. Two of these, a Hercon laminated plastic and Conrel hollow fiber dispensers, are currently in use. The Conrel method provides better control of dosage while the Hercon is more convenient to use. We anticipate no environmental problems because the absolute amounts of chemicals released are very

---

[1] *Also involved, and doing much of the important work, are Wayne Jones, Technical Assistant, and former graduate students Glenn Pearce, Bill Gore, John Bartels and Ernest Elliott.*

small; for example, within a 800 ha beetle suppression area in Syracuse, New York, the maximum for the mean daily release rate is 0.832 g of the three compounds, in aggregate, or 114.8 g over the 160 day trapping period.

Trap placement is a major consideration. Our most effective trap sites are fully-exposed utility poles. Obscured traps, such as those shaded by foilage, may have the detrimental effect of luring beetles they do not catch, thus increasing the net population of living beetles within the trapping area. Traps should not be placed on elm trees which are not thoroughly treated with an insecticide.

Catches on our traps reflect the emergence rate and ratio of the sexes. This is consistent with the natural situation in which attacking virgin females attract both sexes in equal numbers. Females responding to material under attack initiate new attacks while males scurry over the bark surface searching for boring virgin females. Mating occurs at the entrance of the female's gallery. After copulation is terminated, females commence gallery elongation and oviposition in niches in the sides of these tunnels. Males resume their search for mates. Contrary to most of the literature, this insect is not monogamous; Bartels and Lanier (1974) found that, when offered unlimited opportunity, males inseminated an average of 17 females.

Traps must be installed in late spring before the emergence of the beetles, which overwinter as fully grown larvae. Early trapping is most important because it is during this period that healthy elms are susceptible to infections by the DED fungus spores rubbing off the bodies of beetles feeding in twig crotches. Crotch-feeding is not obligatory prior to breeding, as has been erroneously perpetuated in some literary accounts. Newly emerged beetles readily attack and breed in elm logs. Gut analysis showed that 57% of the trapped beetles emerged the same day they were caught and had not ingested bark and xylem of twig crotches. I believe that twig feeding is an adaptation of the beetles for sustaining life until suitable breeding material is located.

Most beetles undergo a dispersal flight before they are responsive to pheromone. Using rings of traps around brood logs hauled into a large limestone quarry, we found that beetles usually flew 200 to 600 m before landing on pheromone-baited traps. Thus, one should not expect to disseminate a population of beetles by traps in the area where they emerge. Trapping in treeless areas suggests that beetles are capable of flying for several miles. Clearly the capacity of the European elm bark beetle to disperse greatly exceeds the 300 m that has been commonly repeated in the literature.

Mass trapping European elm bark beetles as a strategy for reducing DED was evaluated during 1975 in several localities encompassing a variety of situations (Lanier *et al.*, 1976). In Detroit, Michigan, traps were deployed at 30 to 50 m intervals in a 520 ha plot within a generally infested area. At Ft. Collins, Colorado, a similar grid was used, but it covered the entire area (1,350 ha) within which elms occurred. Almost four million beetles were caught on the Detroit traps; this was about twice the number estimated to have originated within the test area. The pheromone clearly concentrated beetles in the treatment zone and resulted in an increase of the DED rate from 4.3% in 1974 to 7.4%. At Ft. Collins, 1.5 million beetles were killed and the DED rate declined from 3.5% in 1974 to 2.8% in 1975.

Another strategy tested was picketing traps around groves of elms to prevent immigration. At Hamilton College, New York, this technique did not decrease the prevailing epidemic rate of DED. However, at Hinerwadel's Grove, North Syracuse, New York, and the University of Delaware the DED rates dropped from 7.1% and 4.4%, respectively, to zero.

At Syracuse, mass trapping was a byproduct of tests of trapping techniques. Over 800,000 beetles were taken on 79 traps, and the DED rate declined from 22.2% in 1974 to 6.7% in 1975. At this time, we view the positive results experienced in some of our tests as coincident with, but not necessarily caused by, beetle trapping.

During 1976 pheromone-baited traps were being evaluated in 12 states. The encirclement tests have increased to 12 and expanded to Evanston, Illinois. In south-central California, Dr. Martin Birch is testing grid and encirclement techniques in three isolated communities. We have increased the minimum distance between traps in the grid system to 100 m, doubled the surface area of the traps, and emphasized their placement on fully exposed tree boles or utility poles. Utility poles treated with methoxychlor are being evaluated as an alternative to sticky traps.

IV. PROSPECTS FOR UTILIZATION OF BEHAVIOR-MODIFYING CHEMICALS

Sex pheromones have recently become commercially available for monitoring populations of certain Lepidoptera damaging to orchards. However, as of the date of this writing, manipulation of insect populations with pheromones, kairomones and allomones has been experimental only. There are two principal reasons for this lack of usage: 1) demonstrations that these systems can be economically effective in suppressing populations or reducing damage has generally been lacking, and 2) protocol for registration of behavior-modifying chemicals has not been established by the U.S. Environmental Protection Agency (EPA). Current EPA standards for registration of these chemicals as pesticides require tests which are prohibitively expensive. Operational use of behavior-modifying chemicals for population management is dependent upon the development of simple systems for their use and the adoption of more appropriate registration standards.

V. SUMMARY AND CONCLUSIONS

The potential of damage by bark beetles in urban areas is generally increasing due to the popular planting of conifers for landscaping, the advancing age of existing shade tree systems and the encroachment on forests by suburban sprawl. Because bark beetles are dependent upon chemical communication for breeding and tree killing, because both sexes of the damaging stage are affected by these chemical messengers, and because the aesthetic values of urban shade trees are high, it is probable that behavior-modifying chemicals can be economically used to reduce damage by these pests.

Disruption of bark beetle aggregation by adaptation of their olfactory sense to high levels of attractant pheromones has not yet been accomplished. This technique would require the continuous dispersal of relatively high concentrations of attractant pheromone at a relatively high cost. On the other hand, disruption with low concentrations of anti-aggregating compounds has been operationally successful. These chemicals may prove useful in both protecting individual trees and manipulating insect populations.

The technical feasibility of using synthetic allomones to prevent attack on individual trees is unknown. Although allomones have been identified, attempts to utilize them for tree protection are lacking.

Mass-trapping bark beetles using attractant pheromones has received considerable attention. The effectiveness of this method in reducing damage was demonstrated for the western pine beetle and current results of tests of pheromones for the control of DED by trapping European elm bark beetles appear favorable. However, mass-trapping is not for the private home owner, because it is not suited to the protection of individual trees. Mass-trapping probably will not, by itself, suppress a bark beetle epidemic; conversely, the impact of mass-trapping probably increases as the population decreases.

In the case of the European elm bark beetle, sticky traps may be effective in reducing incidence of DED by luring beetles to their deaths before they have inoculated healthy elms through twig injuries. However, considerable gains in beetle-killing efficiency might be realized by integrating the use of attractants with insecticides or trap tree techniques. For example, non-elm trees or utility poles carefully treated with insecticide or elms which have been rendered prophylactic by injecting them with systemic fungicides and/or insecticides may be more efficient than sticky traps in killing beetles. Baiting trees which have been injected with the silvicide cacodylic acid could reduce beetle populations as well as decrease the burden of sanitation and diminish the reservoir of inoculum threatening ornamental elms.

The operational use of behavior-modifying chemicals will depend upon the development of simple and economic delivery and their registration for use. Current technology for trapping and for controlled release of pheromones, anti-attractants, etc., seems sufficient for commercialization. However, safety data required for registration and analysis of cost-effectiveness are lacking.

REFERENCES

Anderson, R. F. 1948. Host selection by the pine engraver. *J. Econ. Entomol. 41*: 596-602.

Baker, W. L. 1972. Eastern forest insects. USDA Misc. Publ. No. 1175. U.S. Gov. Print. Off., Washington, D. C.

Bartels, J. M., and G. N. Lanier. 1974. Emergence and mating in *Scolytus multistriatus* (Coleoptera: Scolytidae). *Ann. Entomol. Soc. Am. 67*: 365-370.

Beaver, R. A. 1967. The regulation of population density in the bark beetle *Scolytus scolytus*. *J. Anim. Ecol. 36*: 435-451.

Bedard, W. D., and D. L. Wood. 1974. Management of pine bark beetles - a case history. The western pine beetle. *In* Southern pine beetle symposium (T.L. Payne, R.N. Coulson, and R.C. Thatcher, eds.). Texas A&M Univ. and South. For. Exp. Sta., USDA For. Serv., pp. 15-20.

Beroza, M. 1970. Chemicals controlling insect behavior. Academic Press, New York.

Berryman, A. A. 1973. Population dynamics of the fir engraver, *Scolytus ventralis* (Coleoptera: Scolytidae). I. Analysis of population behavior and survival from 1964 to 1971. *Canad. Entomol. 105*: 1465-1488.

Birch, M. C. (Ed). 1974. Pheromones. North American Elsevier, New York.

Birch, M. C., and D. L. Wood. 1975. Mutual inhibition of the attractant pheromone response by two species of *Ips* (Coleoptera: Scolytidae). *J. Chem. Ecol. 1*: 101-113.

Bright, D. E., and R. W. Stark. 1973. The bark and ambrosia beetles of California. Coleoptera: Scolytidae and Platypodidae. Bull. Calif. Insect. Surv., Vol. 16, Univ. Calif. Press, Berkeley, Los Angeles, London.

Buffam, P. E., and F. M. Yasinski. 1971. Spruce beetle hazard reduction with cacodylic acid. *J. Econ. Entomol. 64*: 751-752.

Copony, J. A., and C. L. Morris. 1972. Southern pine beetle suppression with frontalure and cacodylic acid treatments. *J. Econ. Entomol. 65*: 754-757.

Coulson, R. N., J. L. Foltz, A. M. Mayassi, and F. P. Hain. 1975. Quantitative evaluation of frontalure and cacodylic acid treatment effects on within-tree populations of the southern pine beetle. *J. Econ. Entomol. 68*: 671-678.

Cuthbert, R. A., and J. W. Peacock. 1976. Attraction of *Scolytus multistriatus* to pheromone-baited traps at different heights. *Environ. Entomol. 4*: 889-890.

Doskotch, R. W., S. K. Shatterji, and J. W. Peacock. 1970. Elm bark derived feeding stimulants for the smaller European elm bark beetle. *Science 167*: 380-382.

Furniss, M. M., L. N. Kline, R. F. Schmitz, and J. A. Rudinsky. 1972. Tests of three pheromones to induce or disrupt aggregation of Douglas-fir beetles (Coleoptera: Scolytidae) on live trees. *Ann. Entomol. Soc. Am. 65*: 1227-1232.

Furniss, M. M., G. E. Daterman, L. N. Kline, M. D. McGregor, G. C. Trostle, L. F. Pettinger, and J. A. Rudinsky. 1974. Effectiveness of the Douglas fir beetle antiaggregative pheromone methylcyclohexenone at three concentrations and spacings around felled host trees. *Canad. Entomol. 106*: 381-392.

Gaston, L. K., and H. H. Shorey. 1974. Cotton - the pink boll worm. *In* Pheromones (M.C. Birch, ed.). North American Elsevier, New York, pp. 425-426.

Gilbert, B. L., and D. M. Norris. 1968. A chemical basis for bark beetle (*Scolytus*) distinction between host and non-host trees. *J. Insect Physiol. 14*: 1063-1068.

Gore, W. E., G. T. Pearce, G. N. Lanier, J. B. Simeone, R. M. Silverstein, J. W. Peacock, and R. A. Cuthbert. 1977. Aggregation of the European elm bark beetle, *Scolytus multistriatus:* Production of individual components and related behavior. *J. Chem. Ecol. 3*: 431-448.

Hostetler, B. B., and J. W. Brewer. 1976. Translocation of cacodylic acid in Dutch elm-diseased American elms and its effect on *Scolytus multistriatus* (Coleoptera: Scolytidae). *Canad. Entomol. 108*: 893-896.

Jacobson, M. 1972. Insect sex pheromones. Academic Press, New York.

Knopf, J. A. E., and G. B. Pitman. 1972. Aggregation pheromone for manipulation of the Douglas-fir beetle. *J. Econ. Entomol. 65*: 723-726.

Lanier, G. N., R. M. Silverstein, and J. W. Peacock. 1976. Attractant pheromone of the European elm bark beetle (*Scolytus multistriatus*): isolation, identification, synthesis, and utilization studies. *In* Perspectives in forest entomology (J.E. Anderson and H.K. Kaya, eds.). Academic Press, New York, pp. 149-175.

McCambridge, W. F. 1975. Scotch pine and mountain pine beetles. *Green Thumb 32*: 87.

Norris, D. M. 1970. Quinol stimulation and quinone deterrency of gustation by *Scolytus multistriatus* (Coleoptera: Scolytidae). *Ann. Entomol. Soc. Am. 63*: 476-478.

Payne, T. L., J. E. Coster, and P. C. Johnson. 1977. Effects of slow release formulation of synthetic *endo*- and *exo*-brevicomin on southern pine beetle flight and landing behavior. *J. Chem. Ecol. 3*: 133-141.

Pearce, G. T., W. E. Gore, R. M. Silverstein, J. W. Peacock, R. A. Cuthbert, G. N. Lanier, and J. B. Simeone. 1975. Chemical attractants for the smaller European elm bark beetle, *Scolytus multistriatus* (Coleoptera: Scolytidae). *J. Chem. Ecol. 1*: 115-124.

Pitman, G. B. 1971. *trans*-verbenol and alpha-pinene: their utility in manipulation of the mountain pine beetle. *J. Econ. Entomol. 64*: 426-430.

Pitman, G. B. 1973. Further observations on douglure in a *Dendroctonus pseudotsugae* management system. *Environ. Entomol. 2*: 109-112.

Rasmussen, L. A. 1972. Attraction of mountain pine beetle to small-diameter lodgepole pines baited with *trans*-verbenol and alpha-pinene. *J. Econ. Entomol. 65*: 1396-1399.

Renwick, J. A. A., and J. P. Vité. 1969. Bark beetle attractants: mechanism of colonization by *Dendroctonus frontalis*. *Nature (London) 224*: 1222-1223.

Rexrode, C. O. 1974. Effect of pressure-injected oxydemeton-methyl, cacodylic acid, and 2, 4-D amine on elm bark beetle populations in elms infected with Dutch elm disease. *Plant Dis. Rep. 58*: 382-384.

Rudinsky, J. A., M. M. Furniss, L. N. Kline, and R. F. Schmitz. 1972. Attraction and repression of *Dendroctonus pseudotsugae* (Coleoptera: Scolytidae) by three synthetic pheromones in traps in Oregon and Idaho. *Canad. Entomol. 104*: 815-822.

Schwerdtfager, F. 1973. Forest entomology. *In* History of entomology (R.F. Smith, T.E. Mittler, and C.N. Smith, eds.). Ann. Reviews Inc., Palo Alto, California, pp. 361-386.

Shorey, H. H. 1976. Animal communication by pheromones. Academic Press, New York.

Stark, R. W., and F. W. Cobb, Jr. 1969. Smog injury, root diseases and bark beetle damage in ponderosa pine. *Calif. Agric. 23*: 13-15.

Vité, J. P., P. R. Hughes, and J. A. A. Renwick. 1976. Southern pine beetle: effect of aerial pheromone saturation on orientation. *Naturwissenschaften* *66*: 44.

THE POTENTIAL FOR BIOLOGICAL CONTROL IN URBAN AREAS: SHADE TREE INSECT PESTS

William Olkowski
Helga Olkowski
Alan I. Kaplan
Robert van den Bosch

Department of Entomological Sciences
University of California
Berkeley, California

ISBN 0-12-265250-9

I. INTRODUCTION: CONTEXT FOR A BIOLOGICAL CONTROL APPROACH

This paper presents a conceptual framework for, and a preliminary assessment of, the potential for applying biological control technology to urban pest problems, particularly insect problems on shade trees, through incorporating biological controls as key agents in integrated pest management programs. Biological control is defined as "the action of parasites, predators and pathogens in maintaining another organism's density at a lower average than would occur in their absence" (DeBach, 1964). The intent here is to aid in taking the focus of urban pest control away from the predominant reliance on synthetic chemical tools (Anonymous, 1975a; Appleby, 1976). Because of its hazard to people, pets, wildlife and property, pesticide use in urban areas should be minimized. Pesticides are derived from finite and increasingly costly fossil hydrocarbon supplies, and these raw materials can frequently be put to better use in industry, medicine and agriculture, where losses from insect damage are more serious.

II. URBAN ECOSYSTEM MANAGEMENT

In the process of manipulating pest populations the pest manager actually affects an entire ecosystem. This manipulation is most easily seen in the drastic changes in faunal elements, the side effects on plants (e.g., phytotoxicity), and the ramifications of the toxicants as they biomagnify or degrade to still other toxic compounds or other materials of largely unknown effect. In actual practice, the concept of an ecosystem manager provides a more inclusive context for pest management operations (Olkowski *et al.*, 1976). Practices such as fertilization, irrigation, pruning, species selection and human behavior (e.g., vandalism) all affect the ecosystem in question as well as the pest population and need to be included in decision making. When this perspective is taken, more options for pest suppression become available. This is particularly relevant to the urban ecosystem.

An ecosystem by definition has biotic and abiotic elements, interrelated as energy flows and nutrient cycles. The idea of an urban ecosystem is useful because it recognizes the human element as the dominant characteristic of a contrived or man-made system. It is an important concept for biologists since it aids in the study and redesign of that which is probably the earth's most important area from the viewpoint of environmental impact. Urban complexes are composed of industrial, residential, recreational life supportive (shopping centers, utilities, air and water ports, etc.) and corridor (streets and roadways) areas. Besides the uniqueness of particular pest problems, each area shares general types of problems. For example, areas of the city consisting of wooden structures share the potential for termite problems. The mosquito, on the other hand, would tend to be more general, occurring wherever breeding areas exist. From a managment viewpoint each problem's solution requires a delivery system which has a certain social interface. In urban areas, private operators are largely responsible for pest control in many such delivery systems. Special problems, like excessive mosquito populations, are handled by particular agencies, e.g., mosquito abatement districts, vector control units, public health, public works, and recreation and park departments.

Insects are not the only urban pests; other species include rats, mice, and other vertebrates, (Anonymous, 1975b), weeds, and plant pathogens (Hepting, 1971). The larger framework offered by ecosystem management is needed to encompass all such pests. It is essential for solving problems that are interrelated and defy solutions based on a discipline-limited view. Consider the urban fly problem. Large populations of muscoid and calliphorid flies, especially *Musca domestica* L., *Phaenicia sericata* (Meigen) and *Fannia canicularis* L., are produced in organic refuse and readily enter homes (Greenberg, 1971, 1973). One thousand or more flies per week may be produced in the average garbage can (Ecke and Linsdale, 1967). Poorbaugh and Linsdale (1971) reported on flies emerging from dog feces in California. *M. domestica* and *P. sericata* as well as other fly species feed on aphid-produced honeydew. When honeydew production goes up, flies that congregate in shade trees increase visibly in number. In Berkeley, California, this has caused citizen complaints and requests for treatments. Subsequent reduction of the aphid populations by introduced parasites reduced the fly congregation in the trees (Olkowski *et al.*, 1976) through reducing honeydew production.

The solution to this problem, however, is not limited to biological control efforts (see Legner and Brydon, 1966). Although fly traps and parasitoid releases help alleviate the problem locally, the basic solution lies in the waste management field, particularly in developing recycling systems that reuse discarded resources. Quite possibly combined waste recycling and fly management programs could reduce these populations and radically change the situation that causes the problem, but this approach goes beyond the usual training and thinking of existing pest management personnel. This is one of many examples illustrating why the larger framework of ecosystem management is needed.

III. A TAXONOMY OF URBAN PEST CONTROL

Urban pest management encompasses seven traditional fields of pest management (see Table I). Ideally, the urban pest manager would be trained in all these areas. However, the complexity of operating as a pest manager in urban areas is obvious because of the degree and diversity of education required. For example, problems with house dust mites, *Dermatophagoides* spp. (Keh, 1973; Lang and Mulla, 1977), lie between the architectural and medical fields since these mites inhabit older homes and have been incriminated in bronchial asthma cases.

A number of problem organisms, such as cockroaches, are usually considered within the purview of medical entomology. However, because of a lack of evidence to implicate them in a disease transmission cycle, they are best relegated to psychological concerns; that is, phenomena whose nature is defined through subjective rather than objective perceptions (see Ebeling, 1975). Shade tree-related psychological pest problems are recorded in Olkowski and Olkowski (1976). An example of a medical entomological-tree relation is provided by *Aedes sierrensis* Ludlow, which develops in tree cavities and attacks humans.

The separation between silviculture (the cultivation of forests) and ornamental horticulture theoretically is based on the purpose for which the trees are maintained. Where a tree is raised for its product, pest problems fall into silvicultural areas. If the tree is cultured for its amenity values, its pest problems are those of ornamental horticulture. However, urban plants, including trees, are used or can be used for: screens, fence rows, wind breaks, noise buffers (Cook and Haverbeke, 1971), erosion controls, shade production, snow controls, auto headlight glare reduction (Tunnard and Pushkaren, 1963), fire buffers (Hamilton, 1972), deceleration barriers, bird attraction, animal repellents and barriers, sources of nectar and pollen for bees, sources of aesthetic pleasure (from fragrance and flowers, tree shape and form, leaf shade and color) (Williamson, 1968), and as insectaries for beneficial insects (Stary, 1970). Less recognized benefits obtained from urban plants include the production of oxygen, absorption of carbon dioxide, dust adsorption and emission of water vapor (Root and Robinson, 1959). Nevertheless, aesthetic reasons predominate in justifying the culture of urban plants and leaving urban silvicultural entomology essentially undeveloped, as well as the practical aspects of ornamental horticulture.

Many insects that are forest pests also are problems in urban areas, particularly where forests are contiguous with urban areas. For example, during outbreaks the spruce budworm, *Choristoneura fumiferana* (Clem.), may attack ornamental coni-

TABLE I Areas of Urban Pest Management

Pest management sub-fields	Areas of concern	Examples of pest problems
I. Medical	Human physical health, veterinary health problems	Mosquitoes, flies, fleas, ticks, rats, cockroaches
II. Psychological	Human mental health and comfort	Nuisance pests, including those that trigger hysteria (see Olkowski and Olkowski, 1976)
III. Architectural	Structural problems including artifacts like furniture, fabric, paper	Termites, ants, carpenter bees, wood beetles, wood wasps, carpet beetles, clothes moths, furniture beetles, book lice
IV. Agricultural	Fresh and stored foods such as vegetables, fruits, nuts, grains, animals, spices, oils, beverages, etc.	Tomato hornworms, cucumber beetles, cabbage aphids, armyworms, earwigs, codling moth, fruit flies, flower beetles, raisin moths
V. Floricultural	Plants used for decorations (e.g., cut flowers, house plants)	Chrysanthemum aphids, whiteflies, spider mites
VI. Silvicultural	Trees grown for economic value; i.e., lumber, Christmas trees, etc.	Douglas-fir tussock moth, white pine weevil
VII. Horticultural (ornamental)	Plants as landscaping and recreational elements (street and park plantings, turf areas, plants in travel corridors), special garden areas (arboreta and botanical gardens) and nursery plants	Elm leaf beetle, Japanese beetle, gypsy moth, fall webworm, linden aphid, rose aphid

fers. Polyphagous forest pest species, such as the fall cankerworm, *Alsophila pometaria* (Harris), may easily become urban shade tree pests.

On the other hand, an innocuous insect in the forest may be regarded as a pest in urban areas. For example, *Cinara curvipes* Patch, an aphid that feeds on atlas and deodar cedars (*Cedrus atlantica* Manetti and *C. deodara* (Roxb.) Loud. is under good control by parasites in the Sierra Nevada of northern California (Voegtlin, 1976). However, in the City of Palo Alto, where climatic conditions differ and where these parasites are absent, we have observed this aphid in high numbers producing nuisance amounts of honeydew. Table II, modified from a list of the principal forest pests in the United States (Anonymous, 1975a) is a compilation of forest pests that are also problems on ornamentals.

Many agricultural orchard pests also may occur on urban shade trees. Insects such as the blue green sharpshooter, *Graphalocephala atropuntata (Hordnia circellata)* Baker, which is a problem to grape growers in northern California because it vectors Pierce's disease, is regarded as a pest on shade trees (e.g., on *Liquidambar styraciflua* L.) in the San Francisco Bay Area because it produces honeydew and annoys tree workers during pruning operations. The San Jose scale, *Quadraspidiotus perniciosus* (Comst.), can damage streetside shade trees. Black scale, *Saisettia oleae* (Olivier), a serious pest of citrus and olive, can be found on many ornamentals, especially ash *(Fraxinus velutina* Torr.) in California. *Aphis spiraecola* Patch, a major problem on citrus, occurs on yellow poplar trees, *Liriodendron tulipifera* L., but is a relatively minor pest species in comparison to the recently invaded aphid, *Illinoia (Macrosiphum) liriodendri* (Monell).

In many areas, wood, leaves and prunings are a major part of the urban solid waste stream. There have been some recent attempts to begin recycling programs to compost leaves and use chips for mulches, roadways, paper pulp, particle board, erosion control, livestock feed or bedding, and fuel (Boers, 1976; Swisher, 1976). The City of St. Paul, Minnesota, operates a large-scale chipping program to manage the huge volume of dead tree material resulting from Dutch elm disease. The chipping program is part of the disease management effort because it encourages the collection of dead wood which would otherwise constitute a development source for the bark beetle vector, *Scolytus multistriatus* (Marsham). Mulches can affect pest management, for example, by contributing to increased mortality of the elm leaf beetle, *Pyrrhalta luteola* Müller, when it descends to pupate at the base of elm trees. Although no detailed studies have been conducted, the mulch is believed to increase moisture conditions which enhance the larval pathogen, *Beauvaria* sp. (Broudii, 1973).

TABLE II Principal Forest Pests Which Are Known to Attack Shade Trees in Urban Areas[a]

Common name	*Scientific name*
Western budworm	Choristoneura occidentalis *Freeman*
Gypsy moth	Porthetria dispar *(L.)*
Spruce budworm	Choristoneura fumiferana *(Clem.)*
Douglas fir tussock moth	Orgyia pseudotsugata *(McD.)*
Pitch pine looper	Lambdina athasaria pellucidaria *(G. and R.)*
Fall cankerworm	Alsophila pometaria *(Harr.)*
Jack pine budworm	Choristoneura pinus *Freeman*
Western hemlock looper	Lambdina athasaria lugubrosa *(Walker)*
Saratoga spittlebug	Aphrophora saratogensis *(Fitch)*
Great basin tent caterpillar	Malacosoma fragile *(Strech)*
Pine tussock moth	Dasychira plagiata *(Walker)*
Elm spanworm	Ennomos subsignarius *(Hubn.)*
Western blackheaded budworm	Acleris variana *(Fern.)*
Redheaded pine sawfly	Neodiprion lecontei *Fitch*
Lodgepole needle miner	Coleotechnites milleri *(Busck)*
White fir needle miner	Epinotia meritana *(Heinrich)*
White fir sawfly	Neodiprion abietis *(Harris)*
Arkansas sawfly	Neodiprion taedae linearis *(Ross)*
Eastern hemlock looper	Lambdina athasaria *(Walker)*
Fall webworm	Hyphantria cunea *(Drury)*
Forest tent caterpillar	Malacosoma disstria *(Hubner)*
Larch sawfly	Pristiphora erichsonii *(Hartig)*
Loblolly pine sawfly	Neodiprion taedae *(Ross)*
Lodgepole pine sawfly	Neodiprion burkei *Midd.*
European pine sawfly	Neodiprion sertifer *(Geoff.)*

[a]*Modified from a list by Charles Sartwell, U.S. Forest Service and checked against Johnson and Lyon, 1976.*

The management of urban pest problems, particularly those on shade trees, interfaces with a variety of other applied fields within the urban setting. This should be kept in mind when considering the potential for biological control of shade tree pests, particularly since advances in these related areas could benefit shade tree pest management, and vice versa. For example, municipal shade tree insect spraying operations have been a serious concern to home or community vegetable gardeners because of hazards associated with pesticide drift (Dahlsten, 1970; Walden, 1974).

IV. BIOLOGICAL SETTING FOR URBAN PEST CONTROL

Urban pest control activities take place under private, commercial and public auspices. Obtaining a comprehensive concept of the diversity and size of the user-delivery complex may in itself be a problem to be surmounted before adequate approaches for managing pest problems can be developed. Among users of urban pest control are the homeowner, residence manager, business person (especially in restaurants and other establishments involved with food storage and transport), gardener, the private landscape management firm, and various city, county, state, and federal agencies.

Our approach has been to work on pest problems of ornamental plants and associated vegetation. Since the use of biological control techniques can provide permanent solutions to many of these horticultural pest problems, pesticide use can be reduced where great human density and contact are likely. Our focus on shade trees developed because a) shade tree care accounts for a sizeable portion of the landscape maintenance budget in many cities, b) pesticide applications to shade trees generally pose a drift hazard, c) the shade tree complex constitutes a major, if not the most, stable, productive ecosystem in the urban complex from the standpoint of fixing sunlight and storing biomass, and d) shade trees have concerned, vocal citizen-advocates, e.g., shade tree committees, who might appreciate innovations in pest management technology.

Figure 1 illustrates the complexity of the biological-ecological urban space using the home as a focal point. It provides a general overview of the life support systems for major urban pest groups. The interrelationships are of critical importance. Consider, for example, an aphid problem on the shade tree in front of a home. If the plant is exotic, the aphid also may be foreign and its population size may become excessive because it lacks its natural enemies. As a consequence of its abundance and high honeydew production the aphid is considered a pest and thus engenders a control measure, most likely an insecticide treatment. Honeydew excreted by aphids

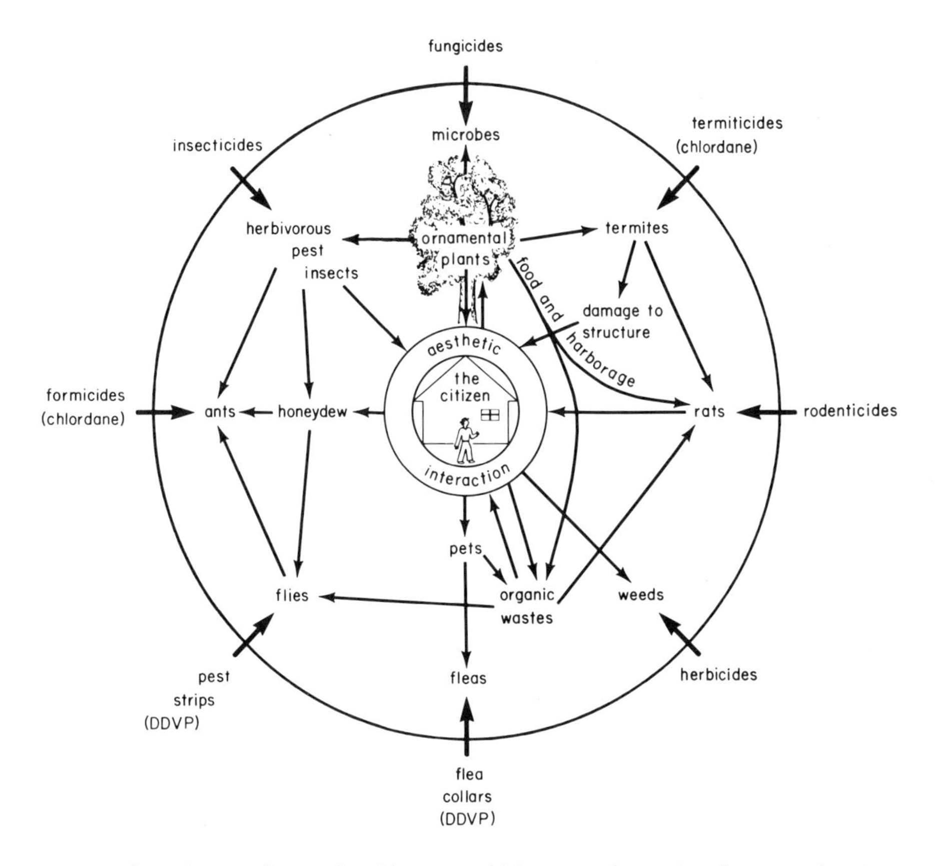

Fig. 1. Schematic diagram illustrating the interrelationships between various urban pests, their major life supports, and the pesticides used against them. (Thin arrows indicate material flow.)

is harvested in large quantities by ants. In the San Francisco Bay area the ant that fits into this niche is the Argentine ant, *Iridomyrmex humilis* Mayr, itself an exotic species. This biological relationship is further enhanced by the fact that the ant frequently prefers to nest under the sidewalk near a shade tree that can supply honeydew and insects as food. When high ant populations occur they are attacked vigorously by homeowners and others. The literature on this ant species and preliminary studies on the relationships described here are reported in Olkowski (1973). Observations in an area of Berkeley, California, suggest that tolerance to the insecticide chlordane may have developed in certain *I. humilis* populations (W. Olkowski, unpub. observation).

Urban rodents, especially the roof rat, *Rattus rattus* L., and the Norway rat, *R. norvegicus* Berkenhout, can gain sustenance from pet droppings and pet foods. In Berkeley, California, a city of slightly over 100,000 people, an estimated 1.4 million kilos of pet feces are produced each year (15% of the population keep pets; 0.25 kg feces/pet/day), a great deal of which is deposited on the streets and public walkways. Many pet owners are quite surprised when they learn of the rodent-pet interrelationship. Flies, especially the synanthropic species (e.g., the green bottle fly, *P. sericata*), also use pet droppings as a food source. Notwithstanding the magnitude of this pet-related problem, urban rodents basically derive their life support from man's physical environment, that is, his structures, debris, and garbage, which provide both shelter and food resources (Davis, 1950, 1972). Date and other palm trees are notorious for harboring rat populations by providing nesting and sheltering sites in southwest cities (Dutson, 1977).

Termites are responsible for damage estimated at about one half billion dollars per year in the United States alone. These insects create structural problems that can be largely eliminated by measures similar to those used to reduce rat problems, i.e., by altering home construction features (Hartnack, 1943; Ebeling, 1975). In the San Francisco Bay area the Argentine ant is an enemy of the western subterranean termite, *Reticulitermes hesperus* Banks (see Beard, 1973). The fact that both are considered structural problems is important, for if the ant helps keep new infestations of termites from becoming established and is itself subject to pesticide treatments, the elimination of it, a relatively minor pest, removes a biological barrier for the termite, a major pest. Poisoning of *I. humilis*, although a common practice in the San Francisco Bay area, may have negative value to homeowners if they understood the ant-termite relationship. Still, the relationship needs to be further studied. Where ants interfere with the biological control of shade tree insects they can be prevented from ascending the tree trunks by barriers of nontoxic sticky materials.

Weeds in the garden are biologically integrated into the complex of pest problems as sources of food and habitat for insects and rodents. Insects that feed upon weeds and help to keep them in check are considered beneficial. Mulches, as alternative tools for managing weeds, also provide habitat for predaceous arthropods such as carabid beetles and spiders which help control certain pest insects. By properly composting the waste debris from shade trees, leaves and chips can be turned from a nuisance with a potential for breeding flies and harboring rats into a useful product for managing weeds and controlling insects. Weed management in general should be approached using integrated control techniques. Examples of the biological control of weeds have been reported by Andres (1971) and Andres *et al.* (1976).

An important aspect of an ecosystem approach is that the pest manager must be made aware of many strategy options including indirect ones, rather than focusing exclusively upon the pest-host relationship. For example, knowing that aphid honeydew supports ants suggests that by applying aphid biological control techniques resulting in the permanent suppression of an aphid population, one can reduce an important ant food source. More difficult to appreciate is the relationship between ornamental plants and rats. In California, food and harborage for roof rats are provided by Algerian ivy (*Hedera canariensis* Willd.), a commonly planted groundcover around residences and highways, and Himalayan blackberry (*Rubus discolor* [=procerus] [Howcel]) which has extensively naturalized (Dutson, 1974a, b). Rats not only feed on the ivy, but also eat invertebrates (e.g., brown garden snail) that live there. The snail, *Helix aspersa* Müller, is an important pest of vegetables, flowers, groundcovers and fruit trees. If this snail constitutes an important part of the diet of rats living in ivy, a reduction in the snail population through use of a natural enemy (e.g., the staphylinid, *Ocypus olens* Muller [Fisher *et al.*, 1976]) might achieve a secondary level of biological control of the rats. Thus, in addition to rodenticides and environmental modifications, a rodent management program must consider ornamental vegetation and its management practices.

Air pollution (see Williamson, 1968) is a factor affecting the whole urban ecosystem including biological controls. In Israel automobile exhaust was believed to kill or otherwise hinder the parasites of the scale *Protopulvinaria mangiferae* (Green) on Java plums (Gerson, 1975). Possibly the mechanism of interference here is the particulate matter deposited from automobile exhaust since dust is known to interfere with natural enemies. Rather than replacing affected trees, a periodic washing with a cleansing agent and/or water might have remedied the situation (see Pinnock *et al.*, 1974a).

V. INTEGRATED PEST MANAGEMENT (IPM)

Several factors are emerging in the field of pest management that portend enormous changes for a science and industry that in the recent past has relied overwhelmingly on chemical controls. First, there is the growing problem of pest resistance to pesticides, as populations of both invertebrate and vertebrate pests develop the capacity to detoxify the compounds used against them (Georghiou, 1972). Pest resurgence, caused by toxic effects to natural enemies, is also a problem. In addition, it is now apparent that the use of synthetic organic pesticides frequently results in secondary pest outbreaks. This occurs because insecticides destroy natural enemies which normally suppress populations of potentially destructive pests. Without control agents these pests build up to injurious levels. Additional factors dictating change are rapidly rising costs of pesticides, new awareness of phytotoxicity problems and increasing concern among the scientific community and the public over man-made, long-term human health hazards and environmental disruptions (Epstein and Grundy, 1974; Pimentel, 1976).

Approximately 14 million kilos of pesticides are used in nonagricultural urban and suburban areas, in industry, in commercial establishments, along streets and highways and in homes and gardens (von Rumker *et al.*, 1972). The dense human populations of these areas require the adoption of safe and effective alternative methods of managing animal and plant pests in urban environments. Further, that pesticide use in ornamental plantings is primarily a response to aesthetic damage rather than a reaction to medical problems or economic losses suggests that use rates can be changed by aesthetic alteration, which is best approached through education. Biological control techniques also offer alternatives to chemical control, but to be maximally effective they must be embedded within the larger framework of integrated pest management (IPM).

IPM, a system of supervised control that reduces pesticide use was originally developed to manage insect pests of certain agricultural crops (Stern *et al.*, 1959; Smith and van den Bosch, 1967). It is an approach which combines cultural, physical, biological, and chemical strategies in a decision-making process having the aim of suppressing pest populations below the level of unacceptable economic or aesthetic injury. An urban IPM program has at least four basic elements, all of which are essential to the realization of the potential for biological control of pests within the system. These are: 1) an entomological research-development component that elucidates the biologies of pest insects and develops new control technologies; 2) a delivery system, which includes monitoring of pest populations, establishment of action thresholds or levels which conserve biological controls, and aid in timing of treatments, incorporating selective pesticides, genetic manipulation and/or physical-cultural processes in target areas; 3) an educational component which includes training of management personnal as well as dissemination of informational materials to the public, and 4) a biological control component responsible for research on the biology of natural enemies and their impact on pest species as well as importation efforts that utilize natural enemies to suppress pest populations.

A pilot program, applying the IPM concept to urban shade trees, has been developed in the San Francisco Bay area over the last eight years. The program currently works with five cities and one school district. Table III summarizes pertinent information about the clients involved and Table IV records the shade tree pest insects of concern along with their origin and current status.

VI. BIOLOGICAL CONTROL COMPONENT OF URBAN IPM

Biological control means population regulation by natural enemies, specifically the manipulation of a pest's parasites, pathogens and predators. This definition is in contrast to the one chosen by Sailer (1975), which includes such approaches as plant resistance to pests, use of growth regulators, attractants, plant or animal-derived toxic substances, autocidal genetic manipulations, and reproductive disruption by use of sex pheromones. Although we encourage the search for and development of all tools necessary for pest management, we also believe that to designate many of these approaches as "biological control" can seriously dilute and misdirect research efforts vitally needed in "traditional" biological control. Categories of natural enemy manipulations considered herein include conservation, colonization, augmentation and importation. Selection of a particular tactic is a function of the available natural enemies and the knowledge to use them to prevent the growth of, or suppress, economically or aesthetically intolerable pest populations.

Availability of natural enemies is a function of their actual existence, knowledge of their existence and ability to find and utilize them. Political realities frequently influence availability. For example, the aphid *Melanocallis* (=*Sarucallis, Myzocallis*) *kahawaluokalani* (Kirkaldy) occurs widely in California, most probably originating in mainland China. Since the Chinese currently are not permitting exploration for parasites by foreign entomologists, there is little possibility of finding effective natural enemies in the immediate future.

Knowledge of the existence of natural enemies is discussed by Delucchi (1976). For the pest manager the ability to find and utilize various natural enemies is a matter of training, interest and motive - all of which are seriously lacking in most areas of the country. One of the most important problems in regard to this concerns the way entomological research is supported. Chemical companies which support university pesticide research and development obtain a product from their investment. In contrast, a method devised to reduce pesticide use, be it an IPM procedure or a natural enemy, provides no product to its investors for subsequent sale. At present the idea of pest control businesses supporting research that devel-

TABLE III Pesticide Reduction Summary of the IPM Shade Tree Component of the Urban Biological Control Project, University of California, Berkeley, at Completion of 1976 Season

City	No. of years under IPM program	City population[a]	Total tree population under management	Total insecticide treatments per year prior to introduction of IPM program[b] (in number of tree treatments)	1976 IPM insecticide treatments (in number of tree treatments): 1976 microbal treatments (Bacillus thuringiensis)	1976 IPM insecticide treatments (in number of tree treatments): 1976 chemical treatments (Diazinon, Malathion, Sevin, Pyrenone, and Meta Systox-R)	1976 IPM insecticide treatments (in number of tree treatments): 1976 total IPM insecticide treatments	% of treatment reduction due to implementation of IPM program
Berkeley	6	107,500	35,000	11,500	32	3	35	99.7%
San Jose	3	557,700	250,000	42,000	4,307	3	4,310	89.7%
Palo Alto	2	54,900	80,000	1,600	369	6	375	76.6%
Modesto	1	85,000	85,000	8,000	8	350[c]	358	95.5%
Davis	1	32,800	12,000	10,000	0	12[d]	12	99.8%
	Total:	837,900	462,000	73,100	4,716	374	5,090	Avg.: 93.0%

[a]Based on census, January 1976.

[b]All treatments were chemical (no microbial agents were used).

[c]4,250 trees were treated with Meta Systox-R (injection) prior to the initiation of IPM program in March, 1976.

[d]805 trees were treated with various synthetic chemicals prior to the initiation of IPM program in April, 1976.

TABLE IV Shade Tree Pest Insects under Study by the University of California Urban Biological Control Project (List Arranged in Decreasing Order of Importance)

Common name	Scientific name	Origin	Status
1. California oakworm	Phryganidia californica *Packard*	*W. U.S.*	*B.t. effective, major parasite* Itoplectis behrensii *(Cresson)*
2. Fall webworm	Hyphantria cunea *(Drury)*	*E. U.S.*	*B.t. effective, importation under study*
3. Elm leaf beetle	Pyrrhalta luteola *(Müller)*	*Europe*	*Releasing parasites - source Iran,* Tetrastichus galerucae *(Fonscolombe) and* T. brevistigma *Gahan*
4. Silver maple aphid	Drepenaphis acerifoliae *(Thomas)*	*E. U.S.*	*Target parasite -* Trioxys americeris *Smith*
5. Tulip tree (yellow poplar) aphid	Illinoia liriodendri *(Monell)*	*E. U.S.*	*Releasing parasites - source E. U.S.,* Aphidius liriodendri *Liu,* Ephedrus incompletus *Provancher,* Praon *sp.*
6. Fruit tree leaf roller	Archips argyrospilus *Walker*	*E. U.S.*	*T.t. effective, importation under study*
7. Green fruit worm	Lithophane attenata *(Walker)*	*N.E. U.S.*	*B.t effective, under study*
8. Modesto ash aphid	Prociphilus fraxinifolii *Riley*	*W. U.S.*	*Pruning and washing are effective if done early enough (before the leaves curl). Importation under study*
9. Green birch aphid	Euceraphis punctipennis *Zett.*	*Europe*	*Target parasite -* Trioxys compressicornis *Ruth*
10. Striped birch aphid	Callipterinella calliptera *Hartig*	*Japan?*	*Unknown parasite, washing effective*
11. Redhumped caterpillar	Schizura concinna *(J.E. Smith)*	*W. U.S.*	*B.t. effective, pruning may also be effective,* Apanteles schizura *Ashmead and* Ophion *sp. present*
12. Holly oak aphid	Tuberculatus *sp.*	*W. U.S.*	*Washing is effective*
13. English oak aphids	Myzocallis castanicola *Baker*	*Europe*	*Under attack by imported parasite* Trioxys pallidus *(Kaltenbach)*
	Tuberculatus annulatus *(Hartig)*	*Europe*	*Under attack by imported parasite* Trioxys pallidus *(Kaltenbach)*
14. Western tussock moth	Hemerocampa vetusta *Boisd.*	*W. U.S.*	*B.t. effective, hand picking of egg masses in dormant season*
15. Mealy plum aphid	Hyalopterus pruni *(Geoff.)*	*Europe*	*Under study for importation, washing effective*

Common names	Scientific name	Origin	Status
16. European elm scale	Gossyparia spuria *(Modeer)*	*Europe*	*Under study for possible parasite importation*
17. Western tent caterpillar	Malacosoma distria *Hubner*	*W. U.S.*	*B.t. and foliage clipping effective*
18. Norway maple aphid	Periphyllus lyropictus *(Kessler)*	*Europe*	*Releasing parasites - source N. Europe,* Ephedrus *sp.,* Aphidius setiger *MacKaver*
19. Linden aphid	Eucallipterus tiliae *L.*	*Europe*	*Under biological control by imported parasite,* Trioxys curvicaudus *MacKaver*
20. Ash bug	Tropidosteptes illitus *(Van Duzee)* T. pacificus *(Van Duzee)*	*W. U.S.*	*Under study*
21. Black Scale	Saissetia oleae *(Bernard)*	*Africa*	*Parasites present, usually kept under control*
22. Western box elder bug	Leptocoris rubrolineatus *(Barber)*	*W. U.S.*	*Spot treatment, washing*
23. Shot hole borer	Scolytus rugulosus *Ratz.*	*Europe*	*Only attacks unhealthy trees*
24. Elm aphid	Tinocallis platani *(Kaltenbach)*	*Europe*	*Under biological control by imported parasite,* Trioxys hortorum *Story*
25. Crepe myrtle aphid	Melanocallis kahawaluokalani *(Kirkaldy)*	*China*	*Washing is effective*
26. Blue-green sharpshooter	Graphalocephala atropuntata *(Baker)* (Hordnia circellata)	*Mexico*	*Under study for possible importation*
27. European fruit lecanium	Lecanium corni *Bouche complex*	*Europe*	*Under study for possible importation*
28. Beech aphid	Phyllaphis fagi *(L.)*	*Europe*	*Importation feasible, washing effective*
29. Citricola aphid	Aphis citricola (=spiraecola) *Patch*	*?*	*Under study for importation*
30. Bowlegged fir aphid	Cinara curvipes *Patch*	*W. U.S.*	*Washing, change of tolerance level*
31. Spruce aphid	Elatobium abietinum *(Walker)*	*Canada ?*	*Spot treatment, parasite present but often effective too late in season to prevent serious damage*
32. Japanese maple aphid	Periphyllus californiensis *(Shinji)*	*Japan*	*Under biological control by imported parasite,* Aphidius areolatus *Ashmead*

ops IPM programs, or sponsoring the importation of a specific natural enemy, appears unlikely but may become a reality.

Selection of other tactics until natural enemies are imported and successfully eliminate the problem, either as interim measures or as substitutes for insecticidal treatments, is essentially a function of the ability of managers to manipulate the pest and existing natural enemy complex. This is related to the competence of the particular people involved, what they know is possible and what they are capable of accomplishing and developing. The following discussion is an effort to acquaint shade tree pest managers with some aspects of biological control technology. Interested readers should consult Huffaker and Messenger (1976) for more detailed information on this topic.

VII. MAJOR SHADE TREE INSECT PROBLEMS IN THE UNITED STATES

Most of the pest problems in the San Francisco Bay area are caused by exotic pests (Table IV). As a consequence, our efforts in this area have a large importation component.

Table V lists what we consider to be the major pest problems on shade trees in the United States and some of the available literature which mentions a natural enemy component. The list was compiled by recording the number of citations in the USDA Cooperative Economic Insect Report (Vols. 23, 24, 25) and Cooperative Plant Pest Report (Vol. 1 through June, 1976) and the number of times each insect was mentioned in cooperative extension publications distributed by various states. The final list was checked against Johnson and Lyon (1976). We do not know the actual distribution of each problem nor its severity, but we anticipate developing a more extensive pest list, by region, and a more complete mapping of the available natural enemies.

This list of shade tree pest problems can only be of use in a general way to pest managers by indicating the major problems that could occur in a particular area. Because of lack of study and documentation on injury level determination, the specific pest problems of most regions are unknown.

VIII. THE CONSERVATION TACTIC

Conservation includes any manipulation of the environment and/or predator and parasite populations which prevents or reduces losses of natural enemies of actual or potential pests. Minimizing the use of insecticides, especially through establishment of aesthetic tolerance levels and economic injury levels, can prevent decreases of populations of parasites and

TABLE V Preliminary List of the Major Shade Tree Insect Pests in the United States

Common name	Scientific name	Origin	Natural enemy references
Elm leaf beetle	Pyrrhalta luteola (Müller)	Eurasia	Luck and Scriven, 1976
Tent caterpillars	Malacosoma spp.	Native	Stehr and Cook, 1968
Bagworm	Thyridopteryx ephemeraeformis (Haworth)	Native	Balduf, 1937; Barrows, 1974a, b; Berisford and Tsao, 1975; Bishop, 1973; Kulman, 1965
Fall webworm	Hyphantria cunea (Drury)	Native	Morris, 1972
Pine needle scale	Chionaspis (=Phenacaspis) pinifoliae (Fitch)	Native	Herrick, 1931; Houser, 1908; Martel and Sharma, 1975
Mimosa webworm	Homadaula anisocentra Meyrick	China	Webster and St. George, 1947
Douglas fir tussock moth	Orgyia pseudotsugata (McD.)	Native	Balch, 1932; Stelzer et al., 1975
Smaller European elm bark beetle	Scolytus multistriatus (Marsham)	Europe	Sitkowski, 1930
European elm scale	Gossyparia spuria (Modeer)	Europe	Ceianu, 1968; Griswold, 1927
Cooley spruce gall aphid	Adelges cooleyi (Gillette)	Native	Cumming, 1959
Fall cankerworm	Alsophila pometaria (Harris)	Native	Fedde, 1973; Schaffner and Griswold, 1934; Larson and Ignoffo, 1971; Wallner, 1971
Spring cankerworm	Paleacrita vernata (Peck)	Native	Schaffner and Griswold, 1934; Wallner, 1971
Nantucket pine tip moth	Rhyacionia frustrana (Comstock)	Native	Cushman, 1927; Lashomb and Steinhauer, 1975; Kearby and Taylor, 1975; Lewis et al., 1970; McGraw and Wilkinson, 1974; Yates, 1967; Baker, 1972; Yates and Beal, 1962; Dick and Thompson, 1971
Redheaded pine sawfly	Neodiprion lecontei (Fitch)	Native	Middleton, 1921; Benjamin, 1955
White pine weevil	Pissodes strobi (Peck)	Native	Harman and Kulman, 1967
Gypsy moth	Lymantria dispar (L.)	Europe	DeBach, 1974; Clausen, 1956
Yellownecked caterpillar	Datana ministra (Drury)	Native	Schaffner and Griswold, 1934
Pine tortoise scale	Toumeyella parvicornis (=numismaticum) (Pettit and McDaniel)	Native	Bradley, 1973; Orr, 1931; McIntyre, 1960
Cottonwood leaf beetle	Chrysomela (=Lina) scripta Fabricius	Native	Head, 1973
None	Archips negundanus Dyar	Native	Parker and Moyer, 1972
Calico scale	Lecanium cerasorum Cockerell	Native	Johnson and Lyon, 1976

Common name	*Scientific name*	*Origin*	*Natural enemy references*
Spruce needle miner	Taniva albolineana *(Kearfott)* (=Olethreutes abietana)	*Native*	*Felt, 1928*
Locust leaf miner	Odontota dorsalis *(Thunberg)* (=Chalepus, Xenochalepus)	*Native*	*Johnson and Lyon, 1976; Martin, 1927*
Spruce budworm	Choristoneura fumiferana *(Clemens)*	*Native*	*Miller, 1963; Simmons* et al., *1975; Tothill, 1922*
Orangestriped oakworm	Anisota senatoria *(J.E. Smith)*	*Native*	*Hitchcock, 1961; Kaya, 1974; Schaffner and Griswold, 1934*
Redhumped caterpillar	Schizura concinna *(J.E. Smith)*	*Native*	*Pinnock* et al., *1974b*
Periodical cicada	Magicicada septendecim *(L.)*	*Native*	*Johnson and Lyon, 1976*
Obscure scale	Melanaspis obscura *(Comstock)*	*Native*	*Stoetzel and Davidson, 1971*
Oak lecanium	Lecanium quercifex *Fitch*	*Native*	*Dozier, 1936; Peck, 1963; Williams and Kosztarab, 1972*
None	Diaspidiotus liquidambaris *(Kotinsky)*	*Native*	*Stoetzel and Davidson, 1974*

predators. Better timing of pesticide applications, spot treatments and the strategy of leaving a pest population residue to support natual enemy populations are ways to increase survival of parasites and predators. Reduction or elimination of fungicide use may allow insect pathogens to survive, thus enhancing any existing microbial mortality factor. Other aspects of the conservation tactic are discussed by Bartlett (1964).

Field observations in the San Francisco Bay area, made after extensive long-term pesticide treatment programs were discontinued, showed what horticultural personnel call a "host tree" phenomenon. This phrase refers to individual tree specimens that invariably each year host excessive pest populations. This phenomenon, which is believed to be due to a combination of environmental, genetic and plant-parasite factors, was observed on the following tree species (see Table IV for pests and other components): Japanese maple (*Acer palmatum* Thunb.), birch (*Betula* spp.), silver maple (*Acer saccharinum* L.), Norway maple (*Acer platanoides* L.), tulip tree (*Liriodendron tulipifera* L.), ash (*Fraxinus velutina* Torr.), elm (*Ulmus* spp.), plum (*Prunus* spp.), linden (*Tilia* spp.), live oak (*Quercus agrifolia* Née) and holly oak (*Q. ilex* L.). That this phenomenon has been observed for such a diverse group of tree and associated pest species in widely different bioclimatic areas within the San Francisco Bay area suggests that it may occur in other areas. The host tree phenomenon could form an important cornerstone for implementation of integrated pest management programs, since it provides a way to create a transition from exclusively chemical controls. Obviously, treating just selected host trees would be a good first step away from treating whole stands.

The following examples were taken from the literature to illustrate how pesticides can upset natural enemy populations. A well documented case of unsuspected pesticide interference occurred at South Lake Tahoe, California. An outbreak of the scale, *Chionaspis pinifoliae*, on lodgepole and Jeffrey pines, caused by malathion fogging against snow mosquitoes (a group of *Aedes* species) took three seasons to recover after malathion treatments were suspended in 1969 (Dahlsten *et al.*, 1969; Roberts *et al.*, 1973). Large scale mosquito fogging campaigns probably upset many urban ecosystems similarly. We have observed such interference in *Tetrastichus galerucae* (Fonscolombe), an egg parasite of *P. luteola*, in northern California.

Nielsen and Johnson (1972), working in New York on the pine needle scale, point out how properly timed insecticide applications can conserve natural enemy populations. They consider spring insecticide applications to have little impact on parasites (*Prospaltella bella* Gahan and *Aphytis* sp.) which are protected within or beneath their host. However, the cocinellid predators *Chilocorus stigma* (Say) and *Microwesia misella* LeConte are probably killed by spring sprays. They suggest that the use of oxydemetonmethyl after the last crawlers emerge would protect more natural enemies. They also observed scales infesting lower branches first and suggested that the lower subdominant laterals and isolated twigs could be pruned without loss of aesthetic value. Although this technique could eliminate the need for insecticide treatments no follow up work was reported. These authors also provide a short review of earlier insecticide efforts using oils against this scale. Generally, oils are less toxic to parasites and predators than traditional chemical insecticides.

Yanin (1975), working on the fir scale, *Diaspidiotus abietis* Schr., and *Carulaspis juniperi* Bouché, suggests that the best time to apply treatments in parks is the second and third ten-day period in November. Such treatments would minimize damage to entomophages because predators are already in overwintering sites and parasites are within resistant host stages. He reports *Aphytis mytilaspidis* LeB. destroying up to 30% of the female scales while up to 50% are killed by the coccinellids *Chilocorus bipustulatus* (L.), *C. renipustulatus* (Scriba) and *Exochomus quadripustulatus* (L.).

An important way to conserve natural enemies is to choose selective pesticides. Colbern and Asquith (1973) tested a series of insecticides, miticides and fungicides against the coccinellid mite predator, *Stethorus punctum* (LeC.), in Pennsylvania. This ladybeetle is one of the most important native predators of the European red mite, *Panonychus ulmi* (Koch). Larvae and adults were tested by a laboratory dip method. Bay Hox 2709, CGA 13608, carbaryl, Dursban^R, Mesurol^R and Ortho 15223 were found to be highly toxic to adults. Carbaryl and Ortho 15223 were moderately toxic to larvae. All other test materials, except benomyl, which showed some mixed results, appeared to be nontoxic. Although this mite is predominantly a fruit pest, the above work could be applied to other *Stethorus* and coccinellid species which attack shade tree pests.

Since many insect pathogens are fungi (e.g., *Entomophthora*), fungicides applied for plant disease control may interfere with insect or mite epizootics. Thus, minimizing fungicide use can also be considered a conservation tactic; however, no examples of this phenomenon could be found in the shade tree literature. A similar argument can be developed for other pesticide subgroups as well as other manipulations, i.e., physical, cultural, chemical, etc., that can interfere with natural enemy populations.

IX. THE AUGMENTATION TACTIC

Augmentation refers to any procedure leading to increased natural enemy populations already in an ecosystem. An ideal example exists in an agricultural ecosystem with the use of the food spray, Wheast (Hagen *et al.*, 1971), which is used to feed and thereby enhance lacewing populations. However, there are as yet no published examples on the use of this tactic for the control of shade tree insect pests. One of the basic sources of information concerning the natural enemy augmentation tactic is provided by van den Bosch and Telford (1964). The concepts derived from these agricultural examples should be applied to shade tree insect management.

One of the most useful augmentation tools available is *Bacillus thuringiensis* Berliner, (B.t), which, over the past 20 years, has been developed into a potent and selective microbial insecticide (Harper, 1974). First reported as a silkworm pathogen in 1901, B.t. was commercially developed in the 1950's. Isolation and development of a more potent strain in the late 1960's stimulated manufacturers to develop it as a pest control tool. About 150 insects, predominantly Lepidoptera larvae, show some susceptibility to B.t., and the material is registered for use against more than 30 insect pests of agricultural, horticultural and forest plants (Falcon, 1971). We have successfully used B.t. for five years in programs in various cities in the San Francisco Bay area against the California oakworm since the work of Pinnock and Milstead (1971). In addition, we have used B.t. against the fall webworm very successfully for two seasons. Polles (1974) reports that B.t. is as effective as carbaryl for use against this insect, yet the latter insecticide is used commercially in our area in spite of the fact that it is known to cause mite outbreaks (Huffaker, 1971). Another species against which B.t. is used in California is the redhumped caterpillar (Pinnock *et al.*, 1974b). Preliminary laboratory studies also indicate that the green fruitworm, *Lithophane antennata* (Walker), and the fruit tree leaf roller are highly susceptible to B.t. Both these lepidopterans occur on ash and maple in parts of California. Many homeowners are disturbed by these insects because they move from trees by silken threads and frequently lower themselves to vegetation where they can cause considerable defoliation to other plants, e.g., roses.

During the past 10 years, B.t. has been tried against the spruce budworm with disappointing results, until Smirnoff (1971) and Smirnoff *et al.* (1973) showed that the combination of B.t. and chitinase could be effective. Results of a large scale test using this combination in Quebec in 1973 gave satisfactory results and offers the most promising alternative to strictly chemical treatments for this important insect.

By combining insecticide with B.t., the volume of insecticide can be reduced without altering the effectiveness of the treatment (Szalay-Marzso, 1971). Large-scale field experiments against *H. cunea* in Hungary showed that one-tenth normal concentrations of a series of insecticides, in combination with a lower B.t. rate, were effective.

Some interesting examples of habitat management which could be used to increase populations of predatory phytoseiid mites were mentioned by Poe and Enns (1969). The presence of phytoseiids in burlap sacking used to protect trees in winter (Huffaker and Spitzer, 1950; Chant, 1959) suggests a strategy for also protecting mites to alleviate winter mortality. Trees could be collared to allow for the trapping of overwintering mites. The collars could then be transported to less exposed situations. At the appropriate time during the following season, the collars could be relocated on the original trees.

Pollen and nectar sources could be planned into urban environments to increase survival of beneficial species. Pollens of certain weed species have been shown to be effective in sustaining the predatory mite *Amblyseius swirskii* Athias-Henriot in Israel (Ragusa and Swirski, 1975). High populations of another predatory mite, *A. hibisci* Chant, in California avocado orchards were closely correlated with pollen availability even when prey mites were absent (McMurtry and Johnson, 1965, 1966). Kennett (1977) sprayed pollen solutions and obtained a distinct growth response for *A. hibisci* populations in California, thus suggesting a potential for this technique. However, in the case of pollen-feeding mites, this method requires further study since the mites in this particular group are believed to be relatively ineffective predators. Induction of allergy problems to workers would also need to be checked and evaluated.

Pinnock (1977) has encouraged the California Department of Transportation to plant California myrtle in roadside plantings after laboratory studies showed that redhumped caterpillar parasites, *Apanteles schizura* Ashmead and *Hyposoter fugivitus* (Say) live longer when they feed on myrtle nectar and pollen. Pollen is also extremely important in the diet of syrphids (DeBach, 1964) which are common aphid predators on shade trees. No doubt, numerous possibilities exist for utilizing specific flowering shrubs, introduced into the urban landscape, to enhance oviposition or to improve the survival of specific natural enemies already present.

X. THE PERIODIC COLONIZATION TACTIC

Periodic colonization refers to the rearing in the insectary, and subsequent repeated field inoculation or inundation of specific natural enemies at particular times in the season to reduce pest populations. Dietrick (1973) discusses the use of insectary-produced natural enemies in a commercial enterprise offering management services to farmers, but no one is offering a similar service in urban areas.

Information provided by Benjamin (1955) on the redheaded pine sawfly suggests that its egg parasite, *Closterocerus cinctipennis* Ashmead, could be an affective inundative or inoculative agent. Benjamin reports this species of attacking up to 100% of the egg batches sampled; correlations with low pest populations suggest that this parasite may be highly significant in controlling outbreaks of this sawfly.

Nematodes of potential use fall into four groups; neoaplectanids, sphaerulariids, entaphelenchids, and mermithids. Nickle (1974) provides a review of this literature. The neoaplectanids usually have bacterial associates which, when released into insect hemolymph, cause septicemia. *Neoaplectana dutkyi* (DD-136) has a wide host range and has been found attacking Lepidoptera, Diptera, Hymenoptera, Coleoptera, Orthoptera, Hermiptera, Homoptera and Isoptera. Recently, it has been mass reared (in the greater wax moth, *Galleria mellonella* [L.]) and used successfully against the codling moth, *Laspeyresia pomonella* (L.). Sphaerulariids and entaphelenchids attack bark beetles and other Coleoptera, mayflies, fleas, thrips, wood wasps and other species. The best known species include *Contortylenchus elongatus* (Massey) on *Ips confusus* LeConte), *Neoparasitylenchus rugulosi* (Schuester) on *Scolytus rugulosus* (Ratz.), and *Heterotylenchus autumnalis* Nickle on *Musca autumnalis* De Geer. *Reesimermis nielseni* Tsai and Grundmann can parasitize the larval stages of 30 species. *Perutilimermis culicis* (Stiles) is relatively host specific, parasitizing up to 65% of the adults of *Aedes sollicitans* (Walker). *Hexamermis albicans* (van Sibe) can cause epizootics in grasshoppers, chironomids, blackflies, mosquitoes, walking sticks and lepidopterans. Other mermithids parasitize spiders, ants, crustaceans, leeches, nematodes and other invertebrates. The Soviets have used larval *Hexamermis* to control a weevil on oak. Eggs of *Mermis nigrescens* Dujo have been sprayed on vegetable crops to control grasshoppers. The most complete studies of the mermithids are those attacking mosquitoes. Little known species have been reported to attack tent caterpillars and other lepidopterans. These organisms deserve further study.

XI. THE IMPORTATION TACTIC

For the management of exotic pest populations in relatively stable urban vegetation, natural enemy importation offers the prospect of permanancy and self-maintenance from a long-term economic point of view. DeBach (1974) estimates a return of $30.00 for each dollar expended for importation.

Yet, importation with its obvious advantages and proven capabilities still represents a relatively unused tactic in relation to the predominant pest management tactic - use of pesticides. Although worldwide there are about 10,000 pest species, over the last 100 years only about 500 importations have been attempted or 0.5%. Of these, 120 successful cases have resulted in complete or substantial control as of 1970 (DeBach, 1972). All of these successes occurred in agriculture and forestry prior to the initiation of the authors' urban biological control importation efforts.

DeBach and Rosen (1976) reviewed the highly successful importation efforts concerning armored scales that were conducted throughout the world. They provide a preliminary list of the most effective known natural enemies of these insects. Many of these attack scales on both shade and orchard trees (e.g., California red scale, *Aonidiella aurantii* [Mask.], yellow scale, *A. citrina* [Coquillett], rose scale, *Aulacaspis rosae* [Bouché], and other scale species).

The nigra scale, *Saissetia nigra* (Nietner), was a serious pest of a wide spectrum of ornamental plants early in this century but has since become relatively unimportant. Smith (1944) documents a series of occurrences in California that suggest natural enemies, especially aphelinid parasites, reduced populations below injurious levels. Of particular importance was the parasite *Metaphycus helvolus* (Comp.). He mentions that in the early summer of 1939, 25 to 100 adults and from 1,000 to 2,000 nymphs per leaf were common on *Pittosporum undulatum* Vent. By the middle of August few scales could be found. No mention was made of a deliberate attempt to introduce natural enemies for control of this scale, but *M. helvolus* and other parasites were introduced for control of the black scale (Clausen, 1956, 1958).

Certain other shade tree pest insects also have been successfully controlled by importation, but they are seldom discussed from this viewpoint. Sailer (1972) points out an important and interesting case of unappreciated biological control of the San Jose scale, *Quadraspidiotus perniciosus* (Comstock). He notes that during 1890 to 1930 there was scarsely a backyard or abandoned orchard where the scale could not be found. In recent years the San Jose scale rarely has been encountered. He states that in Europe, about three years after orchards were colonized by the parasite *Prospaltella perniciosi* Tower, the scale populations collapsed (Mathys and Guignard, 1965). More recently in large areas of Switzerland, France and West Germany, the scale problem is disappearing (Benassy *et al.*, 1968). Evidently this parasite, with little help from entomologists, invaded, much like its host, from the

Orient through the United States to Europe. Additional shade tree examples include the woolly apple aphid, *Eriosoma lanigerum* (Hausmann), with its parasite, *Aphelinus mali* (Hald.), and the cottony cushion scale, *Icerya purchasi* Mask., with the vedalia beetle, *Rodolia cardinalis* (Muls.). Another example is the parasite *Aphidius areolatus* Ashmead, believed to have been introduced into the United States from Japan with or after the arrival of the aphid *Periphyllus californiensis* (Shinji), which occurs on Japanese maple, *Acer palmatum* Thunb., in Berkeley, California.

Sailer (1975) provides examples of two nuclear polyhedrosis viruses thought to have been introduced accidentally. *Borralinavirus reprimens,* which produces spectacular epizootics of the gypsy moth, first appeared in New England in 1907. The virus was probably imported accidentally with parasites from Europe (Stairs, 1972). Another related species, *B. hercyniae,* which causes collapse of European spruce sawfly, *Diprion hercyniae* Hartig, populations in eastern Canada and Northeastern United States, was also thought to have been imported with parasites from Europe (Neilson and Morris, 1964). Obviously, if useful pathogens and parasites are known to have been introduced accidentally, surely other more useful ones can be found by deliberate effort.

For shade tree insect pests the deliberate importation effort is confined largely to the authors' work in northern California. Olkowski *et al.* (1976) have reported on the successful use of certain aphidiid wasps against three shade tree aphids: the linden aphid, the elm aphid, and the English oak aphid. Work on the linden aphid showed that a mortality agent of much less impact than that required with an agricultural or floricultural crop can have a great effect.

The elm leaf beetle ranks as one of the most important shade tree pests in the Palearctic region. This judgment is based on the area covered by the species, number of host tree species afflicted, severity of the effect and length of time it has been a pest (it was introduced into the U.S. in 1834). Importation efforts against this insect were conducted first in the early 1900's, and they continue to this date. Early work on the east coast of North America failed to establish the egg parasite *Tetrastichus galerucae* (Fonscolombe) from France. A similar failure occurred with an egg parasite, possibly *T. galerucae*, from Japan. We have obtained from Iran, and are releasing, this egg parasite in northern California under the supposition that earlier releases of this parasite failed because of climatic unsuitability. Recently, we have made our first field recovery of the species. Possibly, this new ecotype will prove more successful than prior releases. We have also obtained the Iranian strain of the prepupal parasite, *T. brevistigma* Graham, and are mass producing it and making releases.

Importation attempts have focused primarily on parasites, but predators also have good possibilities, as was shown in 1888 by the first major successful importation against the cottony cushion scale. More recently, two coccinellids have been imported into the United States from Europe (*Coccinella septempunctata* L. and *Propylaea quatuordecimpunctata* Gangl.). Another species, *Harmonia axyridus* Pallas, was imported into central Europe from the Far East. Hodek (1973) summarizes the control attempts with the coccinellids as well as the biology of the family.

Additional importation efforts need to be made with predators. Against aphids, the Chamaemyidae have been largely neglected. Eichhorn (1969) suggests that because the *Leucopis* spp. which attack *Adelges* spp. in Turkey are highly effective they should be introduced into the United States. Another possibility for importation (into California) is the predaceous mirid, *Deraeocoris nebulosus* (Uhler). This species, known from Pennsylvania, is associated with mites, aphids, scales, white flies, psyllids and lace bugs on more than 50 ornamental trees and shrubs (Wheeler *et al.*, 1975).

In a literature review of the use of viruses for the control of pest insects, Pinnock (1975) documents the importation of a series of nuclear polydedrosis viruses for control of the European pine sawfly, *Neodiprion sertifer* (Geoffroy), European spruce sawfly, *Diprion hercyniae* (Hartig), and the gypsy moth. In each case introduction proved effective in producing high mortalities or could be correlated with reduced host populations, in one case for as long as 20 years. Obviously, the use of viruses and other pathogens in importation projects has only started.

XII. BIOLOGICAL CONTROL NEEDS

There is a general paucity of work with natural enemies of shade tree pests. Most of the research effort is directed toward chemical control. If IPM became the dominant approach against shade tree insect pests, biological control tactics and techniques would become of critical importance. To encourage the development of IPM for shade trees, additional pilot programs in other areas are needed. Although the authors have developed their programs in the San Francisco Bay area largely as a volunteer activity, it is unreasonable to expect this to occur everywhere. Now that the efficacy of such programs has been demonstrated it remains to be seen if support for similar efforts can be obtained in other areas.

In order to create something new a motive is needed. We suggest that the hazards from widespread use of toxicants against shade tree pest insects, where proximity to people, pets and property and drift from pesticide use is so potentially damaging, present motive enough. The fact that a reordering of national priorities is needed in the whole pest control area, but particularly in biological control, should be clear from an inspection of Table VI. This table compares one year's national expenditure for other types of research against the Agricultural Research Service's total 10-year expenditure on biological control of weeds and insects (Anonymous, 1975a).

TABLE VI U.S. National Research Resources Invested for Various Areas Compared to Pest Control[a] in 1972, and ARS Importation Effort for the last 10 to 20 Years

Category	Scientific man years	% of total	Millions of dollars	% of total
National total	549,700	100	26,700	100
Defense			8,067	30
Space			3,597	13.5
Agriculture (federal, state and industry)	10,539	1.9	537	2
Total chemical industry			1,624	6.1
Pest control (agricultural research and pesticide industry	5,039	0.9	180	0.7
USDA-ARS (importation effort for weeds and insects)	140[b]		2[b]	

[a]*Data taken from Anonymous, 1975a.*

[b]*Note that these data are for the last 10 to 20 years in comparison to the other data in the table which are for 1972 only.*

ACKNOWLEDGMENT

This work was partially supported by EPA Grant No. R 804 205-01.

REFERENCES

Andres, L. A. 1972. The suppression of weeds with insects. *Proc. Tall Timbers Conf. Ecol. Anim. Control Hab. Manage.* *3*: 185-195.

Andres, L. A., C. J. Davis, P. Harris, and A. J. Wapshere. 1976. Biological control of weeds. *In* Theory and practice of biological control (C.B. Huffaker and P.S. Messenger, eds.). Academic Press, San Francisco, pp. 481-500.

Anonymous, 1975a. Pest control: an assessment of present and alternative technologies. Vol. I. *In* Contemporary pest control practices and prospects. Nat. Acad. Sci., Washington, D. C.

Anonymous, 1975b. Vertebrate pests: problems and control. *In* Principles of plant and animal pest control. Vol. 5. Nat. Acad. Sci., Washington, D. C.

Appleby, J. E. 1976. Current control of insect pests. *J. Arbor.* *2*: 41-50.

Baker, W. L. 1972. Eastern forest insects. USDA For. Serv. Misc. Publ. 1175.

Balch, R. E. 1932. The fir tussock moth. *J. Econ. Entomol.* *25*: 1143-1148.

Balduf, W. V. 1937. Bionomic note on the common bagworm, *Thyridopteryx ephemeraeformis* and its natural enemies. *Proc. Entomol. Soc. Wash.* *39*: 169-184.

Barrows, E. M. 1974a. Some factors affecting population size of the bagworm *Thyridopteryx ephemeraeformis*. *Environ. Entomol.* *3*: 929-932.

Barrows, E. M. 1974b. Insect associates of bagworm moth, *Thyridopteryx ephemeraeformis* in Kansas. *J. Kansas Entomol. Soc.* *47*: 156-161.

Bartless, B. R. 1964. Integration of chemical and biological control. *In* Biological control of insect pests and weeds (P. DeBach, ed.). Reinhold Publ. Co., New York, pp. 489-511.

Beard, R. 1973. Some ants eat termites. *Crops and Soils Magazine* *25*: 28-29.

Benassy, C., G. Mathys, G. Neuffer, H. Milaire, H. Bianchi, and E. Guignard. 1968. L'utilisation pratique de *Prospaltella perniciosi* Toro., parasite du pou de San José *Quadraspidiotus perniciosus* Comstock. Entomophaga Mémoires Hors-Serié 4.

Benjamin, D. M. 1955. The biology and ecology of the red-headed pine sawfly. USDA For. Serv. Tech. Bull. 118.

Berisford, Y. C., and C. H. Tsao. 1975. Parasitism, predation and disease in the bagworm, *Thyridopteryx ephemeraeformis*. *Environ. Entomol.* *4*: 549-554.

Bishop, E. J. 1973. Control of bagworm with *Bacillus thuringiensis*. *J. Econ. Entomol.* *66*: 675-676.

Boers, R. W. 1976. Recycling brush and logs. *J. Arbor. 2*: 36-37.

Bradley, G. A. 1973. Effect of *Formica obscuripes* on predator-prey relationship between *Hyperaspis congressis* and *Toumeyella numismaticum*. *Canad. Entomol. 105*: 1113-1118.

Broudii, V. M. 1973. Ecological characteristics of *Pyrrhalta luteola* in the Ukrainian SSR. *Dopov. Akad. Nauk. UKR SSR Ser. B Geol. Geofiz. Khim. Biol. 35*: 852-856, 863. (Ukrainian, Russian and English summaries).

Ceianu, I. 1968. Observations on *Coccophagus gossypariae* Gah. (Hym.: Aphelinidae), parasite of elm scale, *Gossyparia spuria* (Mod.) (Hom.: Eriococcidae). *Rev. Roum. Biol. Ser. Zool. 13*: 307-314. (English summary).

Chant, G. 1959. Phytoseiid mites (Acarina: Phytoseiidae). Pt. I. Bionomics of 7 spp. in southeastern England. Pt. II. A taxonomic review of the family Phytoseiidae, with descriptions of 38 new species. Canad. Entomol. Suppl. 12.

Clausen, C. P. 1956. Biological control of insect pests in the continental United States. USDA Tech. Bull. 1139.

Clausen, C. P. 1958. The biological control of insect pests in the continental United States. Proc. 10th Internat. Congr. Entomol. 4: 443-447.

Colburn, R., and D. Asquith. 1973. Tolerance of *Stethorus punctum* adults and larvae to various pesticides. *J. Econ. Entomol. 66*: 961.

Cumming, M. E. P. 1959. The biology of *Adelges cooleyi*. *Canad. Entomol. 91*: 601-617.

Cushman, R. A. 1927. The parasites of the pine tip moth. *J. Agric. Res. 34*: 615-622.

Cook, D. I., and D. H. Haverbeke. 1971. Trees and shrubs for noise abatement. USDA For. Serv. and Univ. Nebraska College of Agric., Agric. Exp. Sta.

Dahlsten, D. L. 1970. Why are you doing that? *In* Pesticides workbook. Scientists' Institute for Public Information, New York, pp. 9-13.

Dahlsten, D. L., R. Garcia, J. E. Prine, and R. Hunt. 1969. Insect problems in forest recreation areas. *Calif. Agric. 23*: 4-6.

Davis, D. E. 1950. The mechanics of rat populations. *In* 15th North American Wildlife Conference, March, 1950. Wildlife Management Institute, Washington, D. C., pp. 461-466.

Davis, D. E. 1972. Rodent control strategy. *In* Pest control strategies for the future. Nat. Acad. Sci., Washington, D. C., pp. 157-191.

DeBach, P. (Ed.) 1964. Biological control of insect pests and weeds. Reinhold Publ. Co., New York.

DeBach, P. 1972. The use of imported natural enemies in insect pest management ecology. *Proc. Tall Timbers Conf.*

Ecol. Anim. Control Hab. Manage. 3: 211-233.
DeBach, P. 1974. Biological control by natural enemies. Cambridge Univ. Press, London.
DeBach, P., and D. Rosen. 1976. Armoured scale insects. *In* Studies in biological control (V.L. Delucchi, ed.). Cambridge Univ. Press, Cambridge, pp. 139-178.
Delucchi, V. L. (Ed.) 1976. Studies in biological control. Cambridge Univ. Press, Cambridge.
Dick, W. C., and H. E. Thompson. 1971. Biology and control of Nantucket pine tip moth, *Rhyacionia frustrana,* in Kansas. Kansas Agric. Exp. Sta. Bull. 541.
Dietrick, E. J. 1973. Private enterprise pest management based on biological controls. *Proc. Tall Timbers Conf. Ecol. Anim. Control Hab. Manage. 4*: 7-20.
Dozier, H. L. 1936. Descriptions of two new encyrtid parasites of non-diaspine scales. *Proc. Entomol. Soc. Wash. 37*: 183-185.
Dutson, V. 1974a. The use of the Himalayan blackberry, *Rubus discolor,* by the roof rat, *Rattus rattus,* in California. *Calif. Vector Views 20*: 60-68.
Dutson, V. 1974b. The association of the roof rat, *Rattus rattus,* with Himalayan blackberry, *Rubus discolor,* and Algerian ivy, *Hedera canariensis,* in California. Proc. 6th Vertebrate Pest Conf., Anaheim, California, March 5, 6 and 7, pp. 41-48.
Dutson, V. 1977. Personal communication. Bureau of Vector Control, California State Dept. Public Health.
Ebeling, W. 1975. Urban entomology. Univ. Calif. Div. Agric. Sci., Berkeley, California.
Ecke, D. H., and D. D. Linsdale. 1967. Fly and economic evaluation of urban refuse systems. I. Control of green blow flies (*Phaenicia)* by improved methods of residential refuse storage and collection. *Calif. Vector Views 14*: 19-27.
Eichhorn, O. 1969. Investigation on woolly aphids of genus *Adelges* and their predators in Turkey. *Commonw. Inst. Biol. Contr. Tech. Bull. 12*: 83-103.
Epstein, S. S., and R. D. Grundy. (Eds.) 1974. Consumer health and product hazards: Cosmetics and drugs, pesticides, food additives. Vol. 2, Legislation of Product Safety. M.I.T. Press, Cambridge, Massachusetts and London.
Falcon, L. A. 1971. Microbial control as a tool in integrated control programs. *In* Biological control (C.B. Huffaker, ed.), Plenum Press, New York, pp. 346-364.
Fedde, G. F. 1973. Delayed parasitism of fall cankerworm eggs in Virginia. *Environ. Entomol. 2*: 1123-1125.
Felt, E. P. 1928. Observations and notes on injurious and other insects of New York State. *N. Y. State Mus. Bull.*

274: 145-176.
Fisher, T. W., I. Moore, E. F. Legner, and R. E. Orth. 1976. *Ocypus olens:* a predator of brown garden snail. *Calif. Agric. 30*: 20-21.
Georghiou, G. P. 1972. The evolution of resistance to pesticides. *Ann. Rev. Ecol. Syst. 3*: 133-168.
Gerson, U. 1975. A soft scale as an urban pest. *Israel J. Entomol. 10*: 25-28.
Greenberg, B. 1971. Flies and disease. I. Ecology, classification and biotic associations. Princeton Univ. Press, Princeton, New Jersey.
Greenberg, B. 1973. Flies and disease. II. Biology and disease transmission. Princeton Univ. Press, Princeton, New Jersey.
Griswold, G. H. 1927. The development of *Coccophagus gossypariae* Gahan, a parasite of the European elm scale. *Ann. Entomol. Soc. Am. 20*: 553-555.
Hagen, K. S., E. F. Sawall, and R. L. Tassan. 1971. The use of food sprays to increase effectiveness of entomophagous insects. *Proc. Tall Timbers Conf. Ecol. Anim. Control Hab. Manage. 2*: 59-86.
Hamilton, D. 1972. Plants (and trees) used in fire protection. Growing points. Agric. Ext., Univ. Calif., Berkeley, California.
Harman, D. M., and H. M. Kulman. 1967. Parasites and predators of the white pine weevil *Pissodes strobi* (Peck). Univ. Maryland Nat. Res. Inst. Contrib. No. 323.
Harper, J. D. 1974. Forest insect control with *Bacillus thuringiensis*. Univ. Printing Service, Auburn University, Auburn, Alabama.
Hartnack, H. 1943. Unbidden house guests. Hartnack Publishing Co., Tacoma, Washington.
Head, R. B. 1973. Studies on the cottonwood leaf beetle, *Chrysomela scripta*. Diss. Abst. Inter. B. 33: 3118.
Hepting, G. H. 1971. Diseases of forest and shade trees of the United States. USDA For. Serv. Agric. Handbook No. 386.
Herrick, G. W. 1931. Some shade tree pests and their control. Cornell Univ. Agric. Exp. Sta. Bull. 515.
Hitchcock, S. W. 1961. Egg parasites and larval behavior of orange striped oakworm. *J. Econ. Entomol. 54*: 502-503.
Hodeck, I. 1973. The biology of the *Coccinellidae*. Prague: Academia, Publ. House Czech. Acad. Sci.
Houser, J. S. 1908. The more important insects affecting Ohio shade trees. Ohio Agric. Exp. Sta. Bull. 194.
Huffaker, C. B. 1971. The ecology of pesticide interference with insect populations. *In* Agricultural chemicals - harmony or discord for food, people, environment (J.E. Swift, ed.). Univ. Calif. Div. Agric. Sci. Symp., pp.

92-104.
Huffaker, C. B., and P. S. Messenger. (Eds.) 1976. Theory and practice of biological control. Academic Press, New York.
Huffaker, C. B., and C. H. Spitzer, Jr. 1950. Some factors affecting red mite populations on pears in California. *J. Econ. Entomol. 43*: 819-831.
Johnson, W. T., and H. H. Lyon. 1976. Insects that feed on trees and shrubs. Cornell Univ. Press, Ithaca, New York.
Kaya, H. K. 1974. Laboratory and field evaluation of *Bacillus thuringiensis* var. *alesti* for control of orange striped oakworm. *J. Econ. Entomol. 67*: 390-392.
Kearby, W. H., and B. J. Taylor. 1975. Larval and pupal parasites reared from tip moths of the genus *Rhyacionia* in Missouri. *J. Kansas Entomol. Soc. 48*: 206-211.
Keh, B. 1973. The common house dust mites of the genus *Dermatophagoides* (Acarina: Pyroglyphidae). *Calif. Vector Views 20*: 37.
Kennett, C. E. 1977. Personal communication. Div. of Biological Control, Univ. Calif., Berkeley, California.
Kulman, H. M. 1965. Natural control of the bagworm and notes on its status as a forest pest. *J. Econ. Entomol. 58*: 863-866.
Lang, J. D., and M. S. Mulla. 1977. Distribution and abundance of house dust mites, *Dermatophagoides* spp., in different climatic zones of southern California. *Environ. Entomol. 6*: 213-216.
Larson, L. V., and C. M. Ignoffo. 1971. Activity of *Bacillus thuringiensis* varieties *thuringiensis* and *galleriae* against fall cankerworm. *J. Econ. Entomol. 64*: 1567-1568.
Lashomb, J. H., and A. L. Steinhauer. 1975. Observations of *Zethus spinipes* (Hymenoptera: Eumeridae). *Proc. Entomol. Soc. Wash.* 77: 164.
Legner, E. F., and H. W. Brydon. 1966. Suppression of dung-inhabiting fly populations by pupal parasites. *Ann. Entomol. Soc. Am. 59*: 638-651.
Lewis, K. R., H. M. Kulman, and H. J. Heikkenen. 1970. Parasites of Nantucket pine tip moth in Virginia with notes on ecological relationships. *J. Econ. Entomol. 63*: 1135-1139.
Luck, R. F., and G. T. Scriven. 1976. The elm leaf beetle, *Pyrrhalta luteola,* in southern California: its pattern of increase and its control by introduced parasites. *Environ. Entomol.* 5: 409-416.
McGraw, J. R., and R. C. Wilkinson. 1974. Hymenopterous parasites of *Rhyacionia* spp. in Florida. *Florida Entomol. 57*: 326.
McIntyre, T. 1960. Natural factors in control of the pine tortoise scale in the Northeast. *J. Econ. Entomol. 53*: 325.

McMurtry, J. A., and H. G. Johnson. 1965. Some factors influencing the abundance of the predaceous mite, *Amblyseius hibisci*, in southern California (Acarina: Phytoseiidea). *Ann. Entomol. Soc. Am. 58*: 49-56.

McMurtry, J. A., and H. G. Johnson. 1966. An ecological study of the spider mite, *Oligonychus punicae* (Hirst), and its natural enemies. *Hilgardia 37*: 363-402.

Martel, P., and M. L. Sharma. 1975. Parasites de la cochenille du pin dars la region Sherbrooke, Quebec. *Ann. Entomol. Soc. Quebec 20*: 11-14.

Martin, C. H. 1927. Biological studies of two hymenopterous parasites of aquatic insect pests. *Entomol. Am. 8*: 105-156.

Mathys, G., and E. Guignard. 1965. Etude de l'efficacite de *Prospaltella perniciosi* Tow. En suisse parasite du pou de San-José. *Entomophaga 10*: 193-220.

Middleton, W. 1921. LeConte's sawfly, an enemy of young pines. *J. Agric. Res. 20*: 741-760.

Miller, C. A. 1963. Parasites and the spruce budworm. *In* The dynamics of epidemic spruce budworm populations (R.F. Morris, ed.). Mem. Entomol. Soc. Canad. 31: 228-244.

Morris, R. F. 1972. Predation by wasps, birds and mammals on *Hyphantria cunea*. *Canad. Entomol. 104*: 1581-1591.

Neilson, A., and D. Morris. 1964. The regulation of European spruce sawfly numbers in the maritime provinces of Canada from 1937 to 1963. *Canad. Entomol. 96*: 773-784.

Nickle, W. R. 1974. Nematode infections. *In* Insect disease. Vol. II (G.E. Cantwell, ed.). Marcel Dekker, New York, pp. 327-376.

Nielsen, D. G., and N. E. Johnson. 1972. Control of pine needle scale in New York. *J. Econ. Entomol. 65*: 1161-1164.

Olkowski, W. 1973. A model ecosystem management program for street tree insects in Berkeley, California. Ph.D. thesis, 192 pp., Univ. Calif., Berkeley, California.

Olkowski, H., and W. Olkowski. 1976. Entomophobia in the urban ecosystem, some observations and suggestions. *Bull. Entomol. Soc. Am. 22*: 313-317.

Olkowski, W., H. Olkowski, R. van den Bosch, and R. Hom. 1976. Ecosystem management: a framework for urban pest control. *BioScience 26*: 384-389.

Orr, L. W. 1931. Studies on natural vs. artificial control of the pine tortoise scale. Minn. Agric. Exp. Sta. Tech. Bull. 79.

Parker, D. L., and M. W. Moyer. 1972. Biology of a leafroller, *Archips negundanus*, in Utah. *Ann. Entomol. Soc. Am. 65*: 1415-1418.

Peck, O. 1963. A catalogue of the Nearctic Chalcidoidea (Insecta: Hymenoptera). Canad. Entomol. Suppl. 30.

Pimentel, D. 1976. World food crisis: energy and pests. *Bull. Entomol. Soc. Am. 22*: 20-26.

Pinnock, D. E. 1975. Pest populations and virus dosage in relation to crop productivity. *In* Baculoviruses for insect pest control: Safety considerations (L.A. Falcon *et al.*, eds.). Am. Soc. Microbiol., Washington, D. C., pp. 145-154.

Pinnock, D. E. 1977. Personal communication. Division of Entomology and Parasitology, Univ. Calif., Berkeley, California.

Pinnock, D. E., and J. E. Milstead. 1971. Biological control of California oakmoth with *Bacillus thuringiensis*. *Calif. Agric. 25*: 3-5.

Pinnock, D. E., R. J. Brand, J. E. Milstead, and N. F. Coe. 1974a. Suppression of populations of *Aphis gossypii* and *A. spiraecola* by soap sprays. *J. Econ. Entomol. 67*: 783-784.

Pinnock, D. E., J. E. Milstead, N. F. Coe, and R. J. Brand. 1974b. The effectiveness of *Bacillus thuringiensis* formulations for the control of the larvae of *Schizura concinna* on *Cercis occidentalis* trees in California. *Entomophaga 19*: 221-227.

Poe, S. L., and W. K. Enns. 1969. Predaceous mites associated with Missouri orchards. *Trans. Missouri Acad. Sci. 3*: 69-82.

Polles, S. G. 1974. Evaluation of foliar sprays for control of two webworms and the walnut caterpillar on pecan. *J. Georgia Entomol. Soc. 9*: 182-186.

Poorbaugh, J. H., and D. D. Linsdale. 1971. Flies emerging from dog feces in California. *Calif. Vector Views 18*: 51-56.

Ragusa, S., and E. Swirski. 1975. Feeding habits, development and oviposition of the predacious mite *Amblyseius swirski* Athias-Henriot (Acarina: Phytoseiidae) on pollen of various weeds. *Israel J. Entomol. 10*: 93-103.

Roberts, F. C., R. F. Luck, and D. L. Dahlsten. 1973. Natural decline of a pine needle scale population at South Lake Tahoe. *Calif. Agric. 27*: 10-12.

Root, I. C., and C. C. Robinson. 1949. City trees. *In* Trees. USDA Yearbook of Agric., pp. 43-48.

Sailer, R. I. 1972. Concepts, principles and potentials of biological control: Parasites and predators. *Proc. North. Cent. Entomol. Soc. Am. 27*: 35-39.

Sailer, R. I. 1975. Future role of biological control in management. *Proc. Tall Timbers Conf. Ecol. Anim. Control Hab. Manage. 5*: 195-209.

Schaffner, J. V., and C. L. Griswold. 1934. Macrolepidoptera and their parasites reared from field collections in the northeastern part of the United States. USDA Misc. Pupl.

188.

Simmons, G. A., D. E. Leonard, and C. W. Chen. 1975. Influence of tree species density and composition on parasitism of the spruce budworm, *Choristoneura fumiferana*. *Environ. Entomol. 4*: 832-836.

Sitkowski, L. 1930. Observations on parasites of scolytids. *Polsk. Pismo Entomol. 8*: 1-2.

Smirnoff, W. A. 1971. Effects of chitinase on the action of *Bacillus thuringiensis*. *Canad. Entomol. 103*: 1829-1831.

Smirnoff, W. A., A. P. Randall, R. Martineau, W. Haliburton, and A. Juneau. 1973. Field test of the effectiveness of chitinase additive to *Bacillus thuringiensis* Berliner against *Choristoneura fumiferana* (Clem.). *Canad. J. For. Res. 3*: 228-236.

Smith, R. F. 1944. Bionomics and control of the nigra scale, *Saissetia nigra*. *Hilgardia 16*: 255-288.

Smith, R. F., and R. van den Bosch. 1967. Integrated control. *In* Pest control: Biological, physical, and selected chemical methods (W.W. Kilgore and R.L. Doutt, eds.). Academic Press, New York, pp. 295-340.

Stairs, G. R. 1972. Pathogenic microorganisms in the regulation of forest insect populations. *Ann. Rev. Entomol. 17*: 355-372.

Stary, P. 1970. Biology of aphid parasites. W. Junk, The Hague.

Stehr, F. W., and E. F. Cook. 1968. A revision of the genus *Malacosoma* Hubner in North America (Lepidoptera: Lasiocampidae): Systematics, biology, immatures, and parasites. Smithsonian Institution, U.S. Nat. Museum Bull. 276.

Stern, V. M., R. F. Smith, R. van den Bosch, and K. S. Hagen. 1959. The integrated control concept. *Hilgardia 29*: 81-101.

Stelzer, M. J., J. Neisess, and C. G. Thompson. 1975. Aerial applications of a nucleopolyhedrosis virus and *Bacillus thuringiensis* against Douglas-fir tussock moth. *J. Econ. Entomol. 68*: 269-272.

Stoetzel, M. B., and J. A. Davidson. 1971. Biology of the obscure scale, *Melanaspis obscura,* in Maryland. *Ann. Entomol. Soc. Am. 64*: 45-50.

Stoetzel, M. B., and J. A. Davidson. 1974. Biology, morphology and taxonomy of immature stages of 9 species in the *Aspidiotini*. *Ann. Entomol. Soc. Am. 67*: 475-509.

Struble, G. R. 1967. Insect enemies in the natural control of the lodgepole needle miner. *J. Econ. Entomol. 60*: 225-228.

Swisher, B. E. 1976. Alternative uses of wood chips. *J. Arbor. 2*: 13-16.

Szalay-Marzso, L. 1971. Effect of fungicides and insecticides on the biological activity of *Bacillus thuringiensis* Berl. preparations. *Acta. Phytopath. Acad. Sci. Hung. 6*: 295-307.

Tothill, J. D. 1922. Notes on the outbreaks of spruce budworm, forest tent caterpillar and larch sawfly in New Brunswick. *Proc. Canad. Entomol. Soc. 8*: 172-182.

Tunnard, C., and B. Pushkaren. 1963. Man-made America: Chaos or control? Yale Univ. Press, New Haven.

van den Bosch, R., and A. D. Telford. 1964. Environmental modification and biological control. *In* Biological control of insect pests and weeds (P. DeBach, ed.). Reinhold Pupl. Co., New York, pp. 459-488.

von Rumker, R., R. M. Matter, D. P. Clement, and F. K. Erickson. 1972. The use of pesticides in suburban homes and gardens and their impact on the aquatic environment. Pesticide Study Series No. 2, EPA Office of Water Programs, Applied Technology Div., Washington, D. C.

Voegtlin, D. 1976. Personal communication. Dept. of Biology, Univ. Oregon, Eugene, Oregon.

Walden, E. 1974. Personal communication. Director, Seattle Municipal Community Gardens, Seattle, Washington.

Wallner, W. E. 1971. Suppression of 4 hardwood defoliators by helicopter application of concentrated and dilute chemical and biological sprays. *J. Econ. Entomol. 64*: 1487-1490.

Webster, H. V., and R. A. St. George. 1947. Life history and control of the webworm, *Homadaula albizziae*. *J. Econ. Entomol. 40*: 546-553.

Wheeler, A. G., Jr., B. R. Stinner, and T. J. Henry. 1975. Biology and nymphal stages of *Deraeocoris nebulosus* (Hemiptera: Miridae) a predator of arthropod pests of ornamentals. *Ann. Entomol. Soc. Am. 68*: 1063-1068.

Williams, M. L., and M. Kosztarab. 1972. Morphology and systematics of the Coccidae of Virginia with notes on their biology. Virginia Polytech. Inst. and State Univ. Res. Div. Bull. 74.

Williamson, J. F. 1968. Western garden book. Lane Magazine and Book Co., Menlo Park, California.

Yanin, V. V. 1975. Coccids as pests of park planting. *Zaschita Rastenii 3*: 43-45.

Yates, H. O., III., and R. H. Beal. 1962. Nantucket pine tip moth. USDA For. Serv., For. Pest Leafl. 70.

Yates, H. O., III. 1967. Key to Nearctic parasites of the genus *Rhyacionia* with species annotations. USDA For. Serv., Ashville, S. E. For. Exp. Sta. (mimeo).

SOME ASPECTS OF URBAN AGRICULTURE

L. E. Ehler

Department of Entomology
University of California
Davis, California

I. INTRODUCTION

Urban agriculture is small scale agriculture in the urban environment. There are good reasons for scientific investment in this type of agriculture. First, urban agriculture is rapidly gaining in popularity and is in need of an inductive base. Second, various aspects of urban agricultural technology may be directly transferable to less technologically advanced countries where it may be relevant to small scale, peasant agriculture. In the present paper, these and related topics are considered as they relate to crop production in urban areas. Examples of entomological problems are discussed.

ISBN 0-12-265250-9

II. THE URBAN ENVIRONMENT

The urban environment is a spectacular result of the activities of man. It is here that man has perhaps had the greatest impact on the physical environment. Historically, urbanization has usually been preceded by a destruction and/or disruption of much of the native vegetation, often with an intervening period of agriculture (Watt, 1973). The vegetation that remained was altered and new vegetation was added. Urbanization has also had a pronounced effect on abiotic components of the environment, especially in larger cities. For example, changes in temperatures and windfields have been recorded and precipitation and runoff conditions altered (Watt, 1973; Levin, 1974). In effect, man superimposed his needs (e.g., buildings, streets) onto the existing physical environment and derived a new setting - the urban environment.

From a biological viewpoint, the term "urban environment" has little significance. Instead, we are concerned with a disturbed environment. In this context, a disturbed environment is one in which the biotic and abiotic components have been significantly modified by man. Insects in the urban environment are, for the most part, adapted (or preadapted) to a disturbed or altered habitat. Thus, such terms as "urban strategist" (Pyle, 1975) or "urban insect" are not especially appropriate. Few, if any, insects occur solely in urban areas.

III. URBAN AGRICULTURE

A. Why Urban Agriculture

The motivations for carrying on agriculture in the urban environment are varied and involve, among others, philosophy, social sciences and natural sciences. Olkowski and Olkowski (1975) discuss a number of reasons for participating in urban agriculture. According to them, urban agriculture can: 1) reduce the impact of the population-land squeeze; 2) correct an apparent flaw in modern civilization, i.e., the urban-rural separation; 3) bring about an interest in nature and reduce the consumptive urban life style; 4) reduce the pesticide load through insect habitat manipulation; 5) allow people to take greater charge of their life support system; and 6) eventually aid in building a new solar civilization.

Indeed, there are other reasons that can be added to the list. Crossland (1975) suggests that family gardening in the urban environment is a means for: 1) saving money on the food bill, 2) utilizing "waste land," and 3) reestablishing a link with the natural environment. Since most arable land is now under cultivation (Pimentel, 1976), and because cities often develop on prime agricultural land, it should be of value to obtain additional production from small scale agriculture in urban areas. For some, urban agriculture is a rewarding hobby while for others it is an art form, for in such activity there is aesthetic experience.

B. Organic Farming

In urban agriculture, methods of crop production are highly varied. For example, approaches to insect control problems vary from "do nothing" to total reliance on chemical insecticides. In this regard, the concept of "organic farming" merits particular attention since it is gaining in popularity and because many professional scientists do not seem to fully understand this concept.

Organic farming is a philosophy of agricultural production. To fully appreciate this, it will help to assume that philosophies are not necessarily right or wrong; instead, they are just different. A central philosophical theme to be considered is the metaphysical concept of man's relation to nature. Variations about the theme include: 1) a traditional Western view that man has rightful mastery over nature and that nature was created to benefit man (White, 1967; Murdy, 1975); and 2) a belief that man is not a specially privileged creature with nature subordinate, but is instead a harmonious part of nature and not her exploiter (Watanabe, 1974). It is suggested that organic growers adhere to the latter philosophy. This is evident in most treatises on organic farming, e.g., "Most gardeners spend entirely too much time fretting over bugs. Much of that time is unnecessary because the same gardeners haven't learned to live with nature ...(instead they try)... to dominate it" (McKillip, 1973); or "... I have viewed all insects with an innate feeling of curiosity, affection and the realization that they are an integral part of our natural ecology" (Philbrick and Philbrick, 1974). In other words, crop production can be achieved in ways which are less exploitive or aggressive toward nature.

"Organic" and "conventional" farming are two points on a continuum, and it is possible to adopt features of each (Lockeretz *et al.*, 1975). It seems that for most organic growers, the essential point is not the product (i.e., *what* is produced); instead, it is the process (i.e., *how* it is produced). Thus, biological pest control is preferable to chemical control and manure is preferable to inorganic fertilizers, etc. Although various techniques may be effective, only those consistent with the central philosophy of organic farming are acceptable. There is little use in debating the definition of organic farming. There is no acceptable, operational definition of "organic farming," just as there is none for "pest management."

With respect to insect control, organic growers promulgate an essentially different *strategy*, i.e., the use of naturally occurring materials and cultural control while excluding synthetic organic insecticides. The tactics include predators (e.g., *Hippodamia convergens* Guerin), parasitoids (e.g., *Trichogramma* spp.), pathogens (e.g., *Bacillus thuringiensis* Berliner), botanical insecticides (e.g., pyrethrum), hand picking, companion planting, and a variety of empirically derived recipes and formulas.

Organic farming is not without phony claims and cultists (Day, 1971) and it has its limitations (MacDaniels, 1975). However, there are good reasons for scientific input at this point. The many empirically derived organic remedies are often lacking in experimental verification. Such "homemade insect controls" (Philbrick and Philbrick, 1974) or "garden legends" (Hills, 1970), when properly investigated, may well yield new toxic materials. Experimentation will be required to distinguish between superstition and genuine knowledge of cause and effect.

IV. COMMUNITY GARDENS

A. Overview

Urban agricultural plots commonly occur in such places as backyards (including former patios), rooftops, balconies and sundecks. Recent developments in northern California include: 1) designated garden areas at apartment complexes, 2) specially designed garden plots for newly constructed houses, and 3) community gardens in previously vacant lots or on university campuses. In this section, the concept of community gardening is used as a model for illustrating some characteristic features of urban agriculture and attendant entomological problems.

An analysis of three community gardening projects in Davis, California, is given in Table I. At each site, a vacant area (e.g., about 1 ha) was subdivided into numerous garden plots. Plot size varied from 21 to 70 m^2. At any point in time, the majority of the gardens were planted while the remaining plots were either in progress or abandoned. Among the planted gardens, the number of crops per plot varied from 6.0 at site 3 to 9.4 at site 2. However, of particular significance is the overall range, i.e., from monoculture (e.g., tomato, strawberry, sweet corn) to as many as 18 crops (including ornamentals) per plot. Thus, a most striking feature of such community gardens is the array of botanic diversity displayed.

TABLE I Analysis of Selected Community Gardens in Davis, California, (1976)

Site	Date	Plot size (m^2)	Gardens sampled	Percent of gardens			Crops/plot	
				Planted	In progress	Abandoned	Mean	Range
1	May 27	21	209	62.2	22.9	14.8	6.6	1-13
2	May 27	70	105	66.7	21.0	12.4	9.4	2-18
3	June 30	37	103	78.6	15.5	5.8	6.0	1-12

B. Entomological Aspects

Perhaps the most challenging entomological problem in community gardening is the relationship between botanic diversity and insect population dynamics. It is a conventional wisdom among many scientists and most urban agriculturalists that a mixed planting (i.e., diversity) is preferable to monoculture. However, with respect to insect populations, the data to support this proposition are generally lacking (van Emden and Williams, 1974; Murdoch, 1975). Some recent investigations illustrate the nature of the problem.

Tahvanainen and Root (1972) observed that a flea beetle, *Phyllotreta cruciferae* Geoze, was more abundant in a collard monoculture compared to collard adjacent to natural vegetation. Subsequent experimentation revealed that chemical stimuli given off by non-host plants (e.g., tomato, ragweed) interfered with host finding and feeding of flea beetles. Such results are most relevant to mixed cropping techniques such as in urban agriculture. The obvious suggestion in this case is that plant diversity may be of value in crop protection.

A second example concerns a comparative study conducted by the author in a community garden in Davis, California. This study involved population dynamics of strawberry whitefly, *Trialeurodes packardi* (Morrill), on strawberry. In a given plot (21 m^2), strawberry was grown either as a monoculture or mixed with six to 10 other crops. Preliminary studies clearly showed that whitefly density was much greater in plots where strawberry was grown in mixed plantings compared to a strawberry monoculture (Fig. 1). Mean density for the season was 1.73 in the mixed planting compared to 0.14 in the monoculture! In this instance, botanic diversity did not beget stability nor decrease pest density; in fact, the inverse was true. Interestingly, strawberry whitefly is not a serious pest in commerical strawberry monocultures (Allen, 1959).

Clearly, there is considerable need for entomological investigation into problems associated with mixed cropping (for example) in urban agriculture. In this case, it should be of benefit to identify what Atsatt and O'Dowd (1976) refer to as: 1) insectary plants, i.e., those which aid in maintenance of herbivore predators and parasites; 2) repellent plants, i.e., those which directly or indirectly cause herbivores to fail to locate (or reject) the normal host; and 3) attractant-decoy plants, i.e., those which cause herbivores to feed on alternative hosts. Entomological technology developed in large scale monocultures is not necessarily transferable to small scale, mixed cropping agriculture in the urban environment.

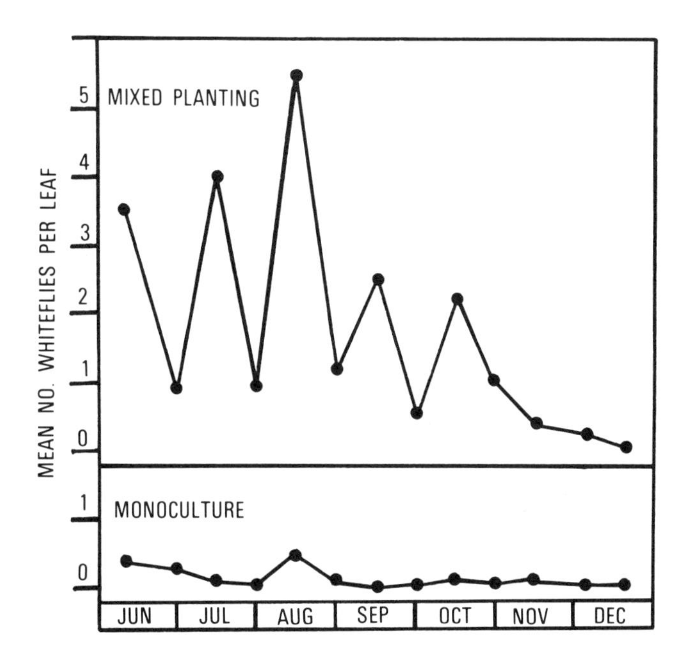

Fig. 1. Population dynamics of Trialeurodes packardi *(Morrill) on strawberry grown in mixed planting versus monoculture in a community garden in Davis, California (1976).*

V. CONCLUDING REMARKS

A. Urban Agriculture

It is suggested that our major goal in urban agricultural research should be to develop predictive theories which we can use to design small scale agro-ecosystems. With the proper knowledge, an agriculturalist or an informed layman should be able to design the system most suited to the occasion.

B. Peasant Agriculture

Agriculture in many less developed countries is characteristically small scale, labor intensive and capital scarce (Way, 1976). Mixed cropping is common. It is clear that the energy and capital intensive food production system of the United States cannot be exported intact to these less developed countries (Steinhart and Steinhart, 1974). Similarly, Green Revolution technology is not always well suited to the small peasant farmer since many of the necessary techniques are too costly, not adapted to his level of education and do not fit the normal scale of operation (Wade, 1974a, b; Greenland, 1975). For the less developed nations, the emphasis shoud be on improved small scale technology (Way, 1976) or "intermediate technology" (Schumacher, 1973). Urban agriculture, like peasant agriculture, is small scale, labor intensive and utilizes multiple and/or sequential cropping. It is suggested that urban agricultural technology, including certain organic methods, may be comparatively easy to export and apply to lesser developed countries. In fact, the reverse may also be true.

REFERENCES

Allen, W. W. 1959. Strawberry pests in California. Calif. Agric. Exp. Sta. Cir. 484.

Atsatt, P. R., and D. J. O'Dowd. 1976. Plant defense guilds. *Science 193*: 24-29.

Crossland, J. 1975. The new city commons. *Environ. 17*: 26-28.

Day, B. E. 1971. Organic gardening...right for wrong reasons. *Calif. Agric. 25*: 2.

Greenland, D. J. 1975. Bringing the green revolution to the shifting cultivator. *Science 190*: 841-844.

Hills, L. D. 1970. Pest control without poisons. Doubleday, Bocking, England.

Levin, M. H. 1974. Commentary - toward an ecology for altered communities. *Ecol. 55*: 225-226.

Lockeretz, W., R. Klepper, B. Commoner, M. Gertler, S. Fast, D. O'Leary, and R. Blobaum. 1975. A comparison of the production, economic returns, and energy intensiveness of corn belt farms that do and do not use inorganic fertilizers and pesticides. Center for the Biology of Natural Systems, St. Louis, Missouri.

MacDaniels, L. H. 1975. Facts about organic gardening. New York State Coll. Agric. Life Sci. Inf. Bull. 36.

McKillip, B. B. 1973. Introduction: Pesticides - who needs them? *In* Getting the bugs out of organic gardening (B.B. McKillip, ed.). Rodale, Emmaus, Pennsylvania, pp. vii-ix.

Murdoch, W. W. 1975. Diversity, complexity, stability and pest control. *J. Appl. Ecol. 12*: 795-807.

Murdy, W. H. 1975. Anthropocentrism: a modern version. *Science 187*: 1168-1172.

Olkowski, H., and W. Olkowski. 1975. The city people's book of raising food. Rodale, Emmaus, Pennsylvania.

Philbrick, H., and J. Philbrick. 1974. The bug book. Garden Way, Charlotte, Vermont.

Pimentel, D. 1976. World food crisis: energy and pests. *Bull. Entomol. Soc. Am. 22*: 20-26.

Pyle, R. M. 1975. Silkmoth of the railroad yards. *Nat. Hist. 84*: 45-51.
Schumacher, E. F. 1973. Small is beautiful. Harper and Row, Hagerstown, Maryland.
Steinhart, J. S., and C. E. Steinhart. 1974. Energy use in the U.S. food system. *Science 184*: 307-316.
Tahvanainen, J. O., and R. B. Root. 1972. The influence of vegetational diversity on the population ecology of a specialized herbivore, *Phyllotreta cruciferae*. *Oecologia 10*: 321-346.
van Emden, H. F., and G. F. Williams. 1974. Insect stability and diversity in agroecosystems. *Ann. Rev. Entomol. 19*: 455-475.
Wade, N. 1974a. Green revolution (I): A just technology, often unjust in use. *Science 186*: 1093-1096.
Wade, N. 1974b. Green revolution (II): Problems of adapting a western technology. *Science 186*: 1186-1192.
Watanabe, M. 1974. The conception of nature in Japanese culture. *Science 183*: 279-282.
Watt, K. E. F. 1973. Principles of environmental science. McGraw-Hill, New York.
Way, M. J. 1976. Entomology and the world food situation. *Bull. Entomol. Soc. Am. 22*: 125-129.
White, L. 1967. The historical roots of our ecologic crisis. *Science 155*: 1203-1207.

INSECT PROBLEMS AND INSECTICIDE USE: PUBLIC OPINION, INFORMATION AND BEHAVIOR

Gordon W. Frankie[1]
Hanna Levenson[2]

Texas A&M University
College Station, Texas

I. INTRODUCTION

Numerous studies dealing with people's attitudes towards their general home environment have been conducted over the past several years. However, relatively few attempts have been

---

[1]*Present address: Department of Entomological Sciences, University of California, Berkeley, California.*

[2]*Present address: Langley Porter Institute, University of California Medical School, San Francisco, California.*

ISBN 0-12-265250-9

designed to assess public opinion and behavior with regard to specific plant and animal elements in their environment. Beginning efforts to document some of these attitudes and behaviors are found in Gerhardt *et al.* (1973), Coster *et al.* (1974), and Keel *et al.* (1974). It is noteworthy that these studies have been planned and carried out by individuals who have apparently been trained primarily in the biological sciences. In this study we have attempted to integrate the thinking and methodology of the social as well as the biological sciences in order to evaluate some of the ways that city and rural people relate to one group of animals, the insects. The study is oriented to explore attitudes and practices of urban dwellers towards insect problems and insecticide use.

Because of new U.S. regulations regarding insecticide use (and pesticides in general), the ecology movement (with all of its consequences), and the necessity for efficiently producing food, there has been much controversy concerning the use of insecticides (see Appendix A for brief historical perspective). Little data have been collected, however, permitting one to estimate how many people in urban areas use insecticides and with what frequency. Are they satisfied? Have attitudes toward insecticide use changed in the last few years? To what extent have people experimented with nonchemical ways of controlling insect pests? Are people aware of the existence of beneficial insects? In fact, do some people even like insects (see 4B, Appendix D)? This survey was designed to answer these and other important questions.

Included in this report are the results from three studies conducted in the summers of 1974, 1975 and 1976. Data were collected in Bryan-College Station, Texas, for each of these years as part of an effort to assess change over time. In addition, people were also interviewed in the Dallas area in 1975 and 1976 in order to make comparisons between two different types of areas - rural (Bryan-College Station) vs. urban (Dallas).

Many of the insect problems of these two cities are similar. Examples of common pests include chinch bugs, army worms, cockroaches and ants. One serious pest, the June beetle, *Phyllophaga crinita* (Burmeister), causes great economic loss in Dallas but very little damage to date in Bryan-College Station.

II. PROCEDURE

A. Location and Subject Selection

In the summer of 1974, data were collected from 147 adults living in the Bryan-College Station area. During the summer months in 1975 and 1976, 100 and 102 adults respectively in this same area were sampled. Bryan-College Station is a rural area (Brazos Co.; population 50,000) in east central Texas between Dallas and Houston. Many of the residents are directly or indirectly affiliated with Texas A&M University (TAMU), the state land-grant institution which is situated in College Station and which has an agricultural college, agricultural experiment station, and extension service. The TAMU System is charged with, among other matters, the responsibility of developing and dispersing state-wide new information regarding insect pest control.

In 1975 and 1976, 102 and 100 adults respectively were interviewed in the Dallas area (Dallas Co.; population 800,000).

In both locations, subjects were selected according to a modified quota-sampling procedure (Cannell and Kahn, 1969). Interviewers were instructed to confine themselves to a specific quadrant of the area to be sampled and to interview the person residing at a certain location on a particular street. (See Appendices B and C for delineation of the area quadrants in Bryan-College Station and Dallas). Interviewers were given a quota of demographic qualities they had to fulfill (see Table I). For example, approximately half of those surveyed had to be male, with 25% 51 years or older. If the interviewer was unable to obtain an appropriate respondent at the first designated address, he/she was instructed to interview the person next door until all quotas were filled. If the person at the assigned address was not home, one call-back at a later time had to be made before going to the next house.

As Table I indicates, approximately the same proportion of people with certain demographic qualities were sampled for each of the years in both locations. The main exception to this equivalence appear to be that: 1) more house owners (compared with apartment renters) were interviewed in 1975 than in 1974 or 1976, and 2) more people with occupations classified as professional were sampled in Dallas in 1975.

In general the type of person interviewed may be described as a white, middle-aged, married homeowner who works in a professional, semiprofessional, or highly-skilled occupation. Selection of the quota variables was made on an a priori basis in order to sample a rather well-educated, home-owning group of people, since the focus of the study was on assessing attitudes of people who would have the resources and information to seek professional help for their pest problems.

B. Interviewers

The inverviewers employed in the study were either graduate or advanced undergraduate students in the Psychology De-

TABLE I Demographic Data: Percentages per Item

Item		Bryan-College Station			Dallas	
		1974	1975	1976	1975	1976
Age Group:	19-25	12%	8%	16%	6%	9%
	26-30	15	10	9	12	16
	31-35	12	7	14	11	13
	36-40	9	13	10	14	18
	41-45	9	16	9	13	6
	46-50	14	13	13	13	6
	51-60	12	13	16	12	22
	61-70	14	8	4	13	5
	>70	3	12	9	6	5
Sex:	Male	42	48	51	50	49
	Female	58	52	49	50	51
Married		82	78	79	82	76
Single		18	21	21	17	24
Residency:	House	79	92	75	91	74
	Apartment	21	8	25	9	26
Home Status:	Own	65	72	67	80	63
	Rent	35	28	33	20	37
Occupation:[a]	I	37	36	32	54	29
	II	16	32	18	28	30
	III	20	19	29	12	23
	IV	22	5	11	-	7
	V	1	-	1	1	5
	VI	3	8	9	5	6

[a]*Occupational prestige ratings (Hodge et al., 1966):*

I. *Professional and administrative*
II. *Semi-professional*
III. *Hi-skill and/or small business*
IV. *Semi-skilled*
V. *Unskilled*
VI. *Student*

partment at TAMU, College Station. Each year two male and two female interviewers were trained by one of the authors (HL) in interviewing techniques and in the use of the structured interview questionnaire. As part of their training, all interviewers met several times prior to the collection of data to discuss problems and to ensure uniformity. In addition, interviewers read and used the concepts outlined in the *Interviewer's Manual*, published in 1966 by the Institute for Social Research of the University of Michigan. Briefly described, this manual discusses interviewing principles and procedures, such as how to build a good interview relationship, how to probe or to stimulate discussion and thereby obtain more information from the respondents, and how to record and edit the interview. Each interviewer was paid a fixed amount for each completed interview. Data collection in Bryan-College Station took three weeks in 1974 and one week in 1975 and 1976. The data collection in Dallas was completed in three days in 1975 and in one week in 1976. All interviews were conducted during the summer of each year.

C. Interview

The content of the interview questionnaire (see Appendix D) was developed by examining several topics that had been discussed by interested citizens at an ornamental pest clinic held in Dallas several months prior to the first year of the study. The clinic is an annual event that is sponsored primarily by TAMU and the county extension service. It is designed to allow homeowners to bring a wide variety of urban plant and animal questions directly before a group of research and extension experts. In addition, questions were added which were considered important by researchers in the Entomology Department at TAMU, College Station.

The format of the questionnaire was designed to permit an evaluation of the "A, B, C's" of attitudes - of how people feel (*A*ffect), act (*B*ehavior) and think (*C*ognition) about insect problems and insecticides (Kretch *et al.*, 1962). One grouping of questions focused on how people *feel* about pests (e.g., "Are there any insects that you like?"); another was constructed to assess specific *behaviors* people might exhibit in dealing with insect problems (e.g., "Do you use professional help in solving your insect problems?"); and a third area dealt with the amount and type of *information* people have concerning insect problems and ways to control them (e.g., "Can you tell me what 'organic gardening' means?"). It was thought that each of the "A, B, C" areas had to be examined in order to obtain a full picture of the attitudes people have toward insect problems. For example, it may be that people voice approval of nonchemical means of controlling pests (Affect) and even have a great deal of information on such means (Cognition), but when actually faced with an insect problem might turn to the most potent insecticide available (Behavior). In fact, social psychological research indicates that behavioral change often lags behind affective and informational change (McGuire, 1969).

All interviewers were given standard instructions for entering and leaving the respondent's home (see Appendix E). The structured questionnaires were read to the subjects and their responses were recorded on the form. The quantitative data were then coded on computer cards, and all analyses were performed on an Amdahl 470-V-6 computer at TAMU, College Station.

Each interview required approximately one-half hour to complete. Almost everyone approached in the Bryan-College Station community agreed to cooperate; participation in the Dallas area was more cautious and therefore resulted in more refusals. In general, however, a large percentage of those selected did agree to be interviewed. All interviews were completed anonymously, but any respondent who wished to know more about the study was asked to self-address an envelope and was eventually mailed relevant information.

III. RESULTS AND DISCUSSION

A. Overview

In general, it appears that for these relatively affluent homeowners and renters in Bryan-College Station and Dallas, insects produce an indoor and outdoor problem which is usually dealt with by professional insecticide applicators. While most people could name nonchemical methods of controlling insects and did know of beneficial insects, they were not knowledgeable of the chemicals being used in their homes and/or yards or of the potentially harmful effects of insecticides. Despite widespread home insecticide use, two-thirds of the respondents were in favor of insecticide use in state and/or national parks only in emergencies.

The first part of the results section describes the demographic differences among samples for each of the years of the study for the two localities. The second part discusses the frequency of indoor and outdoor pest problems. The third, fourth, and fifth sections present data on the informational, behavioral, and affective components of attitudes, respectively.

B. Comparison of Demographic Variables

1. *Overview*

There were surprisingly few differences in the answers to the questionnaire based on the respondent's sex. Females reported significantly more pest problems both inside and outside the home than males (81% vs. 61%; 64% vs. 53%). In addition, they reported almost twice as much personal outdoor insecticide use (60% vs. 33%) than male interviewees. It may be that females are more aware and alert to the presence of "bugs" and take on the responsibility for eradicating such pests. A study to examine these speculations further is underway.

There were several major differences in the way younger interviewees responded as compared to older citizens. However, differences between age groups are confounded by differences between groups in the frequency of homeowners as compared to apartment renters. Younger respondents rarely owned homes and few of the older interviewees rented apartments. Differences based on home ownership will be reported in a following section. In brief, younger interviewees (19-29 years) as compared to middle-aged (30-49) and older (50-80) respondents reported fewer outdoor pest problems (38% vs. 71%, 58%) and had changed to using fewer insecticides (44% vs. 22%, 26%). One age difference which does not appear confounded with marital status and homeownership is that having to do with naming beneficial insects. Older respondents as compared to the two younger groups named significantly fewer beneficial insects (56% vs. 74%, 70%).

2. *Differences between Years*

In Bryan-college Station, the age ranges in both samples were comparable as were the percentages of married to unmarried respondents (see Table I for percentages). In all three years an attempt was made to interview an equal number of males and females, although proportionally fewer males than females were sampled in 1974. Almost all subjects were white and of middle and upper socioeconomic levels (i.e., in professional, semiprofessional, or highly-skilled occupations). In 1974 a few more workers in semiskilled occupations were interviewed than in the other two years, and in 1975 proportionally more white collar workers participated. In all three years the majority owned their own homes, although there were proportionally more homeowners sampled in 1975. In general, the respondents across all three years were quite similar with the exceptions as indicated above. These differences should be noted, since they might have contributed to the results in the analyses that follow.

In Dallas, more homeowners and professional people were sampled in 1975 as compared with 1976.

3. Differences based on Home vs. Apartment

To understand how living in a home versus an apartment might be differentially associated with attitudes toward insects and insecticide use as well as frequency of pest problems, analyses were computed on a number of questionnaire items (see Appendix D) to examine differences based on type of residence. The results from such analyses offer insight on the degree to which other findings can be understood in terms of the greater number of homeowners who were sampled in 1975 as compared with the other two years.

Results indicate that there are many significant differences between house and apartment dwellers (Table II). However, all of these differences involve specific *behaviors*. Respondents who live in a house reported significantly more outdoor insect problems and sought more information on their problems. They said more often than apartment people that they personally and professionally used insecticides outside. While apartment dwellers do not use insecticides as readily, when they do make the decision to personally or professionally use insecticides, they do it with greater frequency. With regard to professional help, it should be mentioned that many apartments are routinely sprayed at the request of apartment managers oftentimes without the consent of tenants. Further, greater personal use by apartment dwellers may, in part, reflect recent interest in potted plants (which may support insect pests that require chemical treatment). It is noteworthy that, contrary to expectation, there are no informational or affective differences between home and apartment dwellers.

C. Pest Problems

In Bryan-College Station the similarities between the 1974, 1975 and 1976 samples are striking with regard to the frequency of indoor pest problems mentioned (Table III). In all three years, approximately three-quarters of the respondents said they had an indoor insect problem. This high percentage in relatively affluent settings revealed how widespread indoor insect problems are in this rural area. Similarly, there were high percentages of outdoor pest problems reported, but less than those mentioned for indoors.

In Dallas, however, differences between the 1975 and 1976 samples were found for both indoor and outdoor pest problems (Table III). While 68% of the 1975 sample replied that they had an indoor problem, only 50% of those in 1976 were so both-

Table II Significant Differences between House and Apartment Dwellers

Question	Percentages House	Percentages Apartment	Chi square	p
Outdoor insect problem	*65%*	*26%*	*53.00*	*.0001*
Obtained info on insect problem[a]	*72*	*52*	*11.85*	*.003*
Personal use of insecticides: outdoors[a]	*54*	*15*	*50.52*	*.0001*
Frequency of personal use: outdoors ($\geq$5/yr)[a]	*28*	*53*	*63.88*	*.005*
Professional use of insecticides: outdoors[a]	*56*	*31*	*19.70*	*.001*
Frequency of professional use: indoors ($\geq$5/yr)[a]	*8*	*13*	*21.29*	*.01*
Frequency of professional use: outdoors ($\geq$5/yr)[a]	*5*	*20*	*31.92*	*.001*

[a]*Of those to whom question applied.*

ered ($p<.01$). Similarly, in 1976 the frequency of outdoor problems was significantly less than those reported in Dallas in 1975 (35% vs. 75%, $p<.001$).

Comparing the two localities, people in Bryan-College Station for all three years reported more indoor pest problems than those in the city of Dallas. In 1975 people in Dallas more frequently reported having outdoor insect problems, but in 1976 they reported having fewer such problems than residents in Bryan-College Station.

D. Information

1. Respondents with Indoor/Outdoor Problems

Specific information. In Bryan-College Station, approx-

TABLE III Percentages of Responses to Behavioral and Attitudinal Questions

Question[a]		Bryan-College Station			Dallas	
		1974	1975	1976	1975	1976
1. Indoor problem[b]	Yes	75%	76%	71%	68%	* 50%
1. Outdoor problem[b]	Yes	60	62	55	75	* 35
1a. Info obtained[bd]	Yes	67	67	71	81	* 54
1b. Info source[d]						
TAMU		26	18	24	8	3
Friends		16	7	8	4	3
TV		1	0	2	0	0
Nursery		9	11	10	14	26
Exterminator		38	49	51	45	40
Printed matter		3	8	2	17	23
Not sure		0	3	0	1	0
Other		5	3	5	10	6
1c. Satisfaction with info[cd]						
Generally		83	85	77	90	81
Occasionally		5	5	15	6	13
Rarely		8	6	3	5	3
2. Personal use of chemicals indoors[cd]						
Yes		73	78	* 63	68	* 47

Question[a]	Bryan-College Station			Dallas	
	1974	1975	1976	1975	1976
2. Indoors[cd] (cont'd)					
No	23%	20%	37%	28%	53%
2. Personal use of chemicals outdoors[cd]					
Yes	49	50	* 43	58	* 33
No	43	47	56	39	66
2a. Chemicals do good[cd]					
Yes	95	88	87	90	88
No	10	9	11	5	2
2b. Chemicals do harm[cd]					
Yes	14	23	11	11	13
No	86	77	85	83	87
3. Professional help[bd] Yes	59	69	68	80	* 62
3a. Professional use of chemicals indoors[cd]					
Yes	89	91	93	89	87
No	11	4	7	10	10

TABLE III (cont'd)

Question[a]	Bryan-College Station			Dallas	
	1974	1975	1976	1975	1976
3a. Professional use of chemicals outdoors[cd]					
Yes	56%	46%	63%	53%	38%
No	37	46	34	43	56
3b. Knowledge of chemicals[cd]					
Yes	4	11	7	8	5
No	95	88	93	91	93
4A. Attitude change					
Yes	29	35	* 46	40	32
No	71	65	54	60	68
4Aa. Attitude change: How[d]					
Don't use	26	6	6	14	6
More cautious	60	91	77	78	85
More use	14	3	11	7	6
4Ab. Attitude change: Why[d]					
Seen negative results	12	15	4	21	31
College course	5	3	4	2	3

Question[a]	Bryan-College Station			Dallas	
	1974	1975	1976	1975	1976
4Ab. Attitude change: Why[d] (cont'd)					
Reading	26%	27%	19%	14%	14%
Publicity/TV	21	21	48	29	41
Friends	7	3	6	0	3
Ecology/Env.Movement	17	29	10	26	3
Need to kill bugs	12	3	8	7	3
4Ac. Nonchemical ways[b] Yes	53	46	54	38	* 21
4Ae. Nonchemical info source[d]					
TAMU	8	9	2	0	5
Friends	28	20	38	31	35
TV	1	2	2	0	5
Intuition	54	49	38	44	25
Nursery	3	0	0	0	10
Exterminator	0	2	0	2	0
Printed matter	4	11	9	20	10
School	1	0	5	0	5
Other	1	7	5	2	5
4Af. Satisfaction with info[cd]					
Generally	62	* 36	35	66	55

TABLE III (cont'd)

Question[a]	Bryan-College Station 1974	Bryan-College Station 1975	Bryan-College Station 1976	Dallas 1975	Dallas 1976
4Af. Satisfaction with info[cd] (cont'd)					
Occasionally	19%	24%	24%	18%	30%
Rarely	16 *	38	36	11	15
4B. Liked insects[b] Yes	-	-	65	-	31
5. Aware of beneficial insects[c]					
Yes	66	64	69	76	64
No	29	28	24	21	33
6. Chemical use in parks					
Never	7	2	5	3	10
Emergencies	71	62	67	60	53
Regular	14	23	20	19	25
Not sure	6	12	9	19	12
7. Organic gardening[b] Yes	80	90	89	98 *	87
7b. Info source on organic gardening[d]					
TAMU	3	0	0	0	0
Friends	33	34	25	23 *	38

Question[a]	Bryan-College Station			Dallas	
	1974	1975	1976	1975	1976
7b. Info source (cont'd)					
TV	7%	7%	4%	6%	5%
Intuition	2	4	1	0	4
Nursery	0	0	1	0	3
Exterminator	2	0	0	0	0
Printed matter	47	40	57	58	* 39
School	3	7	3	4	5
Other	3	8	8	9	8

[a]Refer to questionnaire in Appendix D for exact wording.

[b]Alternate answer: No

[c]Alternate answer: Not sure

[d]Of those to whom question applied.

*Significant difference between respective years ($p<.05$).

imately two-thirds of those with a pest problem sought information and were generally satisfied with this information for all three years sampled (Table III). In all three years, exterminators were the people most frequently consulted with TAMU ranking second. Friends and nurserymen were the next preferred sources of information on indoor and/or outdoor insect problems. Only three people in all three years said that information concerning pest control was obtained from watching television.

In Dallas, over 80% of the 1975 sample obtained information concerning an indoor and/or outdoor pest problem, and almost all of these people were generally satisfied with the information they received. However, in 1976 only half of those with a pest problem sought advice on how to deal with it ($p<.001$), a dramatic difference from the 1975 level. In both years, exterminators were the most frequently mentioned source of information, with nurserymen and printed matter the next most popular. No one in the Dallas sample mentioned obtaining such information from television.

Comparisons between the two localities with regard to information revealed few differences with the exception that, as expected, Dallas residents relied more on printed matter and less on TAMU than those living in Bryan-College Station. It is surprising that so few people in either city mentioned television as an information source despite the fact that thousands of dollars are spent on county extension service TV programs and on TV advertising of insecticides for home and garden use. In a similar questionnaire study in Minnesota, few people cited TV as a source of information for their home gardens (Keel *et al.*, 1974).

Comparisons based on information source. Frequency data were tabulated for both the Dallas and Bryan-College Station samples for all years combined on most questionnaire items as a function of source of information obtained on pest problems. It was thought that such comparisons would permit some preliminary speculation concerning the correlates of a person who would, for example, go to an exterminator as compared with one who would seek advice from friends. The numbers of people who went to specific information sources are as follows: TAMU (N=54), friends (N=27), nurserymen (N=40), exterminators (N=144), and printed matter)N=30).

Several interesting differences emerged from this comparative compilation. For example, while only 8% of those who went to exterminators said they knew what chemicals professionals used in their home and/or yard, 20% of those going to TAMU had such information. Similarly, people going to TAMU were more knowledgeable of nonchemical ways to control insects than those going to exterminators (40% vs. 60%), with 70% of those obtaining advice from friends aware of nonchemical means. Furthermore, while only 48% of those who used exterminators as an information source said they liked some insects, 69% of those using TAMU said there were insects that they liked. One particularly interesting finding emerged from several of the analyses. People who used printed matter (not identified as having originated from TAMU) for their source of information were the least likely to know of nonchemical means for controlling insects (38%) and the least likely to say there were insects that they liked (44%). Further, they were more frequently in favor of routine spraying in parks[3] as compared to people who sought information from TAMU (27% vs. 17% respectively). Unfortunately, we are unable to offer insight on the nature of the printed matter consulted by this group of people.

Several similarities in attitudes were also noted among people who use different sources of information. For example, approximately two-thirds of those who went to any one of the above sources for advice felt that chemical insecticides should be used in a national or state park[3] under emergency circumstances only.

2. *Entire Sample*

General information. All interviewees were asked several questions to assess the amount of information they had concerning beneficial animals or insects and organic gardening. In the rural sample, two-thirds of the respondents said they were aware of beneficial animals or insects that could occur in their garden or yard. There was no significant difference among the three years sampled. Almost everyone said they had heard of organic gardening. However, when asked for a definition of organic gardening, only 62% of those in 1975 could give one compared to 77% in 1974 and 81% in 1976 ($p<.01$). Of those who had said they had heard of the term, almost everyone had learned about it either through printed matter or friends for all three years. TAMU was rarely mentioned in 1974 (3%) and not at all in 1975 or 1976.

In the city sample, approximately two-thirds of the interviewees replied that they were aware of beneficial animals and/or insects. While almost every person in the 1975 Dallas survey had heard of organic gardening, this percentage dropped to 87% in 1976 ($p<.01$). Three-quarters of those who said they had heard of organic gardening could give an accurate defini-

---

[3] *Park question (Appendix D) was designed to determine if people viewed their particular urban environment (residence) differently from a park environment.*

tion of the term. The most popular sources of information for learning about organic gardening were printed matter and friends, although in 1976 people seemed to rely less on printed matter and more on friends than in 1975 ($p<.05$).

There are many similarities between people in Dallas and in Bryan-College Station concerning the amount of their general information on beneficial insects and organic gardening. With regard to organic gardening, it is surprising that so few people consulted either TAMU or nurseries for information. In the future we expect that TAMU will be called upon to a much greater extent since new homeowner publications on organic gardening will soon be available for distribution.

Comparisons. Approximately two-thirds of all persons interviewed were able to name one or more beneficial insects and/or other animals that could occur in their yard. The most commonly cited beneficial insects included: ladybugs, praying mantids and bees. A very few individuals also acknowledged the value of dragonflies and yellowjackets. Commonly cited non-insect beneficial animals included, frogs, toads, lizards, snakes, earthworms, birds and spiders.

In order to understand more about the type of person who is aware of beneficial insects versus one who is not, chi square analyses were computed on a variety of questionnaire items, examining differences between respondents who said they were and were not aware of beneficial insects or other animals. Table IV presents these percentages and significance levels. Results indicate that there are many differences between these two groups. Those who said they were aware of beneficial insects had more knowledge concerning nonchemical ways of controlling pests, organic gardening, harmful effects of insecticides, and specification of such harm. They said they had changed their attitudes to being less in favor of insecticide use and less in favor of regular insecticide use in parks. They reported more liking of insects. In addition, there were several objective differences between the two groups. Those who were more aware of beneficial insects or other animals reported more indoor and outdoor insect problems. They also seemed to seek out different sources of information for their insect problems than people who were not aware of beneficial insects.

E. Behavior

1. *Personal Means to Control Insects*

Frequency. In Bryan-College Station a majority of those who said they had an indoor insect problem personally used a

TABLE IV Significant Differences Between Those Who Were Aware and Not Aware of a Beneficial Insect

Question	*Percentages*		*Chi square*	*p*
	Aware	*Not Aware*		
Indoor insect problem	*73%*	*58%*	*10.89*	*.001*
Outdoor insect problem	*62*	*50*	*6.40*	*.04*
Source of information on pest problem (e.g. friends)	*6*	*1*	*16.59*	*.06*
Personal use of insect-icides outdoors	*51*	*40*	*14.38*	*.003*
Knew of insecticide harm	*11*	*3*	*9.29*	*.03*
Named harmful effects of insecticides	*11*	*4*	*13.56*	*.004*
Changed attitude to less insecticide use	*42*	*22*	*19.49*	*.0001*
Became more reluctant to use insecticides	*33*	*14*	*31.76*	*.0001*
Knew of nonchemical pest controls	*50*	*27*	*23.94*	*.0001*
Liked insects	*23*	*8*	*31.90*	*.0001*
Use insecticides regularly in parks	*17*	*25*	*15.73*	*.004*
Knew of organic gardening	*94*	*75*	*38.07*	*.0001*

chemical insecticide, usually an aerosol spray. In 1976 the number of insecticide users dropped significantly to 63% of the total sample from 73% in 1974 and 78% in 1975 ($p<.02$) (Table III). The frequency of spraying ranged from less than once a year to 100 times annually (Table V). The frequency of personal insecticide usage outdoors was lower than that for indoor usage and appeared to decrease in 1976 to 43% from 50% in the previous two years ($p<.01$) (Table III). Almost everyone who used chemical sprays in the three years felt that they did some good; a relatively lower percentage said that such sprays did harm.

With the exception of 1975, nearly 60% of the rural people

TABLE V Annual Indoor Use of Chemicals by Interviewees in Bryan-College Station and Dallas

	Number of interviewees				
No. times chemicals used/year	Bryan-College Station			Dallas	
	1974	1975	1976	1975	1976
1	16	9	8	11	2
2	16	8	19	11	7
3	8	5	2	1	3
4	8	4	3	8	11
5	8	0	1	4	2
6	8	3	5	2	0
	(3.0)[a]	(2.6)[a]	(2.6)[a]	(2.7)[a]	(3.2)[a]
7	1	0	1	0	0
8	1	2	1	1	0
9	0	0	0	0	0
10	6	1	3	1	6
11	0	0	1	0	0
12	6	16	7	8	14
13-20	6	3	2	3	0
21-30	6	4	2	4	2
31-50	3	2	5	4	0
51-90	8	11	0	2	2
91-100	7	9	4	4	0
Totals	108	77	64	64	49

[a]*Average (weighted) annual insecticide usage among those persons using chemicals one to six times yearly.*

who personally used chemicals indoors did so one to six times yearly (38% in 1975) (Table V);[4] more than 75% of the people who personally used chemicals outdoors did so in the same annual frequency range (Table VI).[4] These people are considered to

[4]*This delineation is arbitrary; however, a rather consistent break between six and seven times annually can be observed in Tables V and VI.*

TABLE VI Annual Outdoor Use of Chemicals by Interviewees in Bryan-College Station and Dallas

No. times chemicals used/year	Number of interviewees: Bryan-College Station 1974	Bryan-College Station 1975	Bryan-College Station 1976	Dallas 1975	Dallas 1976
1	20	9	14	11	5
2	12	8	11	14	8
3	7	7	8	11	2
4	8	8	5	3	7
5	2	1	0	0	1
6	9	4	0	5	4
	(2.8)[a]	(2.9)[a]	(2.1)[a]	(2.6)[a]	(3.1)[a]
7	0	1	0	0	0
8	2	2	0	1	0
9	1	0	0	0	0
10	5	0	1	1	4
11	0	0	0	0	0
12	1	3	1	6	2
13-20	1	2	0	0	0
21-51	3	0	2	3	1
52	1	4	1	2	0
>52	0	0	1	0	0
Totals	72	49	44	57	34

[a]*Average (weighted) annual insecticide usage among those persons using chemicals one to six times yearly.*

be light to moderate users of insecticides. If an index of average annual use is computed for the light to moderate users, an apparent decline in indoor and outdoor applications from 1974 to 1976 is observed (Tables V and VI). In summary, the frequency of personal insecticide use in the rural locality appears to have declined over the three-year study period. Furthermore, in the light to moderate group, chemicals may be receiving less use than in previous years. These two patterns will be reevaluated later in terms of overall residence use.

Approximately half of the Bryan-College Station interviewees in all three years said they had tried other means besides chemicals for controlling insects. The ways mentioned were quite varied, from commonsense means such as using a fly swatter to more sophisticated methods such as planting insect-repellent aromatic herbs. Of those who had used nonchemical means, most had said they used their own intuition to discover such measures or had learned about them from friends. Very few people mentioned exterminators or nurserymen as the source of learning about nonchemical ways of controlling insects. While TAMU was mentioned as a source of approximately 10% of the 1974 and 1975 Bryan-College Station samples, this dropped to only 2% in 1976. Of those who used nonchemical methods, almost two-thirds in 1974 were generally satisfied with the results. This decreased in 1975 and 1976 to one-third with an almost equal number saying they were rarely satisfied ($p<.02$). Thus, while nonchemical methods were tried to an equal extent in all three years, people in Bryan-College Station seem to be more dissatisfied with the results in recent years.

In Dallas personal usage of insecticides was as high as the frequency of indoor pest problems. In 1976 use of such insecticides had dropped from 68% to 47% ($p<.001$), consistent with the lower frequency of indoor problems reported in 1976. Similarly, personal use of chemicals outside had dropped from 1975 levels to 33% ($p<.001$); probably due to the lower numbers of people reporting outdoor problems in 1976. As with the more rural respondents, almost everyone who used insecticides felt that they did some good, and relatively few were aware of any harm from such chemicals.

In contrast to the Bryan-College Station analysis of frequency of personal insecticide use through time, an analysis of the Dallas data revealed that insecticides may have been used more frequently indoors and outdoors in 1976 as compared to 1975, among the light to moderate users (Tables V and VI). This group of people consisted of 50% indoor and about 80% outdoor users. In summary, chemicals were used by a significantly lower number of the Dallas people in 1976; however, the annual frequency of use appears to increase, at least for the people using chemicals one to six times yearly. These patterns will be reevaluated in a later section from the standpoint of overall residence use.

Fewer people in 1976 used nonchemical means of controlling insects than those in 1975 (21% vs. 38%, $p<.02$), again possibly reflecting a lower incidence of problems in general. In both years intuition and friends were the most commonly mentioned sources of learning about nonchemical methods of control. Of those who used such methods more than half were generally satisfied.

People in Dallas and Bryan-College Station who have indoor and outdoor pest problems commonly relied on the personal use of insecticides. It appears that fewer Dallas residents knew of nonchemical ways of controlling insects, and more people in Dallas seemed to depend on printed matter for this information. Although fewer Dallas residents sampled had tried nonchemical ways of controlling insects, they reported that they were more satisfied with these techniques than those in Bryan-College Station.

Comparisons. Separate chi-square comparisons between those who did and did not use nonchemical means of controlling insects on a number of questionnaire items reveal a number of significant differences (Table VII). People who have used nonchemical methods, as compared with those who have not, have more knowledge of beneficial insects, of harmful effects of insecticides, and of organic gardening. These people reported that they had become more reluctant to use insecticides. In addition, people who were more likely to use nonchemical means indicated that they had more indoor and outdoor pest problems.

Based on the responses to questions 2 and 4Ad (Appendix D), we assume that most people were able to distinguish between chemical and nonchemical control methods. Future questionnaire efforts will probably require further qualification since some of the newer methods will not be readily classifiable into chemical or nonchemical categories. For example, packaged microbial products (a few are presently available to homeowners)

TABLE VII Significant Differences Between Those Who Use and Do Not Use Nonchemical Methods

	Percentages			
Question	*Use*	*Not Use*	*Chi square*	*p*
Indoor insect problem	*74%*	*65%*	*5.74*	*.06*
Outdoor insect problem	*64*	*53*	*9.40*	*.05*
Knew of insecticide harm	*14*	*5*	*17.55*	*.01*
Named harmful effects of insecticides	*13*	*5*	*13.13*	*.01*
Became more reluctant to use insecticides	*7*	*2*	*15.67*	*.05*
Knew of beneficial insect(s)	*79*	*60*	*24.63*	*.0001*
Knew of definition of organic gardening	*70*	*62*	*10.31*	*.04*

will be used like chemicals; however, it is clear that microbes act as biological control agents. Other new methods may likely involve behavior-modifying chemicals (synthesized natural products), which will act, e.g., as either attractants or repellents. In addition, synthesized natural products such as juvenile hormone-like compounds may be used in control to alter physiological processes of insects. Although most of the above methods probably would be used by professionals, some would undoubtedly be used personally by homeowners.

2. *Professional Means to Control Insects*

In the rural sample, approximately two-thirds of the respondents with pest problems used professional help in solving their insect problems. Almost all of these professionals used chemicals in the home, anywhere from once to 15 times a year. Approximately half of the sample with problems reported that professionals used chemicals outdoors (Tables VIII and IX).

People who employ professionals to use chemicals one to four times yearly are considered to be light to moderate users

TABLE VIII Annual Indoor Use of Chemicals by Professionals in Bryan-College Station and Dallas

	Number of interviewees				
	Bryan-College Station			Dallas	
No. times chemicals used/year	1974	1975	1976	1975	1976
1	42	21	21	34	18
2	9	8	11	8	18
3	1	4	5	2	2
4	20	20	25	12	7
	(2.0)[a]	(2.4)[a]	(2.6)[a]	(1.9)[a]	(2.0)[a]
5-11	3	1	0	2	2
12	2	1	2	6	7
>12	0	1	0	0	1
Totals	77	56	64	64	55

[a]*Average (weighted) annual insecticide usage among those persons using professionals one to four times yearly.*

TABLE IX Annual Outdoor Use of Chemicals by Professionals in Bryan-College Station and Dallas

No. times chemicals used/year	Number of interviewees				
	Bryan-College Station			Dallas	
	1974	1975	1976	1975	1976
1	33	10	15	21	10
2	3	6	8	6	4
3	2	1	3	1	2
4	7	10	17	7	4
	(1.5)[a]	(2.3)[a]	(2.5)[a]	(1.8)[a]	(2.0)[a]
5-11	3	0	0	1	1
12	1	0	0	0	1
>12	0	1	1	1	2
Totals	49	28	44	37	24

[a]*Average (weighted) annual insecticide usage among those persons using professionals one to four times yearly.*

of insecticides; most people in Bryan-College Station can be placed in this group. Single and quarterly indoor and outdoor insecticide treatments are common in this area. If an index of average annual insecticide use by professionals is computed for the light to moderate group, it appears that indoor and outdoor use is increasing slightly each year. Some exterminators in this area claim that their businesses have expanded substantially in recent years because people have expressed apprehension in personally applying toxicants (Frankie, unpub. data). However, our data reveal that only a relatively low number of people are aware of the possible harm that insecticides may cause (see Table III, questions 2b, 4A and 4Aa).

In the urban sample, 80% of those in 1975 said they used professional help in solving their insect problems (Table III). Consistent with previously reported results, this percentage dropped significantly in 1976 to 62%, reflecting fewer problems in general. Of those people who used professionals in the Dallas sample, approximately 90% said such professionals used chemicals in the home and about 45% said chemicals were used in the yard.

Single and quarterly indoor and outdoor insecticide treatments by professionals were common in Dallas in 1975. These two particular patterns occurred less frequently in 1976 (Tables VIII and IX). The average annual chemical use by professionals for the light to moderate group (1-4 times annually) is approximately the same for both years. In contrast to Bryan-College Station, Dallas professionals are called upon less frequently by people who use their services one to four times annually. However, in the case of indoor situations, professionals in Dallas used chemicals more frequently in the five to >12 use categories.

The patterns of indoor/outdoor insecticide use by professionals in both cities will be reevaluated in the general discussion section.

F. Affective Components

1. Attitude Change

Frequency. In 1974 and 1975 approximately one-third of the interviewees in Bryan-College Station said they had changed their attitudes regarding the use of chemical insecticides; this percentage increased significantly in 1976 to 46% ($p<.03$) (Table III). In 1974 four times as many people said they no longer used insecticides compared with those in 1975 and 1976 (question 4Aa; Table III; $p<.05$). Reasons for attitude changes were similar for the three years sampled, but almost twice as many people in 1976 said they changed their attitudes because of publicity on television.

In both 1975 and 1976 in Dallas approximately one-third of those interviewed said that their attitude toward insecticide use had changed to a more cautious one. Publicity or television was given as the reason for this change. There were few overall differences in the amount and direction of attitude change between rural and urban respondents. However, the reasons given for attitude change do differ for respondents in the two localities. In the rural setting there was more use of reading material and discussions with friends, while the city people were more likely to attribute attitude change to the fact that they had personally seen negative results from insecticides.

Attitude change - comparisons. Table X reveals different percentages on various questionnaire items of the people who said they changed their attitude to less insecticide use versus those who did not experience a change in attitude. In general, it appears that those who have recently changed their attitude concerning insecticide use as compared with nonchangers have

TABLE X Significant Differences Between Those Who Did and Did Not Change Their Attitude

Question	*Percentages*			
	Changers	*Non-changers*	*Chi square*	*p*
Indoor insect problem	*74%*	*65%*	*5.86*	*.05*
Knew of insecticide harm	*13*	*6*	*10.55*	*.03*
Generally satisfied with nonchemical methods	*29*	*18*	*16.49*	*.04*
Knew of beneficial insect(s)	*80*	*61*	*21.95*	*.001*
Use insecticides in park only in emergencies	*70*	*59*	*17.18*	*.03*
Knew of organic gardening	*93*	*85*	*7.98*	*.02*

more knowledge regarding beneficial insects, organic gardening, and harmful effects of insecticides. They are generally more satisfied with nonchemical means of controlling pests and are more in favor of using insecticides in parks only in emergencies. However, one should note that they more frequently report having an indoor insect problem than those people who have not changed their attitudes.

2. Attitude toward Extended Use of Insecticides in Parks

Frequency. In response to a question designed to assess attitudes toward extended use of insecticides (question no. 6; Table III), approximately two-thirds of the rural sample were in favor of using chemical insecticides in a national or state park only under emergency circumstances. A somewhat lower percentage in the Dallas area was in favor of only emergency use, but this difference was not significant. Approximately one-fifth of the residents in both localities was in favor of regular or routine chemical use in parks.

Comparisons. Table XI contains the comparison of percentages on a number of questionnaire items of people who were in

TABLE XI Significant Differences Between Those Who Want to Use Insecticides in Parks in Emergencies and Regularly

Question	Percentages		Chi square	p
	Emergency	Regularly		
Knew of insecticide harm	10%	3%	12.07	.01
Named harmful effects of insecticides	10	3	15.44	.001
Changed attitudes	40	23	10.24	.01
Liked insects	21	14	12.54	.01
Knew of beneficial insect(s)	74	58	12.27	.01

favor of emergency use vs. those who preferred regular insecticide use in parks. Data indicate that people who want insecticides used only in emergency situations as compared with those preferring regular application are more knowledgeable regarding beneficial insects, harmful effects from insecticides, and the specific hazards of insecticides. They also reported significantly more attitude change and liking of insects.

3. *Liking Insects*

Frequency. Although most of the questionnaire was designed to examine techniques and information about pest eradication, a question was asked (in 1976 only) to investigate the degree to which people liked insects (Table III). Two-thirds of those interviewed in Bryan-College Station said that they liked certain insects. In Dallas, however, only one-third of the respondents said they liked insects ($p<.001$). When asked what insects and why these particular groups were liked, different response patterns were recorded for each city. Approximately 60% of the Dallas people liked insects for aesthetic reasons compared to only 25% for the Bryan-College Station people. In contrast, 60% of the Bryan-College Station sample liked insects because of their utilitarian value; 29% in Dallas listed utility as their reason for liking certain insects. These striking differences might reflect a greater awareness of agriculture and associated insects in the more rural Bryan-College Station area.

Comparisons. To understand more fully the type of person who likes insects, separate chi-square analyses were computed on a variety of questionnaire items examining differences between those respondents who said they liked insects and those who said they did not. Table XII presents the percentage differences and significance values. In summary, people who said they liked insects were more able to name nonchemical ways of controlling insects and were more reluctant to use insecticides than those who did not like insects. People who did not like insects were more in favor of regular insecticide use in parks and were less likely to have heard of organic gardening. People who liked insects also reported a greater frequency of outdoor insect problems.

IV. GENERAL DISCUSSION

The patterns described herein provide some indications as to how people relate to insects and insecticides in rural and urban environments. Many of the patterns may not apply, without qualification, to other U.S. cities since the interviewees belonged primarily to one group of people selected on the basis of certain demographic factors. However, overall this ques-

TABLE XII Significant Differences Between Those Who Did and Did Not Like Insects

Question	*Percentages*		*Chi square*	*p*
	Liked	*Not Liked*		
Outdoor insect problem	*55%*	*36%*	*7.91*	*.02*
More reluctant to use insecticides	*54*	*25*	*18.67*	*.001*
Knew of nonchemical ways to control insects	*50*	*25*	*14.30*	*.01*
Knew of beneficial insect(s)	*85*	*49*	*29.60*	*.0001*
Use insecticides regularly in parks	*15*	*29*	*19.03*	*.001*
Never heard of organic gardening	*6*	*18*	*6.80*	*.01*

tionnaire effort proved useful in 1) understanding how one socioeconomic group in a particular location relates to insects and insecticides, 2) revealing attitudes and behavior that suggest and provide justification for new lines of research in urban entomology, and 3) suggesting a series of new questions that should be asked in future efforts which utilize the same or similar questionnaire form (Appendix D).

From the standpoint of research and extension, it is important to recognize how homeowners acquire their information on insect pests. It was surprising to learn that 38 to 51% of all interviewees mentioned exterminators as their primary information source. If the people who seek information from exterminators and nurserymen are combined, this accounts for 47 to 66% of the total to which this question applied. Possible significance of this pattern is realized when the entomological backgrounds of these two groups are examined. Few people in these fields have ever had formal training in entomology. Most of their experience (and subsequent expertise) has been acquired through personal observations, short courses, communication with customers, literature from chemical companies, and limited contact with the extension service. This kind of background apparently does not detract from their perceived effectiveness as reliable information sources as indicated by the positive responses to question 1c on satisfaction (see Table III).

The potential use of nonchemical control methods should be carefully considered by research and extension people. It seems clear that a substantial number of people (21-54%) already use a wide variety of nonchemical means, many of which are based on common sense (Table III, question 4Ac). It is not surprising that the current level of satisfaction among those using these methods is relatively low (35-62% generally satisfied as compared to the relatively high percentage of people who claimed insecticides did some good) since most of the methods have never been adequately researched and implemented. Based on an overall appraisal of the responses to nonchemical methods, there appears to be clear justification and a need for scientifically investigating some of these methods, singly or in combination with chemicals.

It was informative to examine the general relationship between attitude and behavior. The data suggest that affective components of insecticide use and insect problems are most closely linked with cognitive factors or amount and type of information and not necessarily with any behavioral manifestations. Consistent with other psychological research on attitude change (e.g., McGuire, 1969), it appears that in this situation behavioral change lags behind affective and informational change. Further studies over time are necessary to examine the directionality of such change.

V. SUGGESTIONS FOR FUTURE WORK

Future questionnaire efforts that explore similar questions should be directed simultaneously to a wider variety of socioeconomic groups in at least three large U.S. cities, which would allow for more generalizations and improved predictions. Attention also must be focused on developing means for standardizing information that will be comparable from one area to the next.

Questionnaire efforts that are designed to evaluate overall insecticide usage per residence should include more specific questions than were used in this study. First, they should seek to identify the specific kinds of indoor-outdoor insect problems and the respective frequencies of occurrence. Second, the respective types and quantities of chemicals personally used should be documented. In this regard, the numbers and kinds of insecticides annually purchased (e.g., number of aerosol cans of pyrethrins) by homeowners may possibly serve as an index for this variable.[5] Third, types and quantities of chemicals used by professionals should be recorded. Since homeowners generally do not know which chemicals are being used by professionals (Table III, question 3b), inquiry should be made directly with local exterminators for information on insecticides used indoors and outdoors in respective neighborhoods. Indices of use, which are based on chemicals applied against certain consistent insect problems (e.g., cockroaches, termites or ants), may be useful in identifying the materials and respective quantities used. Some effort should also be made to document possible cases of resistance to insecticides. Through appropriate questions, this information may be gathered from urban dwellers and from exterminators.

If properly designed, questionnaires may be used to probe other areas of interest to urban entomologists. For example, the concept of aesthetic injury level (Olkowski, 1974), which is used to describe certain situations in urban environments where insects cause aesthetic rather than economic damage, may be examined and quantified through a questionnaire survey. Since aesthetic injury levels may be roughly translated into how much insect activity people will tolerate, questions could be posed that would allow for a random sample of levels of tolerance to selected pest species. Average levels of tolerance could then be determined from these compilations, which would prove useful for research purposes (see Piper and Frankie, Chapt. 10, this volume for discussion on how tolerance levels

---

[5] *This may represent a crude index since purchasing does not necessarily mean use.*

for cockroaches in Texas are collected and utilized in a cockroach management program).

In conclusion, we believe that the use of questionnaires represents an important tool for gauging trends in attitudes and behavior of urbanites towards insects and insecticides, particularly if these questionnaire efforts are carried out over two to three year periods. Further, we believe that new and meaningful research programs in urban entomology and the social sciences can be designed from results obtained through appropriate questionnaire efforts.

VI. SUMMARY

A questionnaire survey on rural and urban dwellers' attitudes and practices towards insects and insecticides in their respective environments was conducted in two Texas cities from 1974 to 1976. Data were collected in Bryan-College Station (rural city: 50,000 pop.) for each of three years as part of an effort to assess change over time. People were also interviewed in the Dallas area (800,000 pop.) in 1975 and 1976 in order to make comparisons between two different types of areas - rural vs. urban. A total of 551 interviewees were seen (100-147 in each city each year). In general, the type of person interviewed may be described as a white, middle-aged, married homeowner employed in an upper socioeconomic occupation.

Summaries of responses to the major questions asked are presented below.

1. Pests

Problems. In Bryan-College Station, 55 to 76% of the interviewees in all years had an indoor/outdoor insect problem. In Dallas in 1975, 68 to 75% of the interviewees had an indoor or outdoor insect problem respectively; in 1976, a significant decline in indoor (50%) and outdoor (35%) problems was recorded for this city.

Information. Depending on city and year, 54 to 81% of all interviewees who had problems sought information for them. Most people in both cities, who sought information, went primarily to exterminators (38-51%). Nurserymen were also consulted in both cities (9-26%). In Bryan-College Station, 18 to 26% of the people turned to Texas A&M University for information; very few people in Dallas solicited information from Texas A&M. Most people in both cities were generally satisfied with the information they received.

2. Personal Use of Insecticides

From 63 to 78% of the Bryan-College Station interviewees used chemicals indoors during 1974 to 1976; 43 to 50% of the same people used chemicals outdoors during the same period. Depending on year, 47 to 68% and 33 to 58% of the Dallas people used chemicals indoors and outdoors, respectively. There was a significant decline in indoor/outdoor chemical use from 1975 to 1976 for the interviewees of both cities. Most in both cities felt that chemicals did some good; relatively few could describe negative aspects.

3. Professional Use of Insecticides

Depending on city and year, 59 to 80% of the interviewees called upon professional help to solve their insect problems. Chemicals were used indoors in 90% of these cases and outdoors in 38 to 63% of the cases in which professionals were used. Few people in either city knew which chemicals were used by the professionals.

4. Attitude towards Insecticides

Attitude change. A change in attitude towards the use of chemicals was expressed by 29 to 46% of the people in both cities. Most stated they used no chemicals or they used them cautiously. Commonly cited reasons for attitude change included the following: personal experience with negative results, reading, TV, and the ecology/environment movement.

Nonchemical insect control methods. In Bryan-College Station, 46 to 54% of the interviewees said they used nonchemical control methods, while in Dallas only 21 to 38% of those interviewed said they used nonchemical means to suppress insects. In both cities, intuition and friends were cited as the most common sources of information for nonchemical controls. Interestingly, 65% of the Bryan-College Station people and 31% of the Dallas people in 1976 claimed they liked some insects.

5. Aware of Beneficial Insects.

Approximately, two-thirds of the interviewees in both cities were aware of beneficial insects in their urban environment.

Using one-way analysis of variance, differences in attitudes and behavior were examined for people with particular demographic characteristics. Further, correlates of attitude and behavior of persons responding one way to selected questions were examined for differences with persons who responded in a different way to the same selected questions.

Use of the questionnaire as a research tool, shortcomings of the current study and future questionnaire possibilities are discussed.

ACKNOWLEDGMENTS

Partial support for this research was provided by a grant from the Environmental Protection Agency (Research Grant No. R803068-03-1). We thank R. R. Fleet for assisting in the pretesting of the questionnaire form. M. J. Gaylor, V. R. Landwehr and G. L. Piper kindly reviewed the manuscript.

REFERENCES

Cannell, C. F., and R. L. Kahn. 1969. Interviewing. *In* The handbook of social psychology. Vol. II (G. Linzey and E. Aronson, eds.). Addison Wesley, Reading, Massachusetts.

Coster, J. E., B. L. Cunningham, and W. G. Boeer, Jr. 1974. Camper's attitudes toward insect control. *J. For. 72*: 92.

Gerhardt, R. R., J. C. Dukes, J. M. Falter, and R. C. Axtell. 1973. Public opinion on insect pest management in coastal North Carolina. North Carolina Agric. Ext. Serv. Misc. Publ. 97.

Hodge, R., P. Siegel, and P. Rossi. 1966. Occupational prestige in the United States. *In* Class, status and power (R. Bendix and S. Lipset, eds.). Free Press, New York, pp. 324-325.

Keel, V. A., H. P. Zimmerman, and R. A. Wearne. 1974. Communicating home garden information. Phase I and II Reports: The Minnesota-Wisconsin, ES-USDA Home Horticulture Project. Communication research and paper series 1 and 5. Agric. Ext. Serv., University of Minnesota.

Kretch, D., R. S. Crutchfield, and E. L. Ballachey. 1962. Individual and society. McGraw Hill, New York.

McGuire, W. J. 1969. The nature of attitudes and attitude change. *In* The handbook of social psychology. Vol. III (G. Lindzey and E. Aronson, eds.). Addison Wesley, Reading, Massachusetts.

Olkowski, W. 1974. A model ecosystem management program. *Proc. Tall Timbers Conf. Ecol. Anim. Control Hab. Manage. 5*: 103-117.

APPENDIX A

Chronology of Recent and Pertinent Historical Events Relating to Insecticides

1962. Publication of *Silent Spring* by Rachael Carson.

1970. Creation of U.S. Environmental Protection Agency (EPA).

1971. September 7, 1971: Texas Structural Pest Control Board established as an official agency of the state. Major responsibilities include: setting up procedures for the licensing of exterminators who use pesticides.

1971-1972. First state licensing of Texas exterminators who use pesticides. December 7, 1971: "Grandfather clause" promolgated (i.e., those in business less than two years were required to take examination for state license).

1972. Use of DDT restricted by EPA.

1975. Use of aldrin and dieldrin restricted by EPA.

1975. Kepone pesticide incident in Virginia.

1976. Licensing of Texas exterminators who use pesticides and who did not test in 1971 because of exemption through an early "grandfather clause."

APPENDIX B

Quadrants for Quota Sampling in
Bryan-College Station

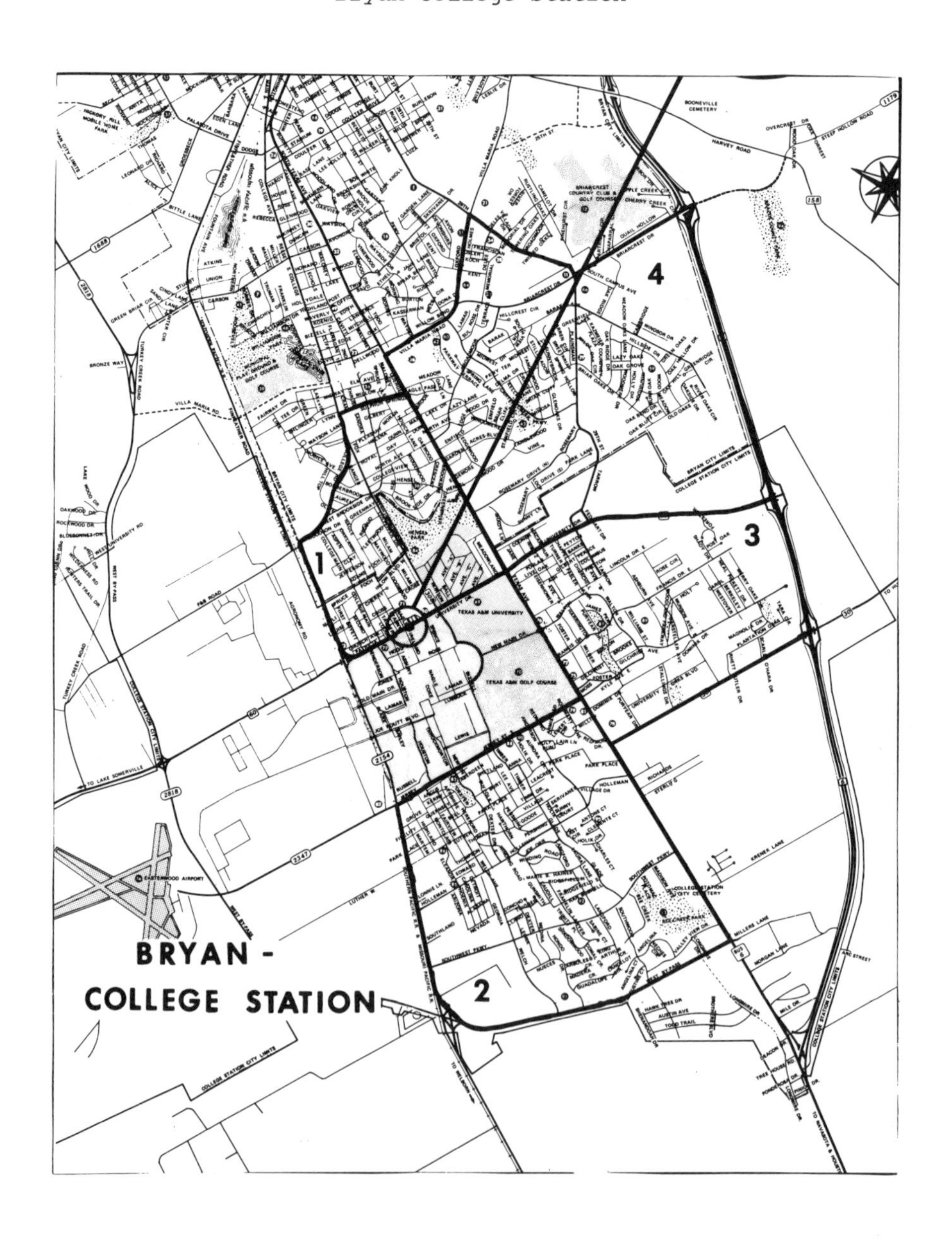

APPENDIX C

Quadrants for Quota Sampling in Dallas

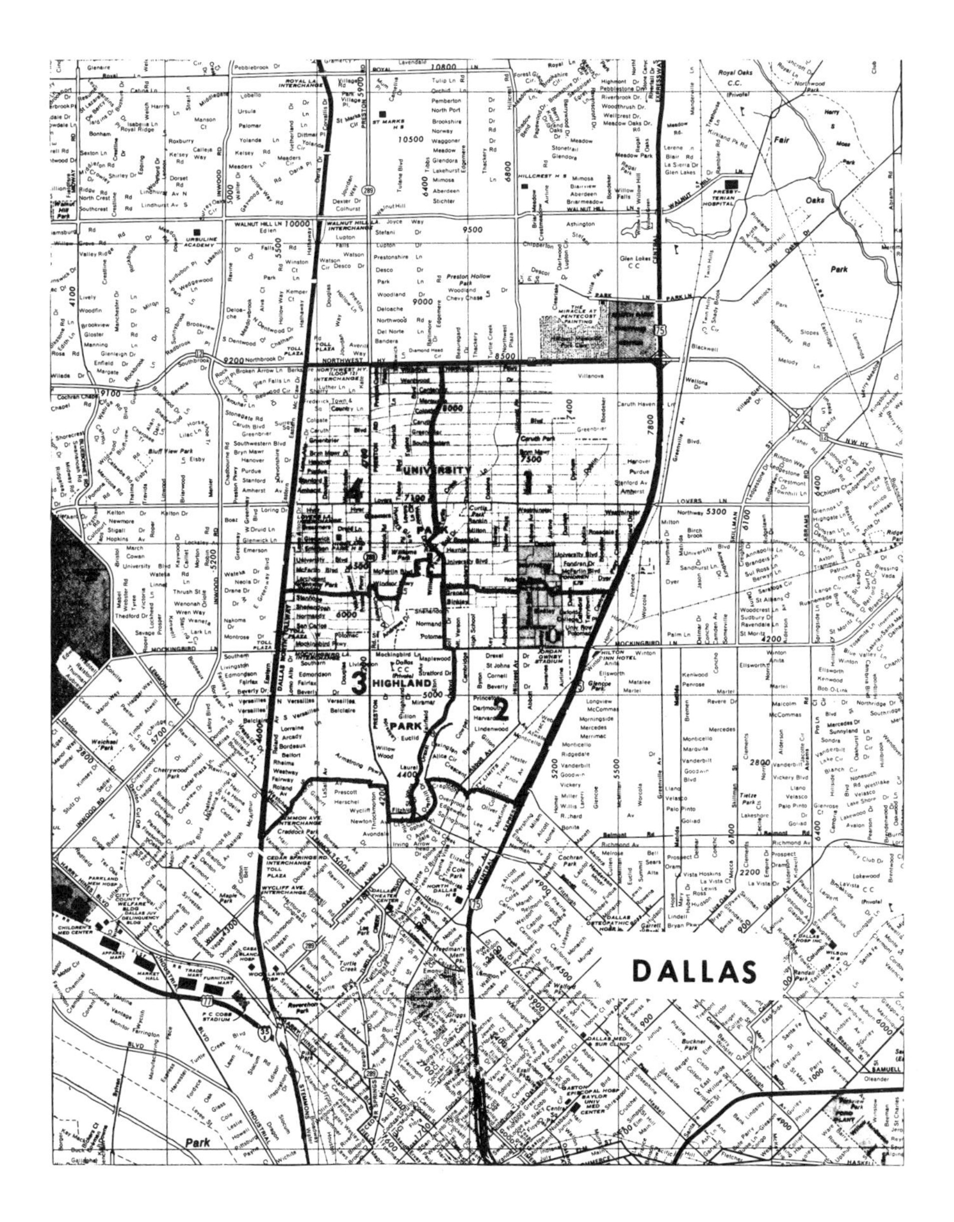

APPENDIX D

Urban Entomology Questionnaire (June 27, 1974)

Age______ House_______ Apt________

Occupation: Married (M F)______ Rent________ Own________

1 2 3 4 5 Single (M F)______ How long lived in this area?

1. Have you ever had an indoor household insect problem? Yes____ No_____

   Have you ever had a yard insect problem? Yes_____ No____

   a. (If yes) Have you ever obtained information about this problem? Yes_____ No_____

   b. (If yes) From whom did you obtain this information?

      Texas A&M — Nurserymen
      Friends — Exterminators
      TV — Printed Matter
      Not sure
      Other____________________

   c. (If yes) Was the information satisfactory?

      Generally — Rarely
      Occasionally — Not sure

2. Do you personally use chemical insecticides for your pest problems (e.g., sprays)?

In the Home		In the Yard	
Yes	times/yr______	Yes	times/yr______
No		No	
Not sure		Not sure	

   a. (If yes) Do they do any good?
      Yes______ No______ Not sure______

   b. Do they do any harm that you are aware of?
      Yes______ No______ Not sure______

   c. (If yes) What harm (list specific examples)?

APPENDIX D (cont'd)

3. Do you use professional help in solving your insect problems (e.g., Exterminators, Gardeners)? Yes___ No___

a. (If yes) Do these people use chemicals?

In the Home		In the Yard	
Yes	times/yr____	Yes	times/yr_____
No		No	
Not sure		Not sure	

b. (If yes) Do you know what chemicals they use? Yes______ No______ (If yes) What chemical__________

4A. Over the years, has your attitude regarding the use of chemical insecticides changed in any way? Yes___ No____

a. (If yes) In what way? ______________________________

b. (If yes) Why did your attitude change?______________

c. Do you now or have you ever used other means besides chemicals for controlling insects (e.g., handpick)? Yes________ No________

d. (If yes) What methods? (List)______________________

e. (If yes) How did you learn about these other methods?

Texas A&M	Nurserymen
Friends, Relatives	Exterminators
TV	Printed Matter
Own intuition	School courses
	Other______________________

f. (If yes) Have you found these methods satisfactory?

Generally	Rarely
Occasionally	Not sure

4B. Are there any insects that you like?

a. (If yes) Which ones?________________________________

b. (If yes) Why do you like them?_____________________

APPENDIX D (cont'd)

5. Are you aware of any beneficial animals or insects that occur or could occur in your garden or yard?
   Yes______ No______ Not sure______

   (If yes) Can you name two of these for me?
   Named two, one, none

6. Under what circumstances would you be in favor of using chemical insecticides in a national or state park?

   a. Never

   b. Under emergency circumstances whenever man, wildlife or vegetation are in danger of widespread damage

   c. As a regular routine procedure

   d. Not sure

7. Have you heard of "organic gardening"? Yes____ No_____

   a. (If yes) Can you tell me what it means?
      Understands____ Does not understand____

   b. (If yes) How did you learn of this technique?

APPENDIX E

Standard Interviewer Instructions

Interviewing Methodology (Entrance and Exit)

My name is __________________, and I am part of a research effort to survey how people feel about certain insect problems. This study is being conducted throughout the community as part of some research supported by the Entomology Department at Texas A&M University. The addresses of the people we are interviewing were chosen by chance, but it is quite important to inverview every person chosen so that the results will not be biased in any way. I am supposed to interview a (female, male) adult at this address. I have some questions to ask you. Of course, all responses are strictly confidential. I'm sure you will find it all quite interesting.

I could either interview you here or I could come in.

APPENDIX E (cont'd)

Which would you prefer?

Thank you so very much for taking the time to help us. Your cooperation will be a big help. If you would like to have a copy of the results of this research, please address this envelope to yourself and I will see that you receive a copy at the completion of this study. Thank you again.

To Whom It May Concern:[6]

This is to identify____________________ who is employed by the Psychology-Entomology Departments as an interviewer in a project designed to assess people's attitudes about insects and gardening.

If there are any questions concerning the above named person's identification, please do not hesitate to contact me at (714) 845-2554.

All interviewees were chosen at random by place of residence. All interviews are anonymous. We appreciate your help. Thank you.

Sincerely yours,

Hanna Levenson, Ph.D.
Associate Professor

---

[6] *This letter was typed on letterhead stationery; Department of Psychology, Texas A&M University.*

INSECTICIDES IN THE URBAN ENVIRONMENT

Frederick W. Plapp, Jr.

Department of Entomology
Texas A&M University
College Station, Texas

I. INTRODUCTION

As an entomologist and insecticide chemist, I have long found it difficult to deal with the subject of insecticide use in the urban environment. As an "expert" on insecticides, I am often asked for chemical advice, i.e., to answer questions as to the specific insecticide to use to control many house and garden insect pests. To make such recommendations is to imply approval of chemical methods of pest control and further, to imply that the insects in question are "bad" and ought to be fought by chemical means. Knowing a little of the possibilities for harm associated with the use of insecticides, making these recommendations is not a simple matter.

"What can I do about the bugs eating up my squash vines?"

ISBN 0-12-265250-9

and "What can I do about the cockroaches in the kitchen?" are examples of questions frequently asked. If I name a chemical that may decimate the insect population, the next question will be, "Is the chemical poisonous to me?" An affirmative answer is resented. People simply don't realize that most chemicals toxic to insects are also toxic to man and most other living animals.

At the same time, non-chemical exhortations such as "Fertilize and water the squash vines!" and "Clean up the kitchen!" are not always well received. People do not wish to share any of their laboriously-grown garden produce with the bugs and they don't like the implication that their housekeeping may be less than perfect. Recommendations for the use of safer, but slower chemicals like boric acid or silica gel fare only slightly better. These methods of insect control, while much safer, are far too slow for the average urbanite. They don't kill the bugs fast enough. We are, as a society, addicted to the use of chemical shortcuts to cure biological problems. Further, we are unwilling to realize that these shortcuts may not represent unmixed blessings, let alone provide adequate solutions to the problems.

It is the purpose of this essay to present some ideas on the role of entomologists, particularly those with expertise in the field of insecticides, as they relate to insecticide use in the urban environment. A more comprehensive treatment of the subject is available (Ebeling, 1975). Among the ideas to be discussed is that much urban insecticide use is unwise, and more of it is unnecessary, i.e., the benefits obtained are not worth the risks involved. Further, the use of insecticides for cosmetic purposes (Walsh, 1976) or because of entomophobia (Olkowski and Olkowski, 1976) can seldom be justified.

The problem is, what is the role the entomologist as a trained professional should play in the solution of urban insect problems? Should he help make available to the urban consumer all the latest insecticides or do his responsibilities include making critical evaluations of these chemicals and warnings against hazards that may be involved in using them? I shall not deal here with alternative methods of pest management. Others writing for this book have greater expertise in this area.

II. INSECTICIDES IN THE MARKETPLACE

In most American supermarkets and grocery stores, lined up on the shelves in close proximity to the canned goods, cosmetics, soaps, etc,. are brightly labeled insecticide containers. Insecticides are marketed under a vast array of brand names and are offered to perform a wide range of services.

They include house and yard space sprays designed to kill all insects, more specific poisons for cockroaches and ants, chemicals that will yield blemish-free produce from the garden and orchard, and preparations designed to keep fleas off the family pets. None of these bears a skull and crossbones label, the universal symbol that indicates the contents are poisonous; rarely is there any intimation outside of the fine print on the label that any of these products are hazardous to anything but insects. After all, one would not expect to find them sold in close juxtaposition to food were they poisonous to humans.

The result of such marketing practices is that many people buy insecticides without realizing that any dangers are involved. People assume that since these products are available for purchase, they must have been approved by someone in authority and must, therefore, be safe.

Insecticide products aimed at the urban consumer are widely advertised on television and by other mass media. This advertising probably relates less to any concern on the part of the supplier for the public health and welfare than to the fact that the sale of home insecticide products is a highly profitable business. For example, a 284-gm can of "Roach Killer," purchased in July, 1976, for $1.69, contains 0.1% pyrethrins insecticide as the active ingredient. This means an actual sales price of $5,948 per kilo for the insecticide. At such prices, it is no wonder that the urban population is constantly reminded of the wondrous advantages of insecticide use.

Detailed figures on the total amount of insecticide used in urban areas are hard to find. Total insecticide sales in the United States in recent years have exceeded 225 million kilos (Fowler and Mahan, 1973). Data from Canada, cited in the same publication, suggest 20 to 25% of the use is urban. If the same figures hold for the U.S., we are dealing with an annual use of 45 million kilos of chemicals in the urban environment.

III. INSECTICIDES IN THE HOME

Although there apparently is a vast array of insecticides offered for home use, the differences are most illusory. In reality, only a few insecticides are widely sold. Like aspirin, they come under a variety of trade names.

By far the commonest insecticide on the market is pyrethrins. This material has been used as an insecticide for several centuries (Matsumura, 1975). It is extracted from flowers of the genus *Chrysanthemum*. The main advantages to pyrethrins (the name for the combined insecticidally active components) are the almost complete lack of toxicity to humans, coupled with a spectacularly rapid knockdown action against many insects. Applied at high enough doses, pyrethrins can cause an almost instantaneous paralysis in insects. This fast action makes the insecticide psychologically attractive to use - a whiff of pyrethrins is almost as good as a bullet for producing fast results.

There are two main disadvantages to using pyrethrins. One is that it is expensive, the second is that it is really quite a poor insecticide. The paralytic effects produced in a target insect are often only temporary and in some cases at least, the victim can recover, apparently none the worse for the experience.

A number of synthetic insecticides related to pyrethrins are now being developed and may eventually replace the natural insecticide. Many of these are more effective than pyrethrins and cheaper to make. Like the natural product, most appear to be fairly safe for humans.

The other insecticides available for home use are mostly organophosphates and carbamates, insecticides related to the highly poisonous nerve gases developed during World War II. Federal law now prohibits the sale of the most highly toxic of these chemicals to the homeowner (Fowler and Mahan, 1973). Insecticides of these types are widely used by commercial pest control operators and to a lesser degree, are available directly to the consumer. Those in use such as the organophosphates diazinon, malathion and chlorpyrifos, and the carbamate propoxur, are generally of moderate toxicity to humans and moderate to high toxicity to insects. They are claimed, with fair accuracy, to be safe for home use when used as directed on the label. Therein, of course, lies the difficulty. Many people believe that if a little insecticide is good, more is better. Besides, these insecticides are slower acting than pryrethrins and therefore may be overused in an attempt to match the speedy knockdown produced by pyrethrins.

Another widely used organophosphate is the insecticide known as DDVP or Vapona[R], the active ingredient in No-Pest[R] strips and related products. This insecticide is of fairly high acute toxicity to humans as well as being highly toxic to insects. It is sold in a slow-release resin formulation, designed to act as a fumigant and kill any insects coming into range. Like other organophosphates, DDVP works by poisoning the insect's nervous system. Since humans have a nervous system that functions similarly to that of insects, the same can happen to them. Indeed, the label on the package containing a DDVP-impregnated strip warns against placing the strips in rooms where food is prepared or served, or in bedrooms where infants, ill or aged persons are confined.

The persistent organochlorine insecticides such as aldrin,

dieldrin, DDT and chlordane were formerly available for direct sale to the urban consumer. Many of these were highly effective household insecticides, being particularly toxic to cockroaches and ants. However, questions as to their safety for humans, as well as their extreme environmental persistence, have led to the ending of their use in most urban situations. Chlordane and dieldrin are still available as termite control agents, designed to be used under and around the house rather than in it. However, it is doubtful this use will continue for long.

IV. INSECTICIDES IN THE YARD AND GARDEN

A wide array of insecticides is available to the urbanite for control of insects outside the home. Most of these are organophosphates and carbamates. In addition, certain of the less persistent organochlorine insecticides such as methoxychlor and lindane are still available.

A major use of these chemicals is to protect garden produce from insect damage. If this is to be done it should be accomplished with chemicals that leave no harmful residues. Labels on garden insecticide preparations specify the number of days that must be allowed between the last application and harvest, and the dose rates that may be used.

Another area of extensive insecticide use is on the lawn for the control of insects such as white grubs, webworms and chinch bugs. The persistent organochlorine insecticides dieldrin, heptachlor and chlordane were formerly used for this purpose. Presently, less persistent organophosphates are used. The treatments may control the pest as intended, but the effect is seldom specific. Insecticides applied to lawns frequently kill off other animal life in the yard, i.e., earthworms and all other detritus feeders of the lawn ecosystem. This means that animal-dependent processes such as soil aeration, dead grass removal and nutrient recycling may be badly disrupted. The net result is that lawn management becomes more difficult, requiring such practices as mechanical aeration and thatch removal and increased fertilizer and chemical inputs. As in agriculture, insecticide use in the lawn may lead to insecticide dependence.

Most garden and yard insecticides, it is argued, are safe when used as directed on the label. The trouble based on personal observation, is that the label directions are widely ignored. Cost of chemical is not a limiting factor in urban insecticide use and so, the urban gardener may reason that if the recommended amount is good, twice as much is better. Time restrictions in relationship to the interval between treatment and harvesting may be ignored. The final result may be unnecessary exposure of the user and his family to residues of poisonous insecticides.

V. INSECTICIDES AND HUMAN HEALTH

It is interesting to note the two widespread, but very divergent attitudes toward insecticide use held by urbanites. Some dislike insects so much they are willing to use all means possible, including chemicals, to obtain an insect-free environment. Such persons, if chemical users, must feel that any dangers associated with exposure to insecticides are less than the dangers represented by insects cohabiting with them. Others distrust chemical insecticides to the extent that they refuse to use them under any circumstances. They seem to assume that insects represent less of a threat than do insecticides.

In the past, health data on insecticide effects on urbanites have dealt primarily with the persistent organochlorine insecticides. Data cited in the Mrak Commission Report (1969) indicated that residues of chlorinated insecticides in body fat averaged between five and 10 parts per million in Americans in the mid 1960s. Where you live is important in determining how much insecticide is stored in your body fat. Those living in the southernmost states averaged twice the residues of those in the northernmost states. There were also racial differences. Residues in the body fat of black Americans, particularly older ones, are significantly higher than in comparable populations of white Americans. In other words, where you live and who you are are important factors in determining the amount of insecticide you accumulate.

Another important study was that of Radomski *et al.* (1968). Working with human fat samples collected during autopsies in Miami, Florida, hospitals, these researchers found a high correlation between the residues of insecticide present in body fat and deaths from certain diseases. In particular, deaths from various types of cancer and from hypertension were associated with persons having greater than average fat residues of insecticides.

At the present time, and with the DDT ban, the importance of DDT as a possible human health hazard is probably decreasing. There seems to be no comparable data relating levels of use of the nonpersistent insecticides with human health problems. There are reports that accidental or repeated human exposure to organophosphate insecticides is associated with both physical (Aldridge *et al.*, 1969) and mental (Metcalf and Holmes, 1969) disturbances in humans. The significance of these findings needs further study.

VI. THE FUTURE OF INSECTICIDE USE IN THE URBAN ENVIRONMENT

The initial optimism of the post-World War II period when the highly effective synthetic organic insecticides were first introduced has been replaced by a more sober understanding that the use of these chemicals involves risks as well as benefits. What then is the future for the use of these chemicals?

A consistent trend in recent years has been toward ever tougher Federal and State restrictions on insecticide use. Insecticides have been classified as being available for "general" or for "restricted" use. General use chemicals are those safe enough that not much harm can result from misuse by the amateur. Restricted use chemicals are those highly toxic to man or the environment whose use will be restricted to trained operators. These operators in theory will be well enough trained to appreciate the hazards involved and consequently, will use insecticides with care. In the middle is a vast array of moderately toxic chemicals. At present these are mostly classified for general use. It seems likely that increasing numbers of these will be placed in the restricted use category as accidents occur and additional information about possible harmful effects is accumulated.

One consequence of tightening regulations is that more and more insecticide use will pass from the do-it-yourselfer to the professional applicator. This will lead inevitably towards more expensive insect control. Hopefully, it will be matched by improvements in the use of insecticides as well.

It also seems reasonable that as costs of control go up, people will rely more on management techniques such as better sanitation and other nonchemical controls. Entomologists will undoubtedly be involved in disseminating information about these methods.

Lastly, it can be hoped that people will gain a little more tolerance for insects in their personal environment, and maybe even some appreciation for them. A major role for entomologists can be to encourage people to appreciate insects rather than to devise ways to exterminate them. Such encouragement can lessen entomophobia and may in consequence lessen human exposure to potentially harmful chemicals.

REFERENCES

Aldridge, W. N., J. M. Barnes, and M. K. Johnson. 1969. Studies on delayed neurotoxicity produced by some organophosphorus compounds. *Ann. New York Acad. Sci. 160, Art. I*: 314-322.

Ebeling, W. 1975. Urban entomology. Univ. Calif. Div. Agric.

Sci., Berkeley, California.

Fowler, D. L., and J. N. Mahan. 1973. The pesticide review 1973. USDA, Washington, D. C.

Matsumura, F. 1975. Toxicology of insecticides. Plenum Press, New York.

Metcalf, D. R., and J. H. Holmes. 1969. EEG, psychological and neurological alterations in humans with organophosphorus exposure. *Ann. New York Acad. Sci. 160, Art. I*: 357-365.

Mrak, E. M. (Chm.) 1969. Report of the secretary's commission on pesticides and their relationship to environmental health. U.S. Government Printing Office, Washington, D. C.

Olkowski, H., and W. Olkowski. 1976. Entomophobia in the urban ecosystem, some observations and suggestions. *Bull. Entomol. Soc. Am. 22*: 313-317.

Radomski, J. L., W. B. Deichman, and E. E. Clizer. 1968. Pesticide concentrations in the liver, brain and adipose tissue of terminal hospital patients. *Food Cosmet. Toxicol. 6*: 209-220.

Walsh, J. 1976. Cosmetic standards: Are pesticides overused for appearances sake? *Science 193*: 744-747.

TECHNOLOGY TRANSFER IN URBAN PEST MANAGMENT

Gary W. Bennett

Department of Entomology
Purdue University
West Lafayette, Indiana

I. INTRODUCTION

The sources of pest control information available to the general public, business and industry are innumerable. As our society has become more urbanized and industrialized, the demand for information and assistance in the management of insects, as well as other arthropod and vertebrate pests, has likewise grown. Governmental agencies, educational institutions, professional organizations, consumer groups and others have responded with information that flows forth in great quantities, prepared and packaged in many and varied forms.

Another basic reason for this increase in need for information is the broadened scope of urban pest management. In the United States and many other countries, the standard of living has continually risen so that man no longer is concerned only

ISBN 0-12-265250-9

with the protection of his health, welfare and food supplies from pests, but also his property, general comfort and aesthetics. Urban pest management is required in homes, businesses, industrial plants, municipal buildings, and outdoor areas frequented by man such as lawns, recreational areas, and other general living and relaxation areas. A wide range of pests and pest situations are of concern here, and include: general pests of man's structures, from silverfish to rodents; structural pests, also known as wood-destroying organisms, such as termites and decay fungi that destroy wood and wood products; lawn and ornamental pests, from sod webworms to plant parasitic fungi; the very large and complex area of public health, which often involves a community-wide pest management effort; pests of food-handling and processing establishments, a complex area of pest management within itself that encompasses everything from the local Coca-Cola bottler to the meat packer to McDonald hamburger restaurants; pests of nonfood industries such as textiles, plastics, etc.; and pests of man in our ever-increasing recreational areas, whether they be a city park or a national forest. The scope of urban pest management is indeed a broad one.

Urban pest management can be simply defined as the creation and maintenance of conditions or situations in which pests of importance in urban environments are prevented from causing significant problems. Its objectives may be achieved by preventing the establishment or spread of pests, controlling established infestations and keeping infestations at levels at which little or no damage or annoyance occurs. As with any type of pest management system, these objectives should be accomplished at the lowest possible costs consistent with minimum risk or hazard to man and the desirable components of his environment.

Pest control is a term that has long been associated with urban and industrial pest management. Although the word control is used, it should not be interpreted as having a different meaning or thrust than pest management. The objectives of urban pest control and urban pest management are the same. The two terms are in fact synonyms. However, whether one calls it by one term or the other, there remain basic differences between the 1950's and today. In addition to the technological advances that have been made, there are new attitudes and philosophies concerning man, his environment and ecological systems, there is greater concern for the risks of using pesticides and other control methods as compared to the benefits of such uses, and there is greater concern for utilizing and integrating all control or management principles and practices into dynamic systems of pest control.

II. SPECIALIZED AREA OF PEST MANAGEMENT

In formulating urban pest managment programs, it must be borne in mind that basic differences exist between this specialized area of pest management and agricultural pest management. The use of economic thresholds is a common practice in agricultural pest management, but in urban pest managment there are few situations in which the economic feasibility of instituting management practices can be determined. For example, how many cockroaches must be present in a home before control is economically justified? There is certainly no single measuring stick for making this determination. To many homemakers, the presence of a single cockroach is ample justification for initiating control measures. How many termites must be present in a structure before control is economically justified? In this pest situation, even the presence of conditions that are conducive to an infestation may be considered sufficient to justify management. As a matter of fact, in areas of high termite incidence, the use of structural features which prevent termite invasion and the use of soil insecticides are recommended during the construction of new buildings. Thus, for most urban and industrial pest problems, there are no economic thresholds.

Another basic difference involves the limitations of biological control, host resistance, genetic manipulation, pest sterilization and many other nonchemical control methods for use in urban and industrial pest management. This does not mean, however, that pesticides are the only tools available for use. Cultural control practices such as sanitation and habitat removal or modification, physical and mechanical control devices such as light traps, air curtains and traps, and chemical attractants and repellents are used extensively in implementing management programs. In some management situations it is possible to use solely nonchemical control measures but in most, pesticides play a major role and will continue to do so in the foreseeable future.

In addition to the limited number of nonchemical control measures, there are a number of other obstacles to effective urban pest management. One that is most frequently encountered is the immediacy of the pest problem. The consumer reacts to a pest problem and wants immediate control results. The only way this can be accomplished in most instances is to use a pesticide. The cost of many management programs is also an obstacle. For example, the initial cost of an air curtain for fly control in a food store is high compared to the use of an insecticide; yet the air curtain will more than pay for itself in the long run. However, when it comes to cash flow, it is human nature to be more concerned about short-term considerations. There is also the obstacle of consumers being unwilling to change their habits. Humans used to living in less than perfect sanitary conditions will continue to do so even though they may be well aware of the fact that appropriate sanitation and the removal of pest feeding and breeding areas is the key to their pest problems. A final roadblock to effective urban pest management is the inadequate supply of suitably trained professionals to do the job that needs to be done. Commercial pest control firms, the Public Health Service, the Food and Drug Administration, the food industry, the Cooperative Extension Service, and other groups, agencies and institutions simply have few sources from which to draw the expertise needed. However, as the demand from the public, industry and government grows, it is quite obvious that more and more talented people are finding their way into the field of urban pest management.

III. COOPERATIVE EXTENSION SERVICE

A frequently used source of information is the Cooperative Extension Service. This agency is the informal educational arm of both the U.S. Department of Agriculture and the respective state land-grant colleges and universities. It takes to people the results of research and practical experience from all pertinent resources available, as well as information on government programs directly affecting people, whether administered by the U.S. Department of Agriculture or by the state and county governments.

As our cities and industries have grown, so has the Extension Service relative to the urban pest management area. It continues to enlarge its educational and informational role in this area through the development of circulars and other literature, the sponsorship of conferences and other types of educational meetings, the development and administration of correspondence courses in pest control technology, and the hiring of additional specialists to work with individuals and groups on urban pest problems. This effort also includes the development of new strategies and programs.

A. Innovations in Technology Transfer

From its inception, the Cooperative Extension Service has channeled information to its audiences through many forms of the written and spoken word. Publications have been designed with both the subject matter and consumer group in mind; thus, the same subject may be developed in several publications in different degrees of depth depending on the target audience. The Extension Service has always been a strong proponent of helping individuals or groups to help themselves in performing certain functions and has accomplished this through meetings (with individuals at their home or place of work, or with groups in public meeting places), workshops, demonstrations, etc. Radio, television, newspapers, magazines and the telephone have also been used extensively in technology transfer.

These traditional media continue to be the backbone of Extension Service technology transfer. However, as educators and communicators, Extension personnel are constantly testing and developing new methods or systems for getting information to consumers in the most efficient and effective manner. Several examples of innovative program packaging are summarized below:

1. Dial-access Taped Messages

Widely used by medical, veterinary, banking and legal associations to provide brief, taped public service messages to interested callers, this system was recently developed for home economics and horticulture use by the University of Wisconsin. Users simply pick up the telephone, dial the tape library service number, and ask to be connected to one of over 300 numbered tapes from a published list of home and garden subjects. The user hears a one to three minute message. Tapes can be changed as necessary to accommodate the changing seasons and their associated problems; this requires that the published list of titles be revised accordingly.

The more significant attributes of the system are: 1) it frees Extension professionals from routine, recurring calls, 2) the system is modular and has no technical limitations as to the number of tapes or telephone lines which can be accommodated, 3) information is available when it is needed, 4) users can request the tape as many times as necessary to better understand the information without burdening anyone, 5) it represents an ongoing popular and visible service to the community, and 6) the system is cost efficient.

2. Indiana's Fast Agricultural Communications Terminal System (FACTS)

FACTS is the acronym for Indiana's Fast Agricultural Communications Terminal System, a computer-based communicaions system linking each county Extension office with Purdue University and each other. When completed in 1979, FACTS will consist of an "intelligent" computer terminal in each county office and in each department in the Purdue agricultural complex.

This system is being developed to meet an urgent need to

place more information, which is complex and current, at the immediate disposal of the county Extension agent. The agent serves as the major communicator and "change agent" between campus and customer, whether on the farm or in the city. He will be able to secure in just a matter of minutes such information as pest management and pesticide use recommendations, and pesticide applicator licensing information to pass along to the inquiring consumer. The system will provide rapid communications with the counties, better data storage, fast retrieval and will result in improved service to clientele.

FACTS is unique because it can perform distributed processing. That is, the campus and county terminals are "intelligent," having the capability to do more than just transmit and receive data. The intelligent terminal is a small computer in its own right, and has local information processing capability. These terminals are linked on telephone lines to a central computer facility (Fig. 1). This computer is called the front-end processor, and its function is to step in when the question or problem at hand is too much for the county terminals. At the request of the user, the terminal automatically calls the campus computer, which will produce the appropriate program and process the information. If there is a message to be sent, the campus computer will automatically call the county and transmit the message. The campus computer also acts as a message-switcher between counties. If the problem or question is one beyond the capability of the front-end processor, the data automatically goes to a dual CDC-6500, which is the host computer. Then it calls the front-end processor, which, in turn calls the county with the appropriate information.

3. *Paraprofessional Gardening Volunteers*

Interested citizens with backgrounds or strong interests in gardening and service to the public are trained for 35 to 50 hours by Extension Service professionals in many aspects of urban horticulture. In return for this training, volunteers are expected to contribute their time, to the extent of about one-half day per week during the growing season. This commitment may involve the staffing of a "plant clinic" at a shopping center or other public place. Regular publicity, through newspapers and other media, is necessary to encourage the public to bring questions and problems relating to plants to the clinic during its hours of operation. Or, the commitment may entail assisting the county Extension agent in handling gardening inquiries from the public, by phone or in person, at the county Extension office.

The important strengths of this type of program, which was initiated by Washington State University, includes several of those cited above for the dial-access information system.

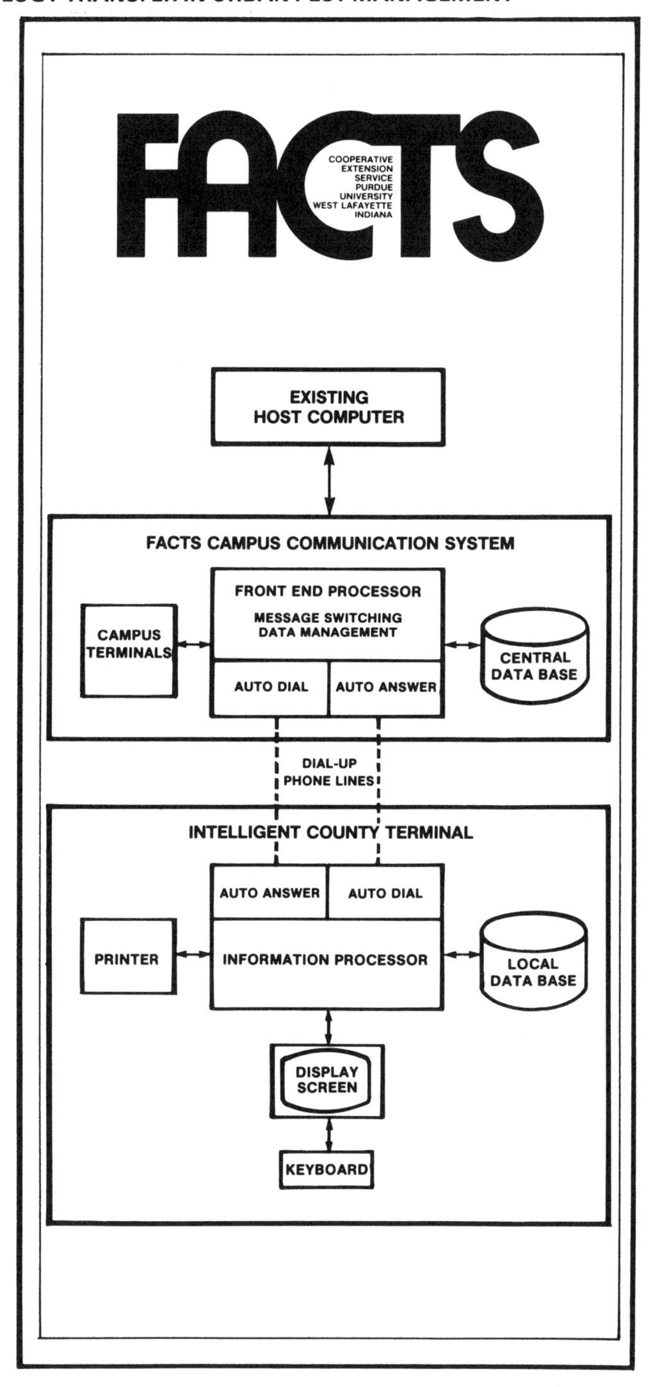

Fig. 1. Indiana's fast Agricultural Communications Terminal System (FACTS).

Additionally, volunteers serve as continuing emissaries of their land-grant institutions since public lectures, and radio and television appearances, are often outgrowths of activity for the interested volunteers.

4. Library Resource Centers

The establishment of home and garden resource centers in public libraries and their branches is a further way to achieve a multiplier effect in dealing with requests for Extension Service information. Such centers are furnished with indexed Extension publications, and encouragement is often given to transferring other appropriate library references to the resource center. Opportunity to order Extension publications through the center may also be provided.

Persons in search of information can be directed to the library resource center by a county Extension office receptionist. Those residing some distance from the county office often find the convenience of a branch library resource center economical in terms of time and fuel, and satisfying in that current information is available the day it is needed.

It is easy to visualize from the innovations discussed above that Extension is viewing new roles - the expansion of adult education in a world where rapid progress challenges all men with obsolescence, and the transmission of technical data in an era when not just the farmer, but all members of society, need up-to-date technical information.

IV. OTHER MEANS OF INFORMATION TRANSFER

The importance of traditional channels through which urban entomology technology is transferred has been discussed earlier in this paper, but they certainly deserve attention for their importance outside the framework of the Cooperative Extension Service. All the media serve an important function in the dissemination of urban pest management information, but the press has been with us longer than any other means. Some of the more effective press channels are naturally those with wider distributions - *Good Housekeeping*, *New York Times*, *United Press International*, *Better Homes and Gardens*, *McCalls*, etc. However, any printed matter, whether one-page circulars or voluminous books, can play an important role.

Radio, television, films, slides, tapes, etc. are packaged and presented in a variety of ways, most of which are to some degree effective means of technology transfer. Information exposure, no matter what the medium, is continuously needed to inform the consumer about urban pest management programs and practices.

V. FUTURE DIRECTIONS

Effective and efficient communication and education relative to urban pest management will be of utmost importance in the years ahead. These should be the primary mission of professionals, whether they are industry, university, government or otherwise employed. Not only must emphasis be placed on communicating with the general public, but also with government and within the various levels of our professional ranks. Urban pest management is facing more rapid and complex changes in technology than ever before; therefore, the mechanisms for disseminating this information must be very carefully designed to ensure the quality of the information being transmitted. The ever-broadening scope of urban pest management will not lessen the magnitude of the job. As urban pest management expands in depth and breadth, so must the methods of technology transfer.

FURTHER READING

Deay, H. O. 1955. Entomology at Purdue. *Proc. Ind. Acad. Sci. 64*: 152-157.

Lin, G. 1939. The earliest bureau of entomology. *J. N. Y. Entomol. Soc. 47*: 307-310.

Parks, T. H. 1955. Extension entomology - its development, present status, and future. Paper presented at A.A.E.E. meeting, Cincinnati, Ohio.

Sanders, H. C. 1972. Instruction in the Cooperative Extension Service. CES, Louisiana State Univ., 152 pp.

Truman, L. C., G. W. Bennett, and W. L. Butts. 1976. Scientific guide to pest control operations. Harvest Publ. Co., Cleveland.

A
B
C 8
D 9
E 0
F 1
G 2
H 3
I 4
J 5